T0281475

THE FRONTIERS COLLECTION

THE FRONTIERS COLLECTION

Series Editors
A.C. Elitzur   Z. Merali   T. Padmanabhan   M. Schlosshauer
M.P. Silverman   J.A. Tuszynski   R. Vaas

The books in this collection are devoted to challenging and open problems at the forefront of modern science, including related philosophical debates. In contrast to typical research monographs, however, they strive to present their topics in a manner accessible also to scientifically literate non-specialists wishing to gain insight into the deeper implications and fascinating questions involved. Taken as a whole, the series reflects the need for a fundamental and interdisciplinary approach to modern science. Furthermore, it is intended to encourage active scientists in all areas to ponder over important and perhaps controversial issues beyond their own speciality. Extending from quantum physics and relativity to entropy, consciousness and complex systems—the Frontiers Collection will inspire readers to push back the frontiers of their own knowledge.

More information about this series at http://www.springer.com/series/5342

For a full list of published titles, please see springer.com/series/5342

Sonya Bahar

The Essential Tension

Competition, Cooperation and Multilevel
Selection in Evolution

 Springer

Sonya Bahar
Department of Physics & Astronomy
  and Center for Neurodynamics
University of Missouri at St Louis
St Louis, MO, USA

ISSN 1612-3018   ISSN 2197-6619   (electronic)
The Frontiers Collection
ISBN 978-94-024-1477-6   ISBN 978-94-024-1054-9   (eBook)
DOI 10.1007/978-94-024-1054-9

Printed on acid-free paper

This Springer imprint is published by Springer Nature
The registered company is Springer Science+Business Media B.V.
The registered company address is: Van Godewijckstraat 30, 3311 GX Dordrecht, The Netherlands

for

Frank Moss

1934–2011

& for C & D

Acknowledgements

There are many people I could acknowledge, but I don't want this to sound like the Oscars, so I will limit myself to the essential thanks that are due to the wonderful Maria Bellantone, for urging me to write this book and for her infinite patience in waiting for the manuscript to finally be completed.

Contents

Introduction

I...suggest that something like 'convergent thinking' is just as essential to scientific advance as is divergent. Since these two modes of thought are inevitably in conflict, it will follow that the ability to support a tension that can occasionally become unbearable is one of the prime requirements for the very best sort of scientific research.

Thomas S. Kuhn

Je juge cette longue querelle de la tradition et de l'invention, de l'Ordre et de l'Aventure.

Guillaume Apollinaire

I MUST begin with an apology to Thomas Kuhn. The title of this book has been borrowed – lifted, liberated, stolen, as you prefer – though my editor assures me that titles cannot be copyrighted and that I commit no actionable offense – from the title essay of a volume of Thomas Kuhn's writings subtitled *Selected Studies in Scientific Tradition and Change*. The essay was initially delivered as a lecture at a 1959 University of Utah conference on, of all things, the "identification of scientific talent." Kuhn was already well known as a self-described "ex-physicist now working in the history of science," though the publication of his masterpiece, *The Structure of Scientific Revolutions*, still lay 3 years in the future. Invited to speak at a conference – run by educational psychologists – on the key characteristics of the creative personality and how to identify such a personality early in the educational process, Kuhn addressed a core issue in scientific education, in terms that clearly show his evolving thought on the themes of creativity, "normal science," and scientific revolutions. Is iconoclasm, he asked, the only characteristic needed for a successful and creative scientist? Kuhn argued that a thorough grounding in tradition is as important to scientific advance as the ability to strike out into the unknown. He said:

> Normal research, even the best of it, is a highly convergent activity based firmly upon a settled consensus acquired from scientific education and reinforced by subsequent life in the profession. Typically, to be sure, this convergent or consensus-bound research ulti-

mately results in revolution. Then, traditional techniques and beliefs are abandoned and replaced by new ones. But revolutionary shifts of a scientific tradition are relatively rare, and extended periods of convergent research are the necessary preliminary to them.

As a result, "the successful scientist must simultaneously display the characteristics of [both] the traditionalist and of the iconoclast" (Kuhn 1977, p. 227).[1]

Kuhn suggested that major scientific advances take place only following the achievement of consensus, when a field has left behind its "natural history stage," filled with a myriad of conflicting results and interpretations. This sequence is necessary because the convergent thought required to build consensus inevitably leads, Kuhn argued, to a period of conservative, incremental research which might at first appear routine, even boring, but is in fact bursting with the seeds of fruitful contradictions. "Work within a well-defined and deeply ingrained tradition seems more productive of tradition-shattering novelties than work in which no similarly convergent standards are involved," Kuhn said. "How can this be so? I think it is because no other sort of work is nearly so well suited to isolate for continuing and concentrated attention those loci of trouble or causes of crisis on whose recognition the most fundamental advances in basic science depend" (Kuhn 1977, p. 234).

When I set out to write a work on the complementary drives that lead to collective behavior in biological systems and how these drives can be illuminated through the lens of physics, I struggled to find a better phrase to represent this delicate balance than "the essential tension". The phrase became the book's provisional title, and, for better or worse, it stuck. Indeed, the tension Kuhn discussed in his 1959 essay has thematic parallels with the tension that will be explored in the following chapters. Kuhn discussed the balance of convergent and divergent thought. This conceptual structure is mirrored in the balance between short-range and long-range forces that leads to clustering of entities in a real or theoretical space and also in the balance between the aspects of nonlinear systems that render them unpredictable, yet constrained, in phase space. Kuhn suggested that an "essential tension" could exist within the mind of a single scientist as well as within a scientific community; the relation between individual and group dynamics will be a recurring theme in the chapters below. Kuhn's model of long periods of "normal science" building up to periods of rapid transition echoes Niles Eldredge and Stephen Jay Gould's mantra that "stasis is data" in their analysis of the fossil record. Indeed, their model of punctuated equilibrium itself reflects the Kuhnian dynamic of consensus followed by revolution.

In short, I swiped the title, it got stuck to my hand, and I couldn't get rid of it.

*** *** ***

---

[1]A note on reference formatting: within the text, citations from books will be given using the author's last name, date of publication or book title, and the page number (e.g., Lannister 1992, p. 205), and articles will be cited using the name of the author(s) or the first author, followed by the year of publication (e.g., Snow et al. 2016; Targaryen and Stark 2016).

Thomas Kuhn analyzed the tension between different styles of thought. In contrast, the tension I will discuss in the present work is a delicate balance between orthogonal and complementary – but not opposing – drives in natural systems. I will trace the development of scientists' understanding of these drives in their various forms, especially as they relate to collective behavior in biological systems and to evolutionary dynamics. These themes all involve the formation of groups of (mostly living) things and lead to the question of how a bunch of individual things comes together to form a group that can be said to exist as a unique entity in its own right. This question has arisen for social scientists investigating crowd behavior and the division of labor in human society, for evolutionary biologists struggling to define what constitutes a species, and for biological physicists investigating the forces that lead fish to swarm and birds to flock. And all these problems are driven by some aspect of the "essential tension" that is the subject of this book. *Short-range birth processes must be offset by global death processes* in order for clusters to form in a reaction-diffusion model (Young et al. 2001). *Balance between competition and cooperation* is critical for the formation of social groups and may hold the key to the process by which a group of organisms begins to function as a collective entity. The *balance between "stretching" of trajectories* in a chaotic attractor as they experience separation at an exponential rate and *"folding"* as the attractor's infinite leaves wrap back on themselves to remain contained in a bounded region of phase space is an essential property of nonlinear dynamical systems. The process of evolution itself is driven by a balance between the noise of random variability and the dual constraints of historical contingency and adaptation to the environment.

In the following pages, I will explore how these themes recur again and again in the study of collective biological systems – groups, swarms, and species. I will also explore how fundamental principles from the physics of complex systems can assist in the struggle to unravel these deep problems. One final tension will emerge from this discussion – a tension between levels of selection, between the individual and the group, and between the needs of the single organism and the species. I will investigate how this delicate balance is driven by the other tensions we have explored and how it relates to fundamental controversies in modern evolutionary theory, such as the role of group selection. Finally, I will propose that the tension between the levels at which natural selection acts is not only essential but that, in fact, it is an inevitable result of the noisy interaction between biological objects at different scales.

One last point needs to be made before we get underway. The concepts that exist in tension here are complementary, not contradictory. They are not dichotomies. *Form* in biology (driven partly by physical constraints and partly by historical contingency) is not antithetical to *function* (driven primarily by adaptation and partly by what Gould and Vrba call "exaptation", the usefulness of a biological character in a role it was not initially selected for). It cannot be too strongly emphasized that concept pairs such as these are *orthogonal*, not *opposing*, axes. Tracing the degree to which each component of such a pair influences a given natural phenomenon becomes, then, a sort of intellectual principal component analysis. In sum, this book will be more than a plea for pluralism. It will be one long argument that pluralism is *essential*.

References

Kuhn TS (1977) The Essential Tension: Selected Studies in Scientific Tradition and Change. The University of Chicago Press, Chicago/London

Young WR, Roberts AJ, Stuhne G (2001) Reproductive pair correlations and the clustering of organisms. Nature 412(6844):328–331

Part I
The History of an Idea

Chapter 1
Crowds

Puisque ces mystères nous dépassent, feignons d'en être les organisateurs.

Jean Cocteau

THE MATHEMATICIAN is tossing fitfully in his bed, talking in his sleep. He was up late talking with his friend the philosopher, and now he is dreaming. Mademoiselle de l'Espinasse watches him, worried. In the morning she calls in Doctor Bourdeu for a consultation. She has noted down her friend's ravings and, looking for reassurance, reads them to the doctor. *"Have you ever seen a swarm of bees leaving their hive? … Have you seen them fly away and form at the tip of a branch a long cluster of little winged creatures, all clinging to each other by their feet? This cluster is a being, an individual, a kind of living creature."* To Mlle. de l'Espinasse's dismay, instead of trying to help the delirious patient, Doctor Bourdeu takes the ball and runs with it. *"Do you want to change the cluster of bees into one individual animal?"* he asks (Fig. 1.1). *"Soften the feet with which they cling to one another, that is to say make them continuous instead of contiguous. Obviously there is a marked difference between this new condition of the cluster and the preceding one, and what can this difference be if not that it is now a whole, one and the same animal, whereas before it was a collection of animals? All our organs… are only distinct animals kept by the law of continuity in a state of general sympathy, unity, identity."* (Diderot, *D'Alembert's Dream*, pp. 168–170).

* * *

The idea of an ensemble of individuals coming together to form a qualitatively different collective entity was not new when Denis Diderot explored it in *D'Alembert's Dream*, a dialogue that touched on ideas so radically new and dangerous that the manuscript was only circulated among a select group of friends during Diderot's lifetime. But the idea, as expressed by earlier thinkers, was more metaphor than theory. In the hands of Diderot, the notion of the emergence of a new collective organism crossed the threshold from analogy to hypothesis.

© Springer Science+Business Media B.V. 2018
S. Bahar, *The Essential Tension*, The Frontiers Collection,
DOI 10.1007/978-94-024-1054-9_1

Fig. 1.1 An engraving of a beekeeper tending his apiary (Lyon, France, 1560). Public domain

With the exception of the poem *Moretum*, attributed to Virgil, from which the phrase *e pluribus unum* derives in the context of the assembly of a salad, early notions of an emergent collectivity come from descriptions of human crowds. Many people assemble, and behave together like a single beast. Plato, who described the crowd as "a large and powerful animal", describes in the *Republic* what happens "when they crowd together into the seats in the assembly, or law courts or theatre, or get together in camp or any other popular meeting places and, with a great deal of noise and a great lack of moderation, shout and clap their approval and disapproval of whatever is proposed, or done, till the rocks and the whole place re-echo, and re-double the noise of their boos and their applause. Can a young man be unmoved by all this? He gets carried away and soon finds himself behaving like the crowd and becoming one of them." (McClelland 1989, pp. 38–39) From the perspective of modern physics, this description – all too real in 2016 as it was in classical Greece – corresponds to local, microscopic interactions between elements of a large ensemble.

The crowd continued to fascinate philosophers in the years following Plato, not least due to its strange manner of emerging as a separate entity from hundreds of individuals. In *Federalist No. 55*, James Madison[1] suggested that

[s]ixty or seventy men may be more properly trusted with a given degree of power than six or seven. But it does not follow that six or seven hundred would be proportionally a better depositary [of the public interests]. And if we carry on the supposition to six or seven thousand, the whole reasoning ought to be reversed. The truth is, that in all cases a certain number at least seems to be necessary to secure the benefits of free consultation and

---

[1] Or possibly Alexander Hamilton; the authorship of this one of the Federalist papers remains uncertain, according to some scholars.

discussion, and to guard against too easy a combination for improper purposes; as, on the other hand, the number ought at most to be kept within a certain limit, in order to avoid the confusion and intemperance of a multitude. In all very numerous assemblies, of whatever character composed, passion never fails to wrest the scepter from reason. Had every Athenian citizen been a Socrates, every Athenian assembly would still have been a mob.

Analyses such as that of Madison, though written in a totally different context (in this case, arguments over the total number of members in the House of Representatives), presage studies of the role of population density in triggering behavioral transitions in flocks and swarms, a topic we will return to in Chap. 8. However, the *inner* dynamics of crowd structure remained relatively unparsed by eighteenth and nineteenth century commentators, who focused on the role of the crowd in the larger flow of history. The Gordon riots in 1780[2] were described by Edward Gibbon in typical elite-versus-rabble style. "Forty-thousand Puritans, such as they might be in the time of Cromwell[,] have started out of their graves, the tumult has been dreadful…that scum which has boiled up to the surface of this huge cauldron [of the City of London]…the month of June, 1780 will ever be marked by a dark and diabolical fanaticism." (McClelland 1989, p. 113)

In his classic history of the French revolution (Fig. 1.2), Thomas Carlyle continued what J. S. McClelland calls a "pre-psychological account of the leaderless crowd". In McClelland's reading, Carlyle painted the crowd as a pure force of history, a collective whirlwind created by fate, by providence, a sort of pre-Marxist sweep of history itself. The crowd was a key protagonist in Carlyle's history, but it did not have an internal structure to be dissected: it simply *was*. In McClelland's analysis, other scholars of the French revolution such as Jules Michelet portrayed the crowd as equally inscrutable, but in a nationalist rather than a metaphysical (or meta-historical) sense: the crowd was "the people", "the nation".

Like Gibbon, Hippolyte Taine's view of the crowd was influenced by his own personal politics: fearfully watching the events of the Paris Commune unfold in 1871 from a sabbatical in Oxford, he saw the crowd as a mob of beasts rather than a collection of Rousseau's noble savages. Taine wrote: "Take women [who] are hungry and men who have been drinking; place a thousand of these together and let them excite each other with their exclamations, their anxieties, and the contagious reaction of their ever-increasing emotions; it will not be long before you find them a crowd of dangerous maniacs." (McClelland 1989, p. 129) Like Plato's, this description centered on the local forces within a group that lead to collective behavior. But Taine's purpose was not to explore the internal structure that binds together elements of a crowd: he was more concerned with describing how the crowd triggers regression to "the animal in us". It is a view driven by Tennyson's simplified Darwinism, red in tooth and claw, exacerbated by an aristocratic (and, by the late nineteenth century, even bourgeois) fear of the working class. Taine appealed to metaphors of society as an organism, in which "the body politic stops functioning properly because the belly tries to usurp the function of the brain…a wild beast,

---

[2] It was during these riots that the phrase "King Mob" was used for the first time, scrawled on the walls of Newgate Prison.

Fig. 1.2 The women of Paris march on Versailles, 1789. Public domain

incompletely domesticated, goes wild again." (McClelland 1989, pp. 134–35) But Taine used these metaphors for emotional effect, not as the starting point of an analysis.

Approaching the turn of the last century, crowd theorists such as Scipio Sighele and Gabriel Tarde struck out in a variety of new directions. They approached the idea of a crowd in a criminological context, but resisted Cesare Lombroso's simplistic identification of criminality (and, by implication, collective behavior of the crowd) with atavism. Sighele argued that, in McClelland's wording, the collective behavior of crowds was

> … genuinely new in a way that no other collective phenomenon is. In the ordinary evolutionary sense, it has no past and no future. This is true phylogenetically and onogenetically. There is no crowd out of which the crowd comes, and there is no crowd-individual before the individual becomes a member of the crowd. When the crowd disperses, crowd-individuals become their ordinary workaday selves. The crowd is in this sense outside ordinary time. (McClelland 1989, p. 164)

The crowd is also different in kind, and is more than simply a sum of its parts. Sighele asked under what circumstances humans' moral and intellectual qualities would be more than simply additive, writing that crowds often behave far better, or far worse, than their constituent individuals would alone. These thoughts suggested the question of how to define a "true individual" in a social group, or in nature itself. In their correspondence, Sighele and Tarde speculated about whether the definition of individuality should be parsed down to the level of the cell, or even to the level of the atom (McClelland 1989, p. 173). McClelland dismisses ideas of this sort as flights of fancy. He notes that Walter Bagehot, in *Physics and Politics* (1872) "makes great leaps from reflections on the evolution of the human body to reflections on the development of human societies as if there were nothing between them… [suggesting] what look like wild analogies to us, analogies between combinations of cells and human groups, or between animal societies and human societies" (McClelland 1989, p. 158). But – as we shall explore over the next few hundred pages together – such crude analogies provide a blunt instrument by which we can begin to excavate the structure of the world. Moreover, these "wild analogies" echo actual, mechanistic, materialist questions asked no less by scholars of Tarde and Sighele's era than by scientists today. Alfred Espinas was not just tossing off wild analogies when he wrote in his 1878 work *Les Sociétés Animales* that

> In fact, we are composed of millions of little entities whose interactions have been compared by the most illustrious physiologists to the work of laborers in a vast factory, to the inhabitants of an immense city, the arteries being like the roads and canals that carry nourishment to different regions, while the nerves resemble telegraphic wires that transmit information and impulses from the parts to the center and from the center to the parts. (Espinas 1878, p. 214, my translation; see also the paraphrased text by McClelland 1989, p. 168)

It was perhaps such metaphors that inspired crowd theorists to inquire – yes, metaphorically – whether the body of a crowd, or indeed any collective, could assemble itself in the absence of a head. Espinas held that a division was essential between the leader and the led; thus, a flock of birds did not constitute a real "society" (McClelland 1989, p. 160). Gabriel Tarde agreed, writing "every mob, like every family, has a head and obeys him scrupulously" (McClelland 1989, p. 185). With perhaps a bit of special pleading, Tarde argued that there are no leaderless crowds: when all else fails, whoever throws the first stone is the leader.[3] Even Freud constructed a theory of the crowd nucleated by a leader. In his view, described in *Group Psychology* (1921), individuals in the crowd projected their own ego-ideal onto the crowd leader. The crowd members then become identified with each other at the ego level, because of the common projection (or, in Freud's terminology, introjection) of their ego-ideal (McClelland 1989, pp. 241–242).

In searching for a mechanism by which a leader could enchant a crowd, some late-Victorian scholars turned to the then-popular phenomenon of hypnotism. Gabriel Tarde took it a step further, suggesting that hypnotism is merely one example of the type of persuasion that society usually exerts on its members, and analyzed social interactions in terms of imitation ("sociability is suggestibility"). Espinas spoke of a "mutual heating-up" between an orator and the audience, a phenomenon we in the United States have witnessed quite clearly in our 2016 election. Many of these interactions can occur between members of a crowd as well as between the crowd members and a leader, so these mechanisms in themselves do not provide an argument that leaderless crowds are an impossibility.

By the end of the nineteenth century, Gustave Le Bon had published a work purporting to show how leaders could manipulate crowds in a popular book largely considered to have been plagiarized from the work of Sighele and Tarde. "The

---

[3] Tarde's investigations were not limited to crowds, but to all of society; he suggested that there is a continuum from a loosely affiliated crowd to what he called a "corporation", a well-defined, durable, organized social structure. The crowd-corporation transition has an important structural resonance with another idea we will explore later: the transition from MLS1 to MLS2 selection.

secret of Le Bon's success," writes McClelland, "was to use science to frighten the public, and then to claim that what science could understand it could also control." (McClelland 1989, p. 196).

Much later, Elias Canetti, writing in *Crowds and Power* (1960), saw the identification between crowd members as far more important than the interaction between the leader and the crowd. He understood crowds as a mass of individuals looking for release from external pressures (the "stings of command"), and from the emotional burdens of separation, through mutual connection ("discharge"). This process could be catalyzed by a demagogue (an example of what Canetti called a "crowd crystal") but was not dependent on him. The demagogue might directly affect a small proportion of individuals in the crowd, who would then spread the excitement horizontally.

* * *

Like the crowd theorists described above, Emile Durkheim (Fig. 1.3) investigated the forces that hold groups of individuals together. In *The Division of Labour in Society* (1893), Durkheim traced the mutual excitations in a crowd to the way like-minded individuals reinforce each other's ideas. Every strong, intense state of

Fig. 1.3 Émile Durkheim. Public domain

consciousness, he argued, is perceived as a strong source of life, since we identify with our feelings.

> It is therefore inevitable that we should react vigorously against the cause of what threatens a lowering of the consciousness…Among the most outstanding causes that produce this effect must be ranged the representation we have of an opposing state…This is why a conviction opposed to our own cannot manifest itself before us without disturbing us. It is because at the same time as it penetrates into us, being antagonistic to all that it encounters, it provokes a veritable disorder. (Durkheim 1997, p. 53)

Hence, right-wingers in the United States rarely watch MSNBC. Likewise, Durkheim continues,

> just as opposing states of consciousness are mutually enfeebling to one another, identical states of consciousness, intermingling with one another, strengthen one another…If someone expresses to us an idea that was one we already had, the representation we evoke of it is added to our own idea; it superimposes itself upon it, intermingles with it, and transmits to it its own vitality…This is why, in large gatherings of people, an emotion can assume such violence. It is because the strength with which it is produced in each individual consciousness is reciprocated in every other consciousness. (Durkheim 1997, p. 55)

Durkheim is describing essentially a resonance effect; in this analysis, shared emotions are *not* simply additive. The interactions Durkheim described can be visualized as parallel

and antiparallel

spin states where the parallel state has lower energy. Reinforcing, parallel interactions generate what Durkheim called *mechanical solidarity*. This is the solidarity of the crowd, the solidarity of the like-minded, which is always weakened by the presence of an anomaly. If Durkheim had ended his analysis here, he would simply have been another member of the crowd of crowd theorists. But he dug much deeper, searching for an explanation of how societies progress along a continuum from fleeting, amorphous crowd to structured society. There is, Durkheim suggested, another type of solidarity, which he called *organic*, because it "resembles that observed in the higher animals. In fact each organ has its own special characteristics and autonomy, yet the greater the unity of the organism the more marked the individuation of the parts." (Durkheim 1997, p. 85).

Organic solidarity derives from the division of labor. Aggregates whose members depend on each other to perform different essential tasks are tied together with qualitatively different bonds than aggregates whose members are connected merely by similarity. Clearly, division of labor exists in human society, as it does among different cell types in multicellular organisms. Durkheim's argument for how division of labor arises in society parallels models of cellular differentiation and, as we shall see in later chapters, of the origin of multiple levels of selection in evolution.

Durkheim began his argument by sketching the development of modular, or segmentary, structures in human societies such as families and clans. These societies "are formed from the replication of aggregates that are like one another, analogous to the rings of annelida worms." (Durkheim 1997, p. 127) Each group might take over a certain role in society, such as herding a particular species of cattle, or performing religious or ceremonial functions. At this point, one type of structure (*social role*) has been superimposed upon another (*social aggregate*). All that remains is for the increasing specialization to overflow the boundaries of the original social aggregate and erode the original structure. "Generally it may be said," Durkheim wrote, "that classes and castes have probably no other origin or nature: they spring from the mixing of the professional organization, which is just emerging, with a pre-existing familial organization." (Durkheim 1997, p. 133) But it is not just mixing: the new structure climbs out of the old one, and sits on it. Explicitly commenting that "the same law governs biological development", Durkheim wrote: "in organic as in social evolution, the division of labour begins by using the framework of segmentary organization, but only eventually to free itself and to develop in an autonomous way." (Durkheim 1997, pp. 139–141) He emphasized that this development happens without any centralized control. The process continues, driven by a positive feedback loop: the old segmentary structure breaks down, rendering "the social substance free to enter upon new combinations" (Durkheim 1997, p. 200). This leads to flow, migration and realignment of human populations, which results in an increase in the division of labor.[4] But just as the bald fact of genetic modularity provides only the raw material, and not the driving force, for nature's experimentation with different numbers of legs and wings, the foregoing does not fully explain *what drives specialization in the first place*. To address this key question, Durkheim invoked an explicitly Darwinian argument.

> If labour becomes increasingly divided as societies become more voluminous and concentrated, it is not because the external circumstances are more varied, it is [rather] because the struggle for existence becomes more strenuous. Darwin very aptly remarked that two organisms vie with each other the more keenly the more alike they are…The situation is totally different if the individuals coexisting together are of different species or varieties. As they do not feed in the same way or lead the same kind of life, they do not impede one another…If therefore one represents these different functions in the form of a cluster of

---

[4] Herbert Spencer had proposed that the simple spread of human populations led to specialization of social roles according to the local environment in which people lived: those who live near the sea make their living as fishermen, etc. (This process can be compared to niche specialization in evolution.) Durkheim argued that this certainly occurs, but not to a sufficient extent to explain the degree of actual division of labor present in human society.

branches springing from a common root, the struggle is least between the extreme points, whilst it increases steadily as it approaches the centre. This is the case not only within each town but over society as a whole. (Durkheim 1997, pp. 208–210)

These divergent societal functions are drawn together through mutual need, resulting in organic solidarity. This evolution is dependent on the prior existence of an initially cohesive, though undifferentiated, group.

In physiology the division of labour is itself subject to this law: it never occurs save with the polycellular masses that are already endowed with a certain cohesion…this integration supposes another sort that it replaces. For social units to be able to differentiate from one another, they must first be attracted or grouped together through the similarities that they display. This process of formation is observed, not only at the origins, but at every stage of evolution. (Durkheim 1997, p. 219)

The development of differentiation from a simple aggregate is indeed a well-worn evolutionary pathway which we will encounter later in these pages, for example, in our exploration of the evolution of the volvocine algae in Chap. 11.

Once established, "*the division of labour unites at the same time as it sets at odds; it causes the activities that it differentiates to converge; it brings closer those that it separates.*" (Durkheim 1997, p. 217, my italics) This is our first glimpse of the essential tension, that delicate balance between things at once drawn together and pulled apart – a structure that endlessly mirrors itself from the ideas of one century to those of another, and from one field of science to another.

Durkheim has been critiqued for a rather selective reading of Darwin's work. Sociologist William Catton (1998, 2002) suggested that Durkheim's assumption that mutualism develops in order to minimize competition is unrealistic in many biological contexts. First, the assumption is roughly equivalent to sympatric speciation (the divergence of co-localized populations), which occurs far less frequently in nature than allopatric speciation (the divergence of populations already separated by some environmental barrier). Second, there is the issue of causality. Darwin himself, Catton argued, never presented population divergence as a *means of* abating competition. Rather, enhanced competition *led to* divergence of populations and extinction of intermediate forms. This does not negate Durkheim's argument, for it is easy to envision a scenario in which someone who works with tin and iron is outcompeted by specialists in each individual metal. Clearly, also, causal factors in sociological evolution are not identical to those at work in the process of speciation. Nonetheless, Catton's critique points to an important issue. Many biological mutualisms, he pointed out, originate in exploitative relationships from which the participants are unable to separate. Here, the ultimate cooperative benefit may be a side-consequence rather than a direct adaptation. Stephen Jay Gould and Elisabeth Vrba called such unintended side-consequences *exaptations*. As Gould noted, Catton explicitly referred to this concept in one of his analyses of Durkheim (Gould 2002, p. 1239). "An adaptation has a function," Catton wrote. "An exaptation has an effect. Once that effect becomes important in the life of an organic type (in its new environment), natural selection may 'improve' the exapted trait, eventually making it an adaptation, and converting the effect into a true function." (Catton 1998) It is entirely likely that many aspects of the division of labor may have arisen as unintended consequences of an initial population divergence, and only later had the added benefit of abating competition, a function for which they were not originally selected. We will revisit the idea of exaptations in later chapters, and we will find that they play a crucial role in mediating the tension and balance between the levels at which natural selection operates.

References

Canetti E (1984) Crowds and Power. Farrar, Straus and Giroux, New York. (Originally published in 1960 as Masse und Macht; trans: Stewart C)

Catton WR (1998) Darwin, Durkheim, and mutualism. In: Advances in Human Ecology, vol. 7. JAI Press, London, pp 89–138

Catton WR (2002) Has the Durkheim legacy misled sociology?. Ch. 5. In: Dunlap RE, Buttel FH, Dickens F, Gijswijt A (eds) Sociological Theory and the Environment: classical foundations, contemporary insights. Rowman and Littlefield, Lanham

Diderot D (1966.) Rameau's Nephew and D'Alembert's Dream (trans: Tancock L). Penguin Books Ltd., London

Durkheim E (1997) The Division of Labor in Society (trans: Halls WD). The Free Press (Simon & Schuster), New York

Espinas A (1878) Les Sociétés Animales, 2eme edn. Libriarie Germer Baillière et Cie, Paris

Gould SJ (2002) The Structure of Evolutionary Theory. The Belknap Press of Harvard University Press, Cambridge, MA/London

Hamilton A, Madison J, Jay J (2015) The Federalist Papers: A Collection of Essays Written in Favour of the New Constitution as Agreed Upon by the Federal Convention September 17, 1787. Coventry House Publishing, Dublin

McClelland JS (1989) The Crowd and the Mob. Unwin Hyman Ltd., London. (Republished by Routledge Revivals, Taylor & Francis, 2011)

Chapter 2
Classification

Only connect.

E. M. Forster

IN THE previous chapter, we explored the problem of how one thing emerges from many. A related intellectual problem, how a multitude of things can be grouped into one, has also preoccupied scholars for many centuries. This is the problem of *classification*. Historically, arguments over classification did not explicitly contain the dynamical axes of differentiation/convergence that we have seen in the work of Durkheim. The tension between differentiation and convergence in biological classification arose only later, in the context of actual evolutionary theory, as we will see in the following chapter. For now, let us explore the conundrum of classification as it occurred before the late eighteenth century, in two interlocking problems: *how to categorize*, and *how to draw boundaries*. The problem addressed by the crowd theorists was a *bottom-up* problem: how individuals form a group. In contrast, classification of animals and plants was the decidedly *top-down* problem of how (and even if) one should draw boundaries. Crowd theory contained the seeds of the essential tension in the question of how groups can be stabilized by counter-balancing forces operating at different levels. In the top-down problem of classification, it is hard to find any hint of a balance between push and pull – things being drawn into a group and yet held separate by orthogonal forces.[1] Instead, we find only a sharp knife slicing from above at the Great Chain of Being.

The history of the Great Chain of Being is masterfully traced in A. O. Lovejoy's groundbreaking 1936 book of the same name; Lovejoy's work virtually inaugurated the study of the history of ideas. The Chain of Being is, in essence, a picture of the world, conceived as a hierarchy extending up toward the "divine". Lovejoy located the origin of the Great Chain in the creation myth in Plato's dialogue *Timaeus*. Here, the world of imperfect beings is formed from a deity's inner abundance, the overflow

---

[1] It is possible that the tension was easier to discern in crowd theory because the opposing forces of push and pull act at the same scales of time and space, while in problem of (biological) classification they do not. Moreover, an understanding of this uneasy balance of forces in biological classification, and the different scales at which they act, requires a vision of classification as extending in time – which was developed only comparatively recently.

© Springer Science+Business Media B.V. 2018
S. Bahar, *The Essential Tension*, The Frontiers Collection,
DOI 10.1007/978-94-024-1054-9_2

of a "best soul" defined by the attributes of goodness and self-sufficiency. Executing what Lovejoy kindly terms a "bold logical inversion", Plato stated that whatever is good must diffuse itself (*omne bonum est diffusivum sui*, in the Medieval Latin formulation used by Thomas Aquinas and the Scholastic philosophers who followed him).

And thus good overflowed, creating a host of imperfect things. How many imperfect things? All of them. For, as Lovejoy put it, "the 'best soul' could begrudge existence to nothing that could possibly possess it and 'desired that all things should be as like himself as they could be.'" (Lovejoy 1964, p. 50). Since "nothing incomplete is beautiful", the created universe must be complete. All things that are possible must become actual. Lovejoy named this the *principle of plenitude* (*lex completio*). According to this "strange and pregnant theorem", Lovejoy wrote, "no genuine potentiality of being can remain unfulfilled, ... the extent and abundance of creation must be as great as the possibility of existence and commensurate with the productive capacity of a 'perfect' and inexhaustible Source, and ... the world is the better, the more things it contains." (Lovejoy 1964, p. 52)[2]

Unlike Plato, Aristotle did not insist upon a world stuffed to the gills with all things, writing that "it is not necessary that everything that is possible should exist in actuality…it is possible for that which has a potency not to realize it." (Lovejoy 1964, p. 55) Instead, Aristotle emphasized a concept that is a consequence (though not a guarantee) of plenitude: *continuity*. The passage from inanimate to animate, as well as from one class of living creatures to another, is so gradual "that their continuity renders the boundary between them indistinguishable" (Lovejoy 1964, p. 56). Lovejoy called this idea the *principle of continuity* (*lex continui*).

The idea of the Great Chain of Being has clear implications for the problem of biological classification (Fig. 2.1). If one accepts that the chain is continuous, is it even possible to define any segment of the chain as a group in its own right? And if so, how? Any segment is to be composed of individual, infinitesimal links on the chain. But how is one to know where to make the cut? As Socrates suggested[3] in

---

[2] This *filled* world is one of all necessity, and no contingency, even for the deity. This caused religious objections to the principle of plenitude, since it could be interpreted as denying the deity free will. Compare this to the sparsely populated, highly contingent world Thoreau pondered in *Walden* (writing in 1854, five years before Darwin published the *Origin of Species*) when he asked "Why do precisely these objects which we behold make a world? Why has man just these species of animals for his neighbors; as if nothing but a mouse could have filled this crevice?" (Thoreau 1982, p. 273)

[3] The actual quote involves a suggestion to divide things "by classes, where the natural joints are, and not trying to break any part, after the manner of a bad carver". I thank my colleague Dan Lehocky for bringing the original quote to my attention. Lehocky notes out that the idea of an "essential tension" may indeed go back to the pre-Socratic philosophers, as for example in Heraclitus's approach to the one-many problem. Heraclitus addressed the problem of division and classification with his famous metaphor of never being able to step into the same river twice. How can something constantly changing remain the same? The same problem arises in parsing a human being's identity: is the self you were as a child the same as the self you are at this moment? Heraclitus suggested that a solution to the problem of constant change lies in an inherent structure in the world that derives from a balance of opposites: the river is made both by the struggle between

Fig. 2.1 The Great Chain of Being, from a 1579 work by Diego de Valades. Public domain

Plato's *Phaedrus*, one must "cut nature at its joints". But if the *scala naturae* is continuous, are there any joints at which to make a section? This is a broad philosophical problem, but it became an increasingly practical one, as we shall see, during the development of seventeenth- and eighteenth-century natural history. A related problem is whether a classification is an actual "thing": does a species have an essence, in the Platonic sense, or is it simply an arbitrary and convenient name for a grouping of individuals[4]? And is a species defined by the collective properties of its members, or are the members of a species so designated because they partake of some universal characteristics that are part of the species essence? Various thinkers took more or less nuanced positions on these questions. John Locke, for example, declared that "the boundaries of species, whereby men sort them, are made by man". (Lovejoy 1964, p. 229). He did hold that species exist in an essential sense, but also that a true understanding of how to parse the components of nature was inaccessible to human knowledge. "In all the visible corporeal world," Locke wrote,

> there are no chasms or gaps. All quite down from us the descent is by easy steps, and a continued series that in each remove differ very little one from the other. There are fishes that have wings and are not strangers to the airy region; and there are some birds that are inhabitants of the water, whose blood is as cold as fishes…There are animals so near of kin to both birds and beasts that they are in the middle between both. Amphibious animals link the terrestrial and aquatic together…and the animal and vegetable kingdoms are so nearly joined, that if you will take the lowest of one and the highest of the other, there will scarce be perceived any great difference between them. (quoted by Lovejoy 1964, p. 184)

Locke further asserted that this structure was part of the "magnificent harmony of the universe", and that "the species of creatures…by gentle degrees, ascend upwards" toward the "infinite perfection" of the universal architect.

Simply extrapolated to the realm of natural history, the idea of a continuous chain of being could lead to the conclusion that separate species did not, indeed *could* not, exist, either in essence or in practicality, and that, therefore, classification was an enterprise doomed from the start. Thus Buffon, who strongly opposed Linnaean classification, wrote in his *Histoire Naturelle* (1749) that

> [t]here will be found a great number of intermediate species and objects belonging half in one class and half in another. Objects of this sort, to which it is impossible to assign a place, necessarily render it vain to attempt to find a universal system…In general, the more one increases the number of one's divisions in the case of the products of Nature, the nearer one comes to the truth; since in reality only individuals exist in Nature. (quoted by Lovejoy 1964, p. 229; Wilkins 2009, p. 75)

Others echoed the same sentiment, but allowed that the establishment of a tentative system of classification was nonetheless of practical use for the naturalist (and for

---

the flow of the water and the pressure of the banks on either side. This also echoes Anaximander's idea of balance and moderation between elements in the universe. Anaximander represented this as a sort of metaphorical form of cosmic justice, writing that things "make reparation to one another for their injustice according to the ordinance of time", as summer succeeds winter. Note, however, that the principles that exist in an "essential tension", as explored in the present work, are *complementary* rather than *opposing*.

[4]This philosophical position is known as species nominalism.

the physician, looking to distinguish herbal cures from poisons!). Bonnet wrote in his *Contemplation de la Nature* (1764) that "[i]f there are no cleavages in Nature, it is evident that our classifications are not hers. Those which we form are purely nominal, and we should regard them as means relative to our needs and to the limitation of our knowledge." (Lovejoy 1964, p. 231; Wilkins 2009, p. 84).

The application of the laws of continuity and plenitude to nature in the seventeenth and eighteenth centuries had an immediate positive result: it spurred observational research, as naturalists struggled to "fill in the gaps". But those laws also posed two clear philosophical problems. The first arose simply from observations of the world. Some data actually did suggest a lack of continuity, and sharp, uncrossable boundaries between species. Indeed, influenced by such observations, Buffon ultimately abdicated his contention that "individuals alone exist in Nature". Convinced by the sterility of hybrid species, he reversed himself totally, claiming in his later writings that "Les espèces sont les seuls êtres de la Nature" and that "an individual, of whatever species, is nothing in the universe"[5] (Lovejoy 1964, p. 230). The second problem arose from a drive that is perhaps as much aesthetic and psychological as it is scientific: the need to order and classify.

An example from recent biological studies serves to highlight the immense conceptual difficulty of defining an individual. One of the most dramatic examples of the porous definitional boundary between group and individual is provided by *Populus tremuloides*, the quaking aspen (Fig. 2.2). The aspen brings home the complexity of defining an individual, and how knowledge of that individual's history may be key to establishing its identity. The quaking aspen, widespread across the northern and western United States and in Canada, varies dramatically in its life cycle according to environmental conditions. Capable of sexual reproduction, seed viability has been measured at greater than 90% (Mitton and Grant 1996). However, seedlings often have trouble germinating in arid environments. A second survival strategy enables aspens to circumvent this difficulty: they are also capable of asexual reproduction, sending out lateral roots in a process called suckering. These lateral roots then send up stems called *ramets*. As discussed in Chap. 13, these could be considered as the "parts" within the "whole" of a *genet*, defined as the "totality of plant tissue that comes from a fertilized zygote" (Okasha 2006, p. 45).

Sustained by a massive underground root system, aspen clones are able to repopulate a forest after most other species, such as conifers, have been decimated by fire, avalanche, or mudslide. As Mitton and Grant (1996) explain, "[p]ersistent root systems allow aspen to colonize, occupy, and even prefer disturbed sites, justifying their general characterization as a successional species. After a fire has removed the conifers, the ramets that sprout from a healthy, mature root system may grow vertically by as much as a meter in a single summer season…The root system of aspen grows so aggressively that adjacent stems can be spaced more than 30 m apart."

The dominance of a clonal aspen grove within an ecosystem is temporary. As other species grow back into a fire-devastated area, conifer competitors may eventually overtake the aspen clone. Nonetheless, giant aspen forests formed from single

---

[5] This stance also led Buffon to conclude that species are fixed and unchangeable.

Fig. 2.2 A forest of a single tree: Pando, a quaking aspen clone, in autumn. Photograph by J. Zapell. http://www.fs.usda.gov/photogallery/fishlake/home/gallery/?cid=3823&position=Promo,   Public Domain, https://commons.wikimedia.org/w/index.php?curid=27865175

clones do persist. The most dramatic of these is a male clone consisting of nearly 50,000 trunks, dubbed "Pando" (from the Latin, *I spread*) by Grant and colleagues. Growing near Utah's Wasatch Mountain Range, Pando – including its root system – is estimated to weigh six million kilograms, making it the most massive known living organism. Its age is difficult to determine, but an upper limit of a million years has been suggested.

Dishearteningly, aspen clones' need for rejuvenation by fire has set Pando's survival at odds with modern society's desire to prevent forest fires. As Grant (1993) explains,

> The quaking aspen gained its name because of the way the tree's leaves tremble in even the slightest breeze. French Canadian woodsmen in the 1600s believed that the trees quaked in fear because the cross Jesus was crucified on was made of aspen. Now giant aspen clones like Pando have a new reason to tremble: human incursions. Several private homes have recently been built within one section of Pando, and another section has been turned into a campground, complete with parking spaces, picnic tables, and toilets. Paved roads, driveways, and power and water lines built to serve these developments dissect this spectacularly beautiful aspen stand. The presence of people has led the U.S. Forest Service to suppress wildfires, and yet Pando's remarkable size and longevity are largely a consequence of the cleansing, rejuvenating power of wildfires. Ironically, ending wildfires could well mean the end of Pando.

Different aspen clones growing in proximity to one another may be distinguishable simply by their physical characteristics. The angle branches make with a ramet's main trunk, the precise time in the spring that a clone comes out of winter dormancy, and the color of fall foliage are all genetically-determined properties of aspen clones. This does not always make for an easy separation between clones by the layperson, however; Grant and colleagues identified two clones, a male and female, whose root systems were intertwined, and who had given other scientists the mistaken impression that a single clone was switching its sex between seasons (something that is entirely possible in plants, although it turned out to be an incorrect hypothesis in this case).

The most complicated aspects of an aspen grove's life take place underground, within its root system. The root system can redirect nutrients and water to deprived areas of the clone (Tew et al. 1969). Individual ramets secrete hormones called auxins, which suppress the growth of nearby suckers (Wan et al. 2006). The clone thus behaves as a multipart system which protects its constituent parts, and also exhibits competitive behavior between those parts, as when one ramet inhibits the growth of other nearby ramets, thereby ensuring sufficient nutrients and light for itself. These properties, as well as the ability of *Populus tremuloides* to switch between sexual and asexual reproduction strategies, make the quaking aspen a tantalizing problem skirting the edge of multilevel selection theory, as pointed out by Clarke (2011). In discussing the philosophical arguments current even now over whether an aspen clone should truly be regarded as a single individual, she notes that Erasmus Darwin was himself involved in an active debate over whether plant buds could be regarded as "offspring" of the parent plant. "A tree," he concluded in 1800, 9 years before the birth of his grandson Charles, "is therefore a family or swarm of individual plants."

* * *

The human mind's need to impose order and structure on the world, and the degree to which this deep need tells us more about ourselves, or more about nature, is the gateway to a suite of problems we will encounter again and again in the following pages. The need is expressed deeply in the pervasive, but often unconscious, use of metaphor, even in such "rationalist" enterprises as the sciences. Viewing a curve in space that closes upon itself, like the circumference of the moon, one looks for similar structures elsewhere in the world. Then (and this is the great leap into metaphor) one begins to look for similar structures along other dimensions of experience, such as time. Soon the cycles of the day, the year, the seasons are mapped onto circles. We still do this, even in introductory physics courses, when we explain how the cycle of an oscillating spring can be mapped onto the unit circle: all the way stretched at zero degrees, released and passing its equilibrium point at 90, all the way compressed at 180 degrees, passing its equilibrium point, but now in the opposite direction, at 270, and back at its fully stretched position at three hundred and sixty degrees, completing a cycle in time, mapped onto a circle in space. Metaphor can be a blunt instrument for generating insight.

There is clearly something in us that necessitates the *mapping* of relations.[6] There is something innate in the human mind that disposes it toward the projection of perceptions and ideas from one dimension onto another.[7] Diderot speculated on this tendency in his *Letter on the Blind*, where he addressed the problem, first posed by William Molyneux, of whether person born blind, suddenly able again to see, would be able to distinguish a cube and a sphere, placed side by side on a table, without reaching out to touch them (Wilson 1972, p. 98). More recent studies suggest that such "cross-modal" transfer between senses occurs quite naturally. Experimental psychologist Richard Gregory observed this in a study of a blind man, Sidney Bradford, who, upon receiving corneal transplants, was able to tell time visually from an alarm clock. As Gregory recalls, Bradford was accustomed to use "a large watch which had no glass [from which] he told ... time by rapidly touching its hands... So he could see immediately from earlier touch experience." (Gregory 2004). A similar phenomenon, if not in neurological substrate then at least in cross-modal outcome, is synesthesia (Brogaard 2013). Cross-modal transfer of perceptions is perhaps the very origin of metaphor; it may also be an antecedent to the development of conceptual thought itself, in which the juxtaposition of ideas forms a structure that maps onto their ideational relation. The mapping of concepts onto a structured space was well developed in classical "memory palaces", in which ideas would be mentally "placed" on spatial images for ease of recall – a phenomenon borne out by recent neuroimaging studies (Wollen et al. 1972; Bergfeld et al. 1982; Nyberg et al. 2003). Memory palaces, initially used as an aid to rhetoric, were eventually adapted by thinkers such as Giordano Bruno as models for the structure of knowledge itself, of the universe and all it contains (Yates 1966). Some of these were spun into vast circles within circles, but more traditional models took the form of a ladder with increasing "complexity" represented as movement along the linear span, upward toward the heavens. Reimagining Plato's plenitude and Aristotle's continuity as a chain was a natural outgrowth of the innate mental tendency to place ideas into a structure. This structure, however, immediately found itself in conflicting with another innately human habit of mind.

In addition to our fundamental need to map relations, there is a fundamental need to identify things, to tag them with names. In a continuous universe, how could that be done? How could one say that *these* individuals will be grouped together as *one* thing, as a collectivity, but these *other* individuals, so close to them on the continuous chain, will be excluded, and placed in a different group? Evidently, one way to do this is to consider the possibility that nature is not, in fact, continuous. Buffon and others eventually moved closer to that realization, but various interpretations of what constitutes a species were considered before sparseness and contingency were accepted as characteristics of nature. (The problem of dividing a continuous set into groups remains critically important for modern biology even now, as we shall

---

[6] Ultimately, this can reveal causal, or at least historical, relations between the juxtaposed elements, in a process of mapping space into time that we will encounter in the next chapter.

[7] We will return to this idea in the following chapter, in the context of the temporalizing of the chain of being.

explore below: the most fundamental problems of biology lie in the blurred regions where one species truly does shade into another).

Before the ideas of plenitude and continuity had hardened into the idea of a linear chain, Plato had suggested a form of dichotomous classification, which parsed each category into two subcategories. Aristotle relaxed this restriction, arguing that categories could be split into more than two parts. This type of structural division, sometimes referred to as Porphyry's "comb" or "tree", in honor of its post-Aristotelian proponents, breaks substance into incorporeal and corporeal, corporeal into animate and inanimate, animate into insensitive and sensitive, sensitive into irrational and rational, and so on (Wilkins 2009, p. 29). While any type of division or splitting is inherently difficult to reconcile with a continuous chain, a more pressing problem faced by the Scholastic philosophers was whether divisions exist in fact or only in the mind. The latter position, species nominalism, was adopted by William of Occam and others. As J. S. Wilkins points out, thinkers who favored the continuity of the "chain" tended generally toward this nominalist position. Locke, for example, wrote that species are "made for communication [and are] an artifice of the understanding…the mind makes those abstract complex ideas to which specific names are given. And if it be true, as it is, that the mind makes the patterns for sorting and naming of things, I leave it to be considered who makes the boundaries of the sort or the species." (quoted by Wilkins 2009, pp. 63–64).

Not all thinkers were comfortable with the nominalist position, however. While John Ray objected to the idea of extinction because he held that it contradicted the principle of plenitude, he sought a definition of species as something more concrete than simply an artifice of the understanding, writing that "no surer criterion for determining species has occurred to me than the distinguishing features that perpetuate themselves in propagation from seed. Thus, no matter what variations occur in the individuals or the species, if they spring from the seed of one and the same plant, they are accidental variations and not such as to distinguish a species." (quoted by Wilkins 2009, p. 66) This definition foreshadows the current understanding of a "biological species", defined by reproductive isolation. Ray adhered to the principle of continuity, however, when it came to the idea of intermediate species, writing in *Methodus Plantarum* (1682) that "Nature, as the saying goes, makes no jumps, and passes from extreme to extreme only through a mean. She always produces species intermediate between higher and lower types, species of doubtful classification linking one type with another and having something common with both – as for example the so-called zoophytes between plants and animals." (quoted by Wilkins 2009, p. 66). We thus see Ray smoothing over the boundaries he had just drawn.

As the idea of species was further developed in the seventeenth and eighteenth centuries (Fig. 2.3), several distinct themes emerged. The first, already evident in the quotation from John Ray, was an emphasis on *defining species based on reproductive history*. Antoine Laurent de Jussieu, for example, defined species as "the perennial succession of like individuals, successively born by continued generation" (Wilkins 2009, pp. 81–82). In contrast, Michel Adanson rejected a reproduction-based definition of species, in part in order to facilitate the inclusion of asexually-reproducing plants, and also because he wanted to include minerals in his classification scheme.

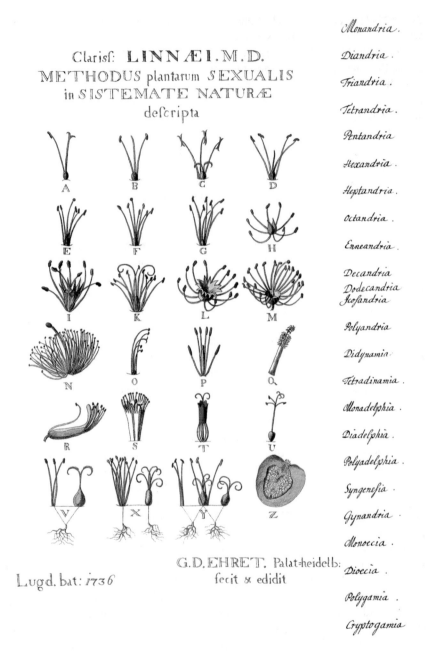

Fig. 2.3 Classification. A page from *Methodus Plantarum Sexualis in Sistemate Naturae Descripta* by G. E. Ehret (Leiden 1736). Public domain

The second theme is a loosening of the idea of arrangement in a linear structure, and the *re-emergence of a branching structure*, which had indeed first appeared in the ideas of Plato and Aristotle, before the hardening of the Great Chain of Being into a predominantly linear form. Some natural philosophers wavered between conceptual schemes. Charles Bonnet, for example, represented nature in a classic ladder scheme in his *Traité d'Insectologie* (1745) He also, however, portrayed nature as branches emanating from a central trunk in a later work, *Contemplation de la Nature* (Foucault 1994, p. 149). Jussieu described the affinities among plants as analogous to those between neighboring regions on a map; he juxtaposed that metaphor, however, with the image of an unbroken chain (Wilkins 2009, p. 82). Other scholars began to consider other structural shapes as well: around 1750, Donati proposed a network model in which every being was "a knot in the web of nature and its resemblance to other forms may be compared to the threads between the knots" (Wilkins 2009, p. 82). Robinet (1763) and Hermann (1783) likewise suggested a three-dimensional lattice model with species situated at the nodes. MacLeay (1819) and Swainson (1834) suggested a model of "osculating circles" in which different groups of species were laid out on adjoining circles. Oken and Kaup suggested pentagram-like models (Wilkins 2009, p. 84) as did Pallas (Foucault 1994, p. 149).

The introduction of models more complex than a simple chain, however, left the problem of continuity unresolved. Lamarck described his branching "Tableau servant a montrer l'origine des différens animaux" as "free from discontinuity, or at least once free from it." (Wilkins 2009, p. 106) However, this branching structure described relations of structure, but not of history: there was no common descent in Lamarck's table, though his model did contain a temporal element in that branches continually arose by spontaneous generation, and then evolved according to a sort of anagenesis. A branching structure did not negate nominalism any more than it did continuity; Lamarck, who had studied with Buffon, held that "all classifications are arbitrary products of thought, and that in nature there are only individuals" (Wilkins 2009, p. 107). Lamarck's phrase "once free from discontinuity" was a reference to extinction. After all, brute observations of nature, even aside from the observations of the infertility of hybrids that made such an impression on Buffon, *did* suggest lack of continuity and plenitude. This led Lamarck to consider the possibility of extinction, as had Adanson before him, though he held that it could only be due to human agency, rather than natural causes.

The possibility of holes in the continuous fabric of nature had a jarring flip side: the notion that new species might arise. While sterile hybrids suggested the idea of uncrossable gaps between species, fertile hybrids told a different story. Studying plants, where fertile hybrids are easier to observe than among animals, Linnaeus saw evidence for the possibility of new species, just as Buffon viewed sterile hybrids as an argument for separation *between* species. Linnaeus ultimately removed the statement that there were no new species from later editions of his *Systema Naturae*, and reportedly crossed out the statement *natura non facit saltum* from his personal copy of his book *Philosophia Botanica* (Wilkins 2009, p. 73).

* * *

Linnaeus was a species nominalist, but wrote that "he was unable to understand anything that [was] not systematically ordered" (Wilkins 2009, p. 71). Classifications are, or may be, arbitrary. But to understand the world, must we order it? This question was central to a radical transformation that occurred in the approach to classification in the seventeenth and eighteenth centuries. Philosopher Michel Foucault argued, in *The Order of Things*, that this transformation went far beyond ladders, branches and osculating circles.

Foucault rejected the standard lens through which the development of natural history is interpreted. The common view asserts that natural history arose when the mechanistic world view of Descartes and Newton failed to explain living creatures. Rather, Foucault wrote, "mechanism from Descartes to d'Alembert and natural history from Tournefort to d'Aubenton were authorized by the same *epistème*." (Foucault 1994, p. 128) The crucial change in perspective, he argued, was from *history* to *natural history*. Of course, we are not using the term "history" to mean the progression of events and interactions through time, the sense of history that transformed the species problem and liberated species into what Foucault called "the irruptive violence of time" (Foucault 1994, p. 132). Rather, this is history in the sense of *historia*, in the sense of tale or story rather than chronological record. Until the mid-seventeenth century, this had encompassed all possible knowledge about an animal, from observations and documents to fables. To a modern eye, such histories could appear as an anarchic jumble. Indeed, Foucault was inspired to begin his "archaeological" investigation of modes of classification by a famous short story by Jorge Luis Borges, which includes a mythical encyclopedia that categorizes animals as follows:

> (a) belonging to the Emperor, (b) embalmed, (c) tame, (d) suckling pigs, (e) sirens, (f) fabulous, (g) stray dogs, (h) included in the present classification, (i) frenzied, (j) innumerable, (k) drawn with a very fine camelhair brush, (l) et cetera, (m) having just broken the water pitcher, (n) that from a long way off look like flies. (Foucault 1994, p. xv).

Foucault identified the historical moment at which authors ceased writing books with titles like "The Admirable History of Plants" and the "History of Serpents and Dragons". In these titles, the word "history" contains "the whole semantic network that connected [the creature in question] to the world." But everything changed, Foucault argued, with Johnson's "Natural History of Plants", published in 1657. "The essential difference," wrote Foucault, "lies in what is missing in Johnson.... The classical age used its ingenuity, if not to see as little as possible, at least to restrict deliberately the area of its experience. Observation, from the seventeenth century onward, is a perceptible knowledge furnished with a series of systematically negative conditions. Hearsay is excluded, that goes without saying; but so are taste and smell, because [of] their lack of certainty and their variability." (Foucault 1994, p. 132)

Everything was stripped away except vision, and vision itself was pared down to structure. Vision presented the naturalist with "screened objects: lines, surfaces, forms, reliefs" (Foucault 1994, p. 133). It was based on vision alone, then, Foucault argued, that Linnaeus elaborated this set of basis vectors for the classification of plants: "[t]he form of the elements, the quantity of those elements, the manner in which they are distributed in space in relation to each other, and the relative magnitude of each element." (Foucault 1994, p. 134).

With nature pried loose from the single axis of the great chain of being, natural historians could now experiment with different means of juxtaposing the beings that make up the world, trying this means, and then that, to obtain a structure that most parsimoniously mirrored that of nature itself (Fig. 2.4). To organize this new structure, to place individuals into species, and species adjacent to one another, naturalists followed one of two possible procedures: the "system", or the "method". In the system, a naturalist would essentially choose a set of basis vectors, as Linnaeus described above, and enumerate the degree of these characteristics in each organism or species under investigation. The problem here, of course, is that the choice of basis vectors is arbitrary, and some characteristics that were important in the description of one species might not be of consequence in the other. Of course, one could simply mentally assign "zeroes" to the characteristics along these "axes" in such a case, but the point was to find a *limited* set of characteristics, and from this perspective a large number of degenerate cases would hardly be helpful. Instead, an alternative approach was proposed: the "method". Here,

> [d]eduction is to be taken … in the sense of subtraction. One begins, as Adanson did in his examination of the plants of Senegal, with a species either arbitrarily chosen or encountered by chance. One describes it in its entirety, leaving out none of its parts and determining all the values that the variables have derived from it. This process is repeated with the next species, also given by the arbitrary nature of representation; the description should be as total as in the first instance, but with the one difference that nothing that has been mentioned in the first description should be repeated in the second. Only the differences are listed. And similarly with the third species in relation to the first two, and so on, indefinitely. So that, at the very end, all the different features of the plants have been listed once, but never more than once. And by arranging the later and progressively more sparse descriptions around the earlier ones, we shall be able to perceive, through the original chaos, the emergence of a general table of relations. (Foucault 1994, p. 142)

Thus everything is described by its relationship to other things. While the method, deriving from a comparison of identities and differences, is uniquely "capable of bringing out vertical relations of subordination", both method and system "are simply two ways of defining identities by means of the general grid of differences." (Foucault 1994, p. 144) These models of nature existed side by side with more traditional "chain-like" characterizations such as that of Bonnet, as well as branching structures, osculating circles, and Linnaeus's system of botanical "calligrams". While models of the world that differed from a simple ladder were not new, the precise and methodological search for such structures, according to defined logical rules and criteria, which would define the relations between natural forms, was a radical departure from previous approaches. These new models coexisted with the questions of continuity and plenitude that had gone hand in hand with the great chain of being; one could just as easily fill in gaps in a continuous plane as on a continuous ladder. However, a multi-dimensional space of possibilities allowed a continuous path to be taken while still leaving some areas unexplored, thus decoupling the *lex continui* from the *lex completio*. The very existence of gaps along

Fig. 2.4 The 2014 sculpture from "Openaries III: Ever open openings/ ever more open openings/ the expanded vessel" by Laura Aldridge and Anna Mayer evokes the arbitrariness, and perhaps the underlying anarchic nature, of classification. Image reproduced with the kind permission of the artists

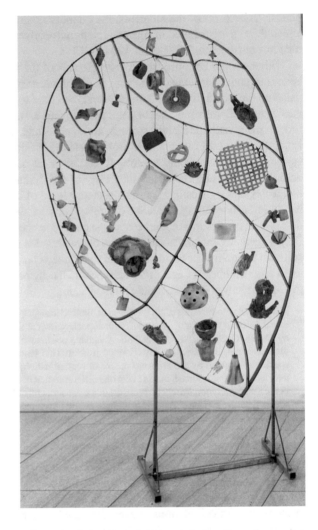

certain directions, of course, also raised the specter of gaps along others, loosening up the soil for a complete abandonment of the *lex continui* altogether. Moreover, the possibility of taking meandering paths through possibility space raised the question of the *means* by which such paths would be traversed. The answer would, of course, be genealogical time. And that realization would at last provide a *mechanism* for the formation of collectivities.

References

Bergfeld VA, Choate LS, Kroll NE (1982) The effect of bizarre imagery on memory as a function of delay: reconfirmation of interaction effect. J Ment Imag 6(1):141–158

Brogaard B (2013) Serotonergic hyperactivity as a potential factor in developmental, acquired and drug-induced synesthesia. Front Hum Neurosci 7:657

Clarke E (2011) Plant individuality and multilevel selection theory. Chapter 11. In: Calcott B, Sterelny K (eds) The Major Tansitions in Evolution Revisited. MIT Press, Cambridge, MA

Foucault M (1994) The Order of Things: An Archaeology of the Human Sciences. Vintage Books (Random House, Inc.), New York

Grant MC (1993) The trembling giant. Discover October issue, pp 83–88

Gregory R (2004) The blind leading the sighted: an eye-opening experience of the wonders of perception. Nature 430:836

Lovejoy AO (1964) The Great Chain of Being: A Study of the History of an Idea. Harvard University Press, Cambridge, MA

Mitton JB, Grant MC (1996) Genetic variation and the natural history of quaking aspen. Bioscience 46(1):25–31

Nyberg L, Sandblom J, Jones S, Neely AS, Petersson KM, Ingvar M, Backman L (2003) Neural correlates of training-related memory improvement in adulthood and aging. Proc Natl Acad Sci U S A 100(23):13728–13733

Okasha S (2006) Evolution and the Levels of Selection. Oxford University Press (Clarendon Press), Oxford

Tew RK, DeByle NV, Schultz JD (1969) Intraclonal root connections among quaking aspen trees. Ecology 50(5):920–921

Thoreau HD (1982) Walden and Other Writings. Bantam Books, New York

Wan X, Landhäusser SM, Lieffers VJ, Zwiazek JJ (2006) Signals controlling root suckering and adventitious shoot formation in aspen (*Populus tremuloides*). Tree Physiol 26(5):681–687

Wilkins JS (2009) Species: A History of the Idea. University of California Press, Berkeley/Los Angeles/London

Wilson AM (1972) Diderot. Oxford University Press, New York

Wollen KA, Weber A, Lowry DH (1972) Bizarreness versus interaction of mental images as determinants of learning. Cogn Psychol 3(3):518–523

Yates FA (1966) The Art of Memory. University of Chicago Press, Chicago

Chapter 3
Time, Just Time: Integrating Up the Great Chain of Being

Mr. Charles Darwin had the balls to ask –

R.E.M.

HOW LOCAL interactions between individuals in a human crowd or in a bee's nest give rise to a group functioning as a single entity – how strong these interactions need to be, whether they are disbursed from a single leader or among equal individuals, whether there is a minimum number of individual organisms needed for the formation of the group – remained unclear, even after the pioneering work of Durkheim. But the problem of collective behavior at least had the advantage of possessing an indisputable underlying mechanism: interaction between individuals within the group. The problem of classification, however, was stymied on that front. Scholars attempting to group organisms in into species had no mechanism by which to organize their classificatory schemes. It is true that classification schemas based on common descent were proposed by Ray and others. But in most cases, common descent was nearly impossible to trace, and all natural historians could do in order to classify plants, for example, was to collect them and order them according to some parsing of their taxonomic properties. The key to classification did ultimately lie in common descent, but applying this concept to the classification of biological species was not potentiated until the discovery of deep geological time and the Darwinian revolution which followed.

*** *** ***

Georges Cuvier (1769–1832) and Étienne Geoffroy Saint-Hilaire (1772–1844) were central figures in European natural philosophy (Figs. 3.1 and 3.2). Widely influential in their own day, their work has been subsequently exploited to promote later intellectual agendas. Their legendary debates held at the Académie des Sciences in Paris (in 1830, while a revolution was about to explode on the streets outside), resulted in a post hoc perceived victory for Cuvier, whose ideas were interpreted as an early form of "adaptationism", and thus sometimes misread as a direct precursor to Darwinian selection (Figs. 3.3 and 3.4). Cuvier's idea of the adaptation of species to their particular "conditions of existence", however, precluded any idea of

© Springer Science+Business Media B.V. 2018
S. Bahar, *The Essential Tension*, The Frontiers Collection,
DOI 10.1007/978-94-024-1054-9_3

Fig. 3.1 Georges Cuvier holding a fish fossil. Credit: Wellcome Library, London. Wellcome Images images@wellcome.ac.uk http://wellcomeimages.org (L0016365, Portrait of Baron Georges Cuvier) (Copyrighted work available under Creative Commons Attribution only licence CC BY 4.0, http://creativecommons.org/licenses/by/4.0/)

evolution; he held that new organisms arose de novo in the aftermath of catastrophic geological events.

Cuvier's idea of geological catastrophes formed one conceptual axis of the catastrophism vs. uniformitarianism dichotomy. Along these axes, Cuvier was interpreted (primarily by Charles Lyell, whose first volume of the *Principles of Geology* appeared in the same year as the Cuvier-Geoffroy debate) as highly unscientific, in contrast to Lyell's own uniformitarianism, which allowed only currently-observable forces as possible agents of change in geology and in nature.

Cuvier and Geoffroy approached the same problem – the use of the structure of living organisms as a means of classification – from fundamentally divergent perspectives. Both men were driven by the desire to find a more "natural" classification for plants and animals than the Linnaean system (Appel 1987, p. 12). But they took radically different approaches. Geoffroy was attracted to the idea of inherent form and structure in nature, writing in 1795 that "[i]t seems that nature has enclosed herself within certain limits, and has formed all living beings on only one unique plan … All the most essential differences which affect each family within the same class come only from another arrangement, in short, from a modification of these same organs" (Appel 1987, p. 28). Both Cuvier's catastrophism and Geoffroy's

Fig. 3.2 Étienne Geoffroy
Saint-Hilaire. Public
domain

emphasis on form rather than function were to be revisited by modern evolutionary theory.

In contrast to Geoffroy's emphasis on form and structure, Cuvier held such grand, sweeping interpretations in great suspicion; he came from a more conservative political background, and "his belief that ideas were at the root of the Revolution led him to fear the consequences of unbridled generalizations in the sciences, as well as in political and social thought" (Appel 1987, p. 30). Nevertheless, influenced by Jussieu and other colleagues, Cuvier did come to believe that a "natural" taxonomic classification of the entire living world could be achieved. He proposed grouping creatures based on their functional systems rather than their external similarities of form. What, he asked, is most important for any animal? The internal systems that allow for generation, respiration, circulation, and other functions.[1]

In Foucault's interpretation, Cuvier's focus on the functional needs of each animal in navigating its environment was in itself a radical shift of perspective. It was the first time, Foucault contended, that organisms were truly viewed as wholes unto

---

[1] Cuvier did not at first include the nervous system among these critical processes, but he added it to the list in 1812 and eventually came to see it as dominant over the other systems (Appel 1987, pp. 43–46).

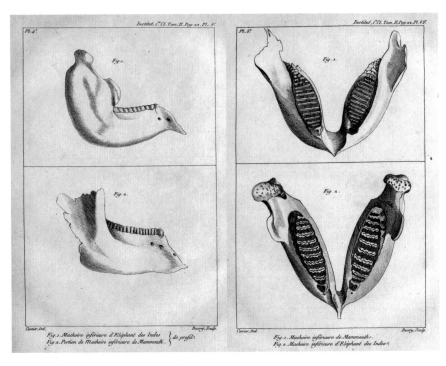

Fig. 3.3 Comparison of jawbones of the Indian elephant and the mammoth, from a paper by Cuvier published in 1799 (the year VII, according to the revolutionary calendar). Public domain

themselves, the first time in centuries that the axis of classification had passed through a living being. Before Cuvier, "the plant and animal [were] seen not so much in their organic unity as by the visible patterning of their organs. They are paws and hoofs, flowers and fruit, before being respiratory systems or internal liquids. Natural history traverses an area of visible, simultaneous, concomitant variables, without any internal relation of subordination or organization" (Foucault 1994, p. 137). This is why, Foucault suggested, botany had enjoyed an "epistemological precedence" in the eighteenth century – it was easier to classify plants, rather than animals, according to such patterning. After Cuvier, "the fundamental arrangement of the visible and the expressible [once again] passed through the thickness of the body" (Foucault 1994, p. 137). In this interpretation, the radical shift precipitated by Cuvier cannot be underestimated, and was, perhaps ironically for a man who distrusted revolutions, profoundly revolutionary. Foucault imagined Cuvier's "iconoclastic gesture" in dramatic fashion.

> One day, toward the end of the eighteenth century, Cuvier was to topple the glass jars of the Museum, smash them open and dissect all the forms of animal visibility that the Classical age had preserved in them. This iconoclastic gesture … was the beginning of what, by substituting anatomy for classification, organism for structure, internal subordination for visible character, the series for tabulation, was to make possible the precipitation into the old flat world of animals and plants, engraved in black on white, a whole profound mass of time … (Foucault 1994 pp. 137–138)

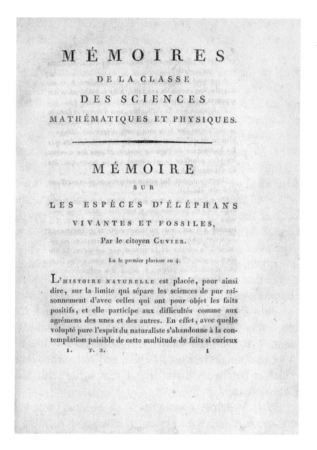

Going even further, Foucault made the radical claim that the idea of life itself, as
interpreted by the last two centuries of science, originated with Cuvier's classificatory emphasis on the functional systems necessary for survival. Foucault suggested
that this marked the beginning of a transition from *natural history* to *biology*. Prior
to this shift, "[i]f biology was unknown, there was a very simple reason for it: that
life itself did not exist. All that existed was living beings, which were viewed
through a grid of knowledge constituted by natural history" (Foucault 1994,
pp. 127–128).

   In Foucault's view, then, Cuvier stands as the founder of modern biology. But
this interpretation must be read with caution, for if Cuvier *made possible* the "pre-
cipitation of a whole profound mass of time" into the "old flat world of animals and
plants", it was not he who actually dropped the seed crystal into the graduated cylinder. Cuvier may have pictured a series more than a table, but so did Lamarck, and
so did others who interpreted the natural world according to the great chain of being.
Cuvier himself viewed the great chain as his *bête noire*, writing that "[t]he pretended

scale of beings is an erroneous application of partial observations to the totality of creation ... and this application has, in my opinion, harmed to a degree that can scarcely be imagined the progress of natural history" (Appel 1987, p. 51). Along with the great chain, Cuvier also dismissed the *lex continui*. As Appel writes, "from 1800 he began to reject all attempts to find graded nuances, and to insist on gaps in nature. This change of heart may be connected with the appearance in the same year of Lamarck's theory of evolution[2]... Perhaps Cuvier came to see that the assumption of continuity in nature's productions left the way open to evolutionary speculation" (Appel 1987, p. 50). Cuvier's view of nature omitted all possibility of evolution. For him, adaptation was a static concept, with organisms exhibiting functions ideal for their current conditions of existence; when Cuvier admitted the possibility of extinction, he used this as an argument against evolution: organisms died because they could not adapt to changing conditions (Gould 2002, p. 176). Another central theme in Cuvier's work the "correlation of parts": only certain mutually dependent combinations of organs and structures could coexist in an animal; change one organ, and the whole might no longer be sustainable. Cuvier argued from the correlation of parts to the unviability of intermediate forms, thus providing a simultaneous argument against both the great chain *and* evolution. It also provided society with the image of the anatomist who could identify an animal simply by inspecting a single one of its bones. As Cuvier himself wrote, "[t]oday comparative anatomy has reached such a point of perfection that, after inspecting a single bone, one can often determine the class, and sometimes even the genus, of the animal to which it belonged, especially if that bone belonged to the head or the limbs" (Rudwick 1997, p. 36).

Cuvier's actual classification of animals was based on four categories, or *embranchements*: vertebrates, articulates, mollusks, and radiates. No gradations were possible within each *embranchement*, and there were no links between the four. For his part, Geoffroy focused on structural homologies between species (though he used the term analogies, which has a different meaning in biology

---

[2] It is important to emphasize (see Gould 2002, pp. 183–186) that Lamarck's evolutionary theory was far subtler than the clichéd view most people know from textbooks, with the giraffe lengthening its neck to reach leaves on high branches. While Lamarck did argue for the inheritance of acquired characteristics, an inherent tendency toward increased complexity or perfection – which resulted in series of organisms moving upward along the chain of being throughout time – was an even more important element of his theory. He also held that new "simple" organisms at the base of the chain would arise by spontaneous generation, resulting in organisms mounting upward along the chain in a series of continually renewed "escalators", some of which were farther advanced than others. Lamarck's attempts to classify species, however, led him to conclude that the chain, such as it was, exhibited several branches, whose divergence was caused by a combination of environmental factors and the inheritance of acquired characteristics. Gould presents Lamarck's view as envisioning a hierarchy of causal factors, but one can also interpret it as a tension between a "pulling apart" (branching) and a drawing together (tendency toward increased complexity or perfection). Interestingly, Lamarck also considered a biologically null model, in which the environment remained constant, presaging twentieth and twenty-first century studies of genetic drift and neutral theory in ecology.

today[3]). In contrast to Cuvier's adaptationist perspective, Geoffroy held that an animal's structure determines its mode of life, rather than vice versa. This focus on form and structure led Geoffroy to virtually found the field of comparative anatomy by performing the first bone-for-bone comparisons between various species of vertebrates (others later extended this approach of "philosophical anatomy" to invertebrates). This led to "the disquieting possibility that homologous bones in different animals could be modified to perform entirely disparate functions" (Appel 1987, p. 85). The evolutionary implications of this view appalled Cuvier. Moreover, in allowing the "radical thesis that [a] bone changed its function as it passed from one class [of animals] to another ... an abstract element of organization which could serve multiple functions as it was placed in different circumstances" (Appel 1987, p. 87), Geoffroy opened the door to historical contingency in the development of animal forms. This view is fundamentally history-based in a way that Cuvier's was not; one can easily see Geoffroy's appeal for later scientists like Stephen Jay Gould, who placed a high value on the contingencies of history and the co-opting of "spandrel" structures for different uses over evolutionary timescales, a topic we will explore in much more depth in later chapters. In practice, Geoffroy focused not only on bone-for-bone homologies, but also on the connections between bones. He was essentially mapping *topology*, the juxtaposition of structures.

The clash between Cuvier and Geoffroy that erupted in 1830 had been brewing for years. A decade before, Geoffroy had suggested that the common plan he had identified in vertebrate structures could be applied to insects as well. This threatened the boundaries between Cuvier's *embranchements*. In an 1820 paper, titled "On Insects Reduced to the Embranchement of Vertebrates", Geoffroy mapped the insect exoskeleton piece by piece onto the vertebrate skeleton, going so far as to suggest that insect appendages were homologous to vertebrate ribs. The two scientists had developed clusters of younger followers, and fought over them, with Cuvier often winning out due to his greater access to academic patronage. After years of sparring, both intellectually and via professional politics, in 1828 Cuvier published a direct attack on one of the specific structural homologies proposed by Geoffroy: the homology between the operculum in fishes and the mammalian middle ear. Two years later, things finally came to a head when two young naturalists, Pierre-Stanislas Meyranx and a colleague of his named Laurencet,[4] attempted to join the molluscan and vertebrate *embranchements*. Geoffroy spun this paper to his advantage, reading a report on Meyranx and Laurencet's paper to the Académie in February 1830, in which he sarcastically quoted a statement by Cuvier from 1817 to the effect that cephalopods were "not a transition to anything" (Appel 1987, p. 146). Cuvier rose and objected. Meyranx later wrote to Cuvier to apologize.

---

[3] The term homology was first used in this sense by Richard Owen in 1843; following Owen's change in usage, homology is still used today essentially to mean what Geoffroy called an "analogy". For that reason, I use the "modern" term here in the description of Geoffroy's ideas. See Gould (2002, pp. 1070ff), for a discussion of this lexical shift.

[4] According to Appel, Laurencet's first name has been lost to history, as has been the original paper by Laurencet and Meyranx (Appel 1987, p. 145).

When the Académie met the following week, Cuvier tore Meyranx and Laurencet's work to pieces, and showed up Geoffroy's lack of knowledge of cephalopods. On March 1, Geoffroy presented his rebuttal but, when pressed by Cuvier that day, and when the Académie reconvened later that month, had difficulty precisely defining his broad theories of "unity of composition" and "unity of plan". Cuvier scoffed, "How can one discuss a question when one cannot pose the terms of it?" (Appel 1987, p. 150). He further tried to show the topological impossibility of some of the structural transformations Geoffroy had proposed between species. Speaking of the change in position of the hyoid bone between species, Cuvier quipped that "[w]ithout a doubt, this is a somersault possible to conceive in a skeleton whose bones are held together only by brass wires ... But I ask of anyone who has the slightest idea of anatomy: is it admissible when one considers all the muscles, all the bones, all the nerves, and all the vessels which are attached to the hyoid bone? It would necessitate – But I stop myself! The very idea would frighten the imagination!" (quoted by Appel 1987, p. 151).

By March 29, the debate had descended into a wrangle over who should speak first. Both Cuvier and Geoffroy tried to return the argument to broad questions. "It is a question of philosophy that divides us," said Geoffroy. Cuvier, for his part, asked what "necessity would have constrained [nature] to employ only the same pieces, and employ them always?" (Appel 1987, p. 154). It was a good question indeed, and Geoffroy had no immediate answer.

While Cuvier came gradually to be seen as the victor in the debate, some initial public reaction in France favored Geoffroy. He was seen as "lead[ing]... a progressive and synthetic school of natural history which was displacing the conservative and outmoded school of Cuvier" (Appel 1987, p. 155). Newspapers sympathetic to the 1830 revolution sided with Geoffroy. Goethe – naturalist[5] as well as poet – wrote a commentary supporting Geoffroy, originally published in German and then later in French translation. The debate was even immortalized in two novels by Balzac, *Louis Lambert* (1835), and *Un Grand Homme de Province a Paris* (1839), with a character named Meyraux standing in for Meyranx.

The gradual scientific eclipse of Geoffroy was driven by a number of factors. Geoffroy's desire to build sweeping systems led him in directions that his colleagues considered outlandish, and he grew alienated from the scientific establishment; his former supporters downplayed his emphasis on evolution; his work was seen as too close to the romantic German school of Naturphilosophiren, from which some French naturalists, including Geoffroy's son, strove to distance themselves. Perhaps most significant, Cuvier's teleological adaptationism was strongly favored by British naturalists raised on William Paley's "Natural Theology" and influenced by William Whewell's 1837 *History of the Inductive Sciences*, which argued that teleological arguments were essential in natural history.

---

[5] Goethe was known in particular for a theory of the archetypal structure of plants (see Dornelas and Dornelas 2005, as well as pp. 283–286 in Gould 2002 for a brief review of his work in plant physiology).

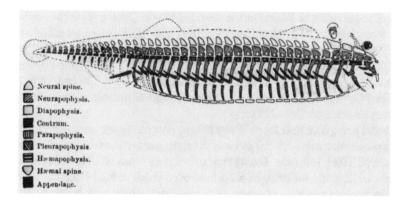

Neural spine.
Neurapophysis.
Diapophysis.
Centrum.
Parapophysis.
Pleurapophysis
Hæmapophysis.
Hæmal spine.
Appendage.

Fig. 3.5 Richard Owen's 1847 conception of an archetypal vertebrate structure. Public domain

Despite the decline in Geoffroy's popularity, the idea of homologies still hovered in the wings of science, however, and Whewell ultimately moderated his position in response to Richard Owen's work on "archetypal" structures in vertebrates (Fig. 3.5). As Appel explains, Owen realized that "[m]any homologies could not be accounted for solely on the basis of functional requirements. One might argue, for example, that the bones of the human cranium were separate in infants and later fused together because parturition was thereby facilitated. But why then were the same bones separated in a bird which picked its way out of a shell, or in a marsupial which was only a fraction of an inch long when born?" (Appel 1987, p. 227). Owen wrote in 1847 that "[t]hese and a hundred such facts force upon the contemplative anatomist the inadequacy of the teleological hypothesis to account for the acknowledged concordances expressed in this work [Owen's] by the term 'special homology'" (quoted by Appel 1987, pp. 227–228). Despite these insights, however, Owen was not an evolutionist, and agreed with Cuvier's strictly separated *embranchements*. However, by this time, Charles Darwin had returned from his voyage on the Beagle. Everything was about to change.

*** *** ***

During his debates with Geoffroy, Cuvier had asked what "necessity would have constrained [nature] to employ only the same pieces, and employ them always?" The answer to this question is nothing less than history itself. In proposing this answer to Cuvier's question, Darwin laid bare the *mechanism* by which organisms could be grouped into species, and by which those species could be grouped in relation to one another. Of course, Darwin also proposed another mechanism, natural selection, by which species evolve from one another. This balance between selection endlessly drawing lineages apart, while history binds them together, is another instance of the essential tension that is the subject of the present book; we have seen already how Durkheim used a similar balance of opposing forces to explain the origin of the division of labor. Darwin described the balancing principles at the core of his own great theory as "great laws", writing that "[i]t is generally acknowledged that all organic beings have been formed on two great laws – Unity of Type, and the Conditions of Existence ... On my theory, Unity of Type is explained by unity of descent. The expression of conditions of existence, so often insisted upon by the illustrious Cuvier, is fully embraced by the principle of natural selection" (Darwin, *Origin of Species*, pp. 261–262).

In order to explore how Darwin was able to provide an answer to Cuvier's question, however, one more crucial piece of the story needs to be told. What is essential for history? Time, just time. Darwin's revolutionary ideas would have been inconceivable without the substrate of deep time over which natural selection could act. We must now turn to the discovery of "deep time", which took the great chain of being and rotated it sideways until, instead of reaching up to heaven, it lay stretched across the earth.

As powerfully recounted by various historians of science, not least by Stephen Jay Gould in his masterful *Time's Arrow, Time's Cycle*, the discovery of the age of the Earth was deeply entwined with a growing understanding that the Earth, its structure, and the life it contains, have not just an age, but a history: that the planet is not a static picture, but an unfolding story. Thus the discovery of the Earth's age was crucial for the development of evolutionary theory on multiple fronts: it provided a field of time across which natural selection would work, and it strengthened a metaphorical picture of life as something unfolding, and therefore capable of change and development.

Early scholars of geology focused on the mechanism by which changes arose in the structure of the planet. Thomas Burnet developed a theory of the Earth's history that attempted to explain biblical processes using currently observable geological processes. Burnet was influenced by the contemporary Newtonian idea that the same laws of nature must obtain on Earth, among the planets, and throughout the rest of the universe. He expressed this view in terms which presage the uniformitarian theory of Charles Lyell, in the following discussion of why natural causes for geological events should be more impressive than "miraculous" causes imposed from above. "If one should contrive," Burnet wrote, "a piece of clock-work so that it should beat all the hours, and make all its motions regularly for such a time, and that time being come, upon a signal given, or a spring touched, it should of its own accord fall all to pieces; would not this be looked upon as a piece of greater art, than if the workman came at that time prefixed, and with a great hammer beat it into pieces?" (Gould 1987, pp. 29–30).

Burnet explored theoretical models for the forces changing the Earth's structure during hypothesized periods of heating and cooling, postulated a mechanism for a universal flood, and used sounding to calculate the total volume of water in the oceans. He argued with Newton over the origin of the current structure of the Earth's surface, contending that its unevenness was a result of the great flood, while Newton held that the Earth's rugged landscape could simply have arisen as the Earth cooled following its formation from an initial chaos; Newton pointed out that even uniform liquids can form irregular solids as they cool (Westfall 1980, p. 391). Assuming initial perfection of structure, both Newton and Burnet interpreted the Earth's current jagged brokenness as a sign of its *history*. As Gould points out (1987, pp. 43ff), this foreshadows the role of "imperfections" in nature as markers of history, a theme of critical importance to Darwin and others, as we shall soon see.

Nicolaus Steno, in his 1669 book *Prodromus*, analyzed the processes driving the breakup of the Earth's crust. In contrast to Burnet, Steno proposed a mechanism for the deposition of new sediment, thus enabling a cyclic process of deposition and collapse. As Gould emphasizes, Steno's system encompassed a series of cyclic events: vacuities created in sediment below the Earth's surface due to shifts in the Earth and movement of underground water, collapse of the crust, deposition of a new layer of sediment on top of the collapsed region, followed by the formation of new vacuities and another collapse. This process followed a conceptual spiral, however, rather than a circle: successive rounds of collapse led to an increasing ruggedness of the Earth's surface; thus, Steno's model contains both an irreversible process ("time's arrow") well as a cyclical one ("time's cycle").[6]

James Hutton made a crucial leap forward in his *Theory of the Earth*, published in 1795 (and presented in a more user-friendly writing style by Hutton's friend John Playfair in 1802). His work has been described as the discovery of "deep time". His theory of geological processes extended the ideas of Steno and others by providing a mechanism for uplift of geological structures rather than just deposition of sedimentary layers. His method of identifying such structures was based on the identification of so-called *unconformities* – groups of layers that existed at a tilted angle, or even orthogonal, to other layers (Fig. 3.6). He identified a process by which such unconformities could arise: upper horizontal layers could form by deposition in an ocean basin, after the underlying tilted layers – which had previously been formed by a similar process – had undergone breakdown, uplift, erosion, and tilt. The observation of many such unconformities in the geological record suggested to Hutton that the process had, in the famous phrase, "no vestige of a beginning, no prospect of an end". In his model, the increasing weight of horizontal strata, formed by deposition, generates enormous heat and pressure (which induced, among other things, the formation of igneous rocks; Hutton identified many rocks that had previously been thought to be formed from sedimentation as igneous in origin). The heat generated by this process would eventually build up and generate sudden expansions and shifts, providing a mechanism of uplift (so that the horizontal layers formed in an ocean basin might emerge above sea level). These uplifted strata would undergo decay and erosion, and the products of this decay would eventually be washed into the oceans, providing a source of new, horizontal, sedimentary strata. In short, it was "a self-renewing world machine" where "[e]ach stage automatically entails the next" (Gould 1987, pp. 65–66).[7]

---

[6] Burnet's system also contains these elements of arrow and cycle, as Gould discusses in depth. Gould's theme in *Time's Arrow, Time's Cycle* is how these two structural motifs have driven (and have been misrepresented in) the history of geology.

[7] Cuvier was strongly opposed to Hutton's arguments, considering them an overly broad "system" in the same category as Geoffroy's formalist taxonomy or the great chain of being.

Fig. 3.6 An unconformity identified by James Hutton near Edinburgh, shown in a 1787 engraving by John Clerk. Public domain

Charles Lyell began his great *Principles of Geology* (1830) with a historical sketch of the progress of the field. As Gould carefully dissects in *Time's Arrow, Time's Cycle*, Lyell presented Hutton's ideas according to his own interpretation, and in many respects the textbook "Hutton" is really Lyell's version of Hutton.[8] Using Hutton's discovery of "deep time", Lyell (Fig. 3.7) developed not only an interpretation of the Earth's geological history, but also a related methodological approach to the subject. He advocated *uniformitarianism*,[9] according to which current geological observations can be ascribed to processes now acting and observable on Earth. Note the conceptual link between this idea and those of Burnet and Newton discussed above; with Lyell, however, these principles applied to a world existing (and undergoing cycles of decay, deposition and uplift) in a limitless span of time.

Lyell associated outdated and unscientific theories of a "young Earth" with hypotheses of geological change driven by large, catastrophic events unlike anything we observe in daily life. With an ancient Earth, there is no need for catastrophes,

---

[8] For example, Lyell presented Hutton as an empiricist champion of fieldwork. In fact, Hutton, though a gentleman farmer for a decade and a half, published the first version of his theory (1785) before seeing a single unconformity. Far from deducing a theory from observations, Hutton wrote that his "theory [was] confirmed from observations made on purpose to elucidate the subject" (Gould 1987, p. 72). Gould also points out that Hutton's system was entirely cyclical, and contained no "arrow" metaphor, no vector of change. "The discoverer of deep time", Gould acidly remarks, "denied history."

[9] The name was actually bestowed by Whewell, in an 1832 review of Lyell's work.

Fig. 3.7 Charles Lyell (1797–1875). Public domain

he argued. Crucially, Lyell identified the scientific approach of rational empiricism with the hypothesis that catastrophic events could *never* take place. He mocked Cuvier for his catastrophism,[10] and presented uniformitarianism as an empirical fruit of the Enlightenment, writing that catastrophists

> felt themselves at liberty to indulge their imaginations, in guessing at what might be, rather than inquiring what is … [preferring to] speculate on the possibilities of the past, [rather] than patiently to explore the realities of the present … Never was there a dogma more calculated to foster indolence, and to blunt the keen edge of curiosity, than this assumption of the discordance between the former and the existing causes of change. (quoted by Gould 1987, p. 111)

Lyell thus accused his intellectual opponents of "a desire … to cut, rather than patiently to untie, the Gordian knot". (It would bear noting here that Lyell seems to be missing the whole point of the Gordian knot story …).

Lyell made use of four distinct meanings of uniformity in his work, and blurred the lines between them.[11] Two of these meanings, however, relate to *methodological techniques of doing science*, and two relate to *substantive claims about the world*.

---

[10] There is a true irony in this, since Lyell's criticism of Cuvier is based on the latter's perceived abdication of empiricism; Cuvier, of course, attempted to dismember Geoffroy's theories using the same tactic.

[11] This argument was originally presented in Gould's *Time's Arrow, Time's Cycle* (Chap. 4). What follows in this paragraph is an outline of Gould's analysis of Lyell's rhetorical tactics.

First, Lyell began with the assumption of *uniformity of law*, as did Burnet, Newton and many others. The assumption that nature follows the same laws at different locations in time and space is essential in order to apply inductive reasoning at all. Lyell also assumed *uniformity of process*: the idea that past phenomena are explicable using currently observable processes. (This could be interpreted simply as an application of the uniformity of law along the axis of time). These were Lyell's two methodological claims. He followed them with two substantive claims, which he made essentially inviolable by conflation with the methodological ones: to doubt the substantive claims would appear to be unscientific. Lyell's postulate of gradualism held that geological processes occurred with *uniformity of rate*: major events (which he did not consider "catastrophic") such as floods, volcanic eruption, and cycles of erosion and uplift, occur with the same frequency at all times in Earth's history. Lastly, Lyell concluded that the natural world exhibited *uniformity of state*. This meant no mass extinction, no progression up the great chain toward increasing complexity. For a large portion of his career, Lyell held that uniformity of state meant that there could be no new species, and no evolutionary process at all. On this final point, Lyell was eventually brought round by his friend Darwin, and the later editions of *Principles of Geology* reflect this change of mind (Fig. 3.8). This forced him to resign one of his initial uniformitarian claims: uniformity of state. But the slow, gradual change driven by natural selection did allow the retention of Lyell's other mechanistic claim: uniformity of rate. He would have been most uncomfortable with punctuated equilibrium.

*** *** ***

The work of Lyell and others on the depth of geological time, as well as the exploration of new styles of classification which introduced the ideas of the organism's essential functions rather than its separate parts as an axis of classification (Cuvier) and the explanation of homologous forms by common descent (Geoffroy), were among the many intellectual developments which loosened up the soil for Darwin's great contribution. But a gradual philosophical shift, described by Lovejoy as the "temporalizing" of the great chain of being, was part of the groundwork as well. This idea was touched on at the end of the previous chapter, and now we must explore it in more detail. Only a century before Darwin published the *Origin of Species*, the Abbé Pluche had written – with the same certainty that Lord Kelvin used to declare physics dead circa 1895 – that

> nothing more…will be produced in all the ages to follow. All the philosophers have deliberated and come to agreement on this point. Consult the evidence of experience; elements always the same, species that never vary, seeds and germs prepared in advance for the perpetuation of everything…so that one can say, *Nothing new under the sun*, no new production, no species which has not been since the beginning. (quoted by Lovejoy 1964, p. 243)

However, scholars were increasingly coming to view the great chain "not as the inventory but as the program of nature" (Lovejoy 1964, p. 244). In Lovejoy's analysis, this shift was driven by various theoretical problems with the principle of plenitude. One set of difficulties derived from moral considerations. Plenitude

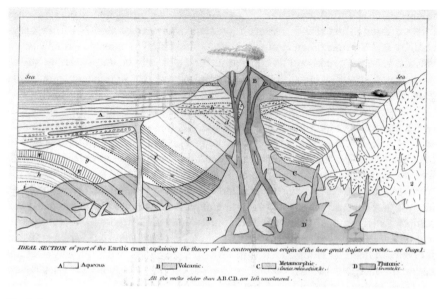

IDEAL SECTION of part of the Earth's crust explaining the theory of the contemporaneous origin of the four great classes of rocks... see Chap.1.

A ☐ Aqueous.   B ☐ Volcanic.   C ☐ Metamorphic. /Gneiss, mica-schist,&c./   D ☐ Plutonic. /Granite,&c./

All the rocks older than A,B,C,D, are left uncoloured.

Fig. 3.8 The frontispiece of an 1857 edition of Lyell's *Principles of Geology*. Public domain

left no room for hope, as Voltaire pointed out in his criticism of Leibniz's "best of all possible worlds" in the aftermath of the Lisbon earthquake of 1755. If the world is stuffed full, how can any being's happiness (or the proximity of a species to "perfection" at the top of the chain) be advanced without a corresponding decrement in another's? There appeared to be no way for any being to advance toward perfection unless some other being were displaced, if one accepted that "each degree of possible difference can have only one representative at a time", as Edmund Law argued (Lovejoy 1964, pp. 245–246). In short, a static chain of being reaching upward, and a world stuffed full of all possible things, seemed to allow no possibility of progress. The great chain had begun to sway in the breezes of the Enlightenment.

A related moral difficulty that further rattled the chain's foundations was the psychological theory that pleasure arises from change. Even Leibniz himself held that

the very law of enjoyment [requires] that pleasure does not have an even tenor, for this begets loathing, and makes us dull, not happy... Our happiness will never consist, and ought not to consist, in a full enjoyment, in which there is nothing more to desire, and which would make our minds dull, but in a perpetual progress to new pleasures and new perfections. (quoted by Lovejoy 1964, pp. 248–250)

Edmund Law echoed these sentiments, writing that "[a] finite being fixed in the same state, however excellent, must according to our conceptions (if we be allowed to judge from our present faculties, and we can judge from nothing else) contract a kind of indolence or insensibility...which nothing but alteration and variety can cure" (quoted by Lovejoy 1964, p. 249). This idea goes back to Renaissance thinkers such as Bruno, and before him, to some Greek philosophers; it also stretches forward through nineteenth century romanticism to current models of neuropsychology and the phenomenon of neural accommodation to repeated stimuli. But it was the renewed focus on this concept in the eighteenth century, even by strong advocates of the great chain of being such as Leibniz, that significantly eroded the great chain of being as a pillar of philosophy.

The second theoretical problem that the great chain encountered was plain evidence against both continuity and plenitude. Despite eighteenth century scholars such as Jean-Baptiste Robinet,[12] who wrote that he had "formed so vast an idea of the work of the Creator that from the fact that a thing can exist I infer readily enough that it does exist" (quoted by Lovejoy 1964, p. 272), and held that division of the world into categories such as organic and inorganic violated the law of continuity, many others focused their attention on the mounting evidence that gaps clearly did actually exist.[13] Philosophers such as Voltaire, Samuel Johnson, and Diderot's friend and fellow Encyclopedist the Baron d'Holbach directly challenged the existence of *lex continui* itself. D'Holbach wrote in 1780 that "[o]f those who ask, why does not nature produce new beings, we inquire in turn how they know that she does not do so. What authorizes them to believe this sterility in nature?" (quoted by Lovejoy 1964, p. 269).

Samuel Johnson was equally dismissive. Attacking the chain of being with empirical arguments (an infinite number of creatures simply does not exist) and using Zeno's paradox to argue for the necessity of gaps, Johnson concluded that "this Scale of Being I have demonstrated to be raised by presumptuous Imagination, to rest on Nothing at the Bottom, to lean on Nothing at the Top, and to have Vacuities from step to step through which any Order of Being may sink into Nihility without inconvenience, so far as we can Judge, to the next Rank above or below it" (quoted by Lovejoy 1964, p. 254).

Leibniz pondered whether other gradations in the broken continuity might be found on other planets or stars, and Maupertuis suggested that breaks in the observed continuity might have been caused by the collision of a comet with the Earth (Lovejoy 1964, p. 255). Compounding the moral and factual discordances of the static great chain was the very fact that the world does unfold in time. Clearly, "something had to be done to fit the postulate of the necessary complete realization of all possibles with the fact that the concrete world *is* temporal" (Lovejoy 1964, p. 255). Gradually, even Leibniz – despite the contradiction with his philosophical stance regarding plenitude and the "best of all possible worlds" – began to flirt with the idea that plenitude may unfold in time rather than being realized all at once. This unfolding, Leibniz wrote, will continue without end, and will never result in a static perfection:

---

[12] Robinet was later mocked by many in the scientific community for his belief in mermaids and mermen.

[13] Lovejoy (1964, p. 276) also raised the question of whether emergent properties are possible in a world defined by continuity and plenitude: does emergence imply discontinuity? (Note his early use of the contemporary buzzphrase "emergent property"!)

...a perpetual and unrestricted progress of the universe as a whole must be recognized...As for the objection which may be raised, that if this is true the world will at some time already have become paradise, the answer is not far to seek: even though many substances shall have attained to a great deal of perfection, there will always, on account of the infinite divisibility of the continuum, remain over in the abyss of things parts hitherto dormant, to be aroused and raised to a greater and higher condition...And for this reason progress will never come to an end. (quoted by Lovejoy 1964, p. 257)

The idea of progress and advancement – essentially, a form of evolution *avant la lettre* – was a dominant theme in eighteenth century literature. Edward Young wrote of stellar evolution in his 1740s poem cycle *Night Thoughts*, concluding that "Nature delights in progress."[14] In a description that echoes Lamarck's continuously progressing evolutionary "escalator", the physician-poet Mark Akenside wrote that "[i]nferior orders in succession rise / To fill the void below." Lovejoy notes that this vision of progress still retains the idea of plenitude: voids must be filled (Lovejoy 1964, pp. 262–265). Kant developed a theory of cosmological evolution similarly based on progress and plenitude. Even Robinet, despite his adherence to the law of plenitude, wrote of nature as a continual unfolding:

Nature has never been, and never will again be, precisely what she is at the moment at which I am speaking...I doubt not that there was a time when there were not yet either minerals or any of the beings that we call animals; that is to say, a time when...not one of them had come to birth...At least it appears certain that Nature has never been, is not, and never will be stationary, or in a state of permanence; its form is necessarily transitory... Nature is always at work, always in travail, in the sense that she is always fashioning new developments, new generations. (quoted by Lovejoy 1964, p. 275)

The great chain had swung sideways, extending forward in time rather than up toward the heavens. The world was ready for Darwin.

*** *** ***

Charles Robert Darwin (Fig. 3.9) was not an evolutionist when he set sail as the ship's naturalist aboard the *HMS Beagle*, but he did carry a copy of Lyell's *Principles of Geology* with him. Lyell, of course, was not an evolutionist either at the time, but his ideas of slow and gradual change informed Darwin's observations during his travels, as well the long intellectual journey upon which Darwin embarked after his return to England. The development of Darwin's ideas has been traced in depth and detail many times, by many scholars, and I will certainly not attempt to compete with them here. For the purposes of the present work, a few crucial points need to be made. The essence of Darwin's idea was this: by analogy with artificial selection by humans in breeding animals, he showed how the differential survival of individual animals could, over time, lead to significant differences between groups of organisms. This was natural selection, acting on the small, random variations between one animal and its siblings, which left some individuals better equipped to

---

[14]This does not mean, of course, that the idea of nature in general, and stars in particular, did not also retain a metaphorical image as fixed and unchangeable: some seventy years later Keats was to write "Bright star! Would I were steadfast as thou art!"

Fig. 3.9 Darwin in his mid-forties. He was working on writing the *Origin of Species* at the time this photograph was taken. Public domain

survive than others. Those that survived might pass their slightly more advantageous traits on to their offspring. The idea of small incremental change, built up over an immensity of time, derives from Lyell's uniformitarianism. The principle of natural selection, acting throughout deep time on variations among individuals, was Darwin's great and radical insight. Through detailed arguments and examples, constructed with painstaking care during the years after his return from voyaging, Darwin laid out the evidence for the role of natural selection in driving speciation (Fig. 3.10). Mendel's genetics was unknown at the time, and one of the principal arguments Darwin faced was the mystery of where the variation came from in the first place (another conundrum was the mechanism of inheritance; it was not known until many years later that the answer lay in the same molecule).

Darwin's critics, including (at first) Lyell, argued that natural selection could act as a "destroyer", but not as a creator (Gould 2002, pp. 139–140). Darwin simply cited observable evidence that variability in nature was, in Gould's phrasing, "small, copious and undirected", and suggested that future research would ultimately uncover the underlying mechanisms of such variation. "I have, he wrote, "hitherto sometimes spoken as if the variations – so common and multiform with organic beings under domestication, and in a lesser degree with those under nature – were due to chance. This, of course, is a wholly incorrect expression, but it serves to acknowledge plainly our ignorance of the cause of each particular variation" (Darwin, *Origin of Species*, p. 172).

Fig. 3.10 Darwin's famous first sketch of an evolutionary tree, drawn in an 1837 notebook, shortly after his return from traveling aboard the Beagle. Public domain

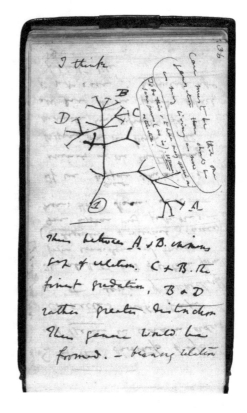

For our purposes, there are a few critically important aspects of Darwin's work that relate to the idea of collectivity and classification. We have discussed at length the struggles of naturalists and philosophers to find a method or system for classifying living organisms. By combining Lyell's uniformitarianism and Hutton's deep time with his own new idea of natural selection, Darwin proposed not merely a method, but a mechanism, for classification: common descent. Closely related organisms should be grouped more closely together, and closely related species likewise. Determining the degree of closeness might be easy, or quite difficult, depending on the biological evidence available. But if a lineage could be determined, then a meaningful classification was possible, based on tangible shared history.

The radical innovation of Darwin's approach may be seen by comparison to Lamarck, for whom adaptation via inheritance of acquired characteristics was far less important than the unseen structural driving forces pushing organisms upward toward greater complexity. Darwin completely removed these unseen forces, laying everything at the doorstep of adaptation. As Gould writes, Darwin

acknowledged Lamarck's implied claim that small scale adaptation to local environment defines the tractable subject matter of evolution. But he refuted the disabling contention that adaptation in this mode only diverted the 'real' force of evolution into side channels and dead ends. And he revised previous evolutionary thinking in the most radical way – by denying that Lamarck's 'real' force existed at all, and by encompassing its supposed results as consequences of the 'subsidiary' force accumulated to grandeur by the simple expedient of relentless action over sufficient time." (Gould 2002, p. 98)

Darwin added another revolutionary idea in introducing natural selection as the mechanism of adaptation to the natural environment. He argued that natural selection alone can generate the range of natural diversity observable in the world, given sufficient time. Darwin eliminated the need for Lamarck's orthogonal forces of adaptation, which increased diversity, and the drive toward greater complexity, which kept all creatures together along an "upward" path. Yet, Darwin's system did contain two drives pulling nature's diversity in different directions. As with Lamarck, adaptation pulled groups of creatures apart as they branched into different ecological niches. But instead of being held together by an indefinable push upward along a chain of being, similarities were maintained by the same mechanism that allowed natural selection to act: common descent. "On my theory," Darwin wrote, "unity of type is explained by unity of descent" (Darwin, *Origin of Species*, p. 261). Descent with modification caused the branching off of different morphologies as species radiated into different ecological niches. But modifications and branching did not erase the traces of common descent. Common patterns derive from common history, even as an accumulation of local differences, given sufficient time, cause phenotypic divergence. Thus Darwin's historical argument contains another essential tension, this time between orthogonal forces derived from a common source – shared history ("Unity of Type") in the presence of natural selection ("Conditions of Existence"). Note the parallel between these forces and the ideas of form (unity of type) and function (adaptation to conditions of existence) propounded, respectively, by Geoffroy and Cuvier.

It is important to note that, while Darwin's work was primarily focused at the level of competition between individual organisms, he did explore some group-level tensions, such as cooperation and competition, mentioned in the previous chapter. This is especially important to remember since the "strict Darwinian" view of evolution is often (incorrectly) reduced to the operation of selection at a single level. For example, in a letter written in September 1856, three years before the publication of the *Origin of Species*, Darwin wrote of human populations that

[t]he advantage in each group becoming as different as possible, may be compared to the fact that by division of labor most people can be supported in each country. – Not only do the individuals in each group strive against the others, but each group itself with all its members, some more numerous, some less, are struggling against all other groups, as indeed follows from each individual struggling. (quoted by Gould 2002, p. 233)

Likewise, in the *Origin*, when Darwin discussed the problem of insects with sterile castes, a situation that seems to fly in the face of organismal selection, he remarked that the difficulty of explaining such a case "is lessened, or, as I believe, disappears, when it is remembered that selection may be applied to the family, as well as to the individual" (Darwin, *Origin of Species*, p. 354). Of bees, he wrote that "if on the whole the power of stinging be useful to the social community, it will fulfill all the requirements of natural selection, though it may cause the death of some few members" (Darwin, *Origin of Species*, p. 257). Returning to the problem of human societies in *The Descent of Man*, Darwin addressed the evolutionary origins of altruism.

> It must not be forgotten that although a high standard of morality gives but a slight or no advantage to each individual man and his children over the other men of the same tribe, yet that an increase in the number of well-endowed men and an advancement in the standard of morality will certainly give an immense advantage to one tribe over another. A tribe including many members who, from possessing in a high degree the spirit of patriotism, fidelity, obedience, courage, and sympathy, were always ready to aid one another, and to sacrifice themselves for the common good, would be victorious over most other tribes; *and this would be natural selection*. (Darwin, *Descent of Man*, pp. 157–158, my italics)

As Gould emphasizes, the "one long argument" Darwin constructed in the *Origin of Species* relies on *methods of historical inference* to support his conclusion that natural selection and the immensity of time are sufficient causes for evolution. The first method used by Darwin is an extrapolation of Lyell's uniformitarianism to the biological realm. Accumulation of small change, given sufficient time, can produce radically divergent morphologies. Gould describes this approach as the "worm principle", in honor of Darwin's last book, a study of the role of worms in processing the soil of Britain. Taking a physicist's perspective, we might alternately call this the "definite integral principle", in analogy to the definite integral as creating something finite by summing infinitesimal parts over some given interval. As he did later in his book on worms, Darwin constructed his argument based on deceptively mundane data. He began the *Origin* with a discussion of artificial selection on domesticated pigeons, and then extrapolated to the process of natural selection, acting over an immense reach of time.

> I look at individual differences, though of small interest to the systematist, as of the highest importance for us, as being the first steps toward such slight varieties as are barely thought worth recording in works on natural history. And I look at varieties which are in any degree more distinct and permanent, as steps towards more strongly-marked and permanent varieties; and at the latter, as leading to sub-species, and then to species. The passage from one stage of difference to another may, in many cases, be the simple result of the nature of the organism and of the different physical conditions to which it has been exposed; but with respect to the more important and adaptive characters, the passage from one stage of difference to another, may safely be attributed to the cumulative action of natural selection… (Darwin, *Origin of Species*, pp. 77–78)

The second method of historical inference that Darwin employed was to *infer history from a series*. Gould refers to this as "the coral reef" principle, in a nod to Darwin's first book, in which he proposed "a single historical process for the formation of coral atolls by recognizing three configurations of reefs – fringing reefs, barrier reefs, and atolls – as sequential stages in the foundering of oceanic islands" (Gould 2002, p. 104). Again taking a physicist's perspective, we could propose another name, and call this the "Fourier series principle", since it takes data from the spatial domain and transforms it onto an axis of time. Note the metaphorical parallel between this and the rotation of the chain of being from a ladder climbing up toward the heavens to a horizontal axis stretching forward in time. As he did with his discussion of the cumulation of small changes, Darwin introduced this method of inference first with observations from the artificial selection of pigeons, before extrapolating out of the pigeon fanciers' clubs of Victorian London and into the vastness of evolutionary time. Darwin wrote that the differences between varieties, and indeed between individuals "blend into each other by an insensible series; *and a series impresses the mind with the idea of an actual passage*" (Darwin, *Origin of Species*, p. 77, my italics).

Inferring history from a series is neither more nor less than inferring causality from correlation.

References

Appel TA (1987) The Cuvier-Geoffroy Debate: French Biology in the Decades Before Darwin. Oxford University Press, New York

Darwin C (2004) The Descent of Man, and Selection in Relation to Sex. Penguin Books Ltd., London

Darwin C (2009) The Origin of Species by Means of Natural Selection. Modern Library, New York. (This edition reprints the second edition of the *Origin*, from early 1860)

Dornelas MC, Dornelas O (2005) From leaf to flower: revisiting Goethe's concepts on the "metamorphosis" of plants. Braz J Plant Physiol 17(4):335–343

Foucault M (1994) The Order of Things: An Archaeology of the Human Sciences. Vintage Books (Random House), New York

Gould SJ (1987) Time's Arrow, Time's Cycle: Myth and Metaphor in the Discovery of Geological Time. Harvard University Press, Cambridge, MA

Gould SJ (2002) The Structure of Evolutionary Theory. The Belknap Press of Harvard University Press, Cambridge, MA

Lovejoy AO (1964) The Great Chain of Being: A Study of the History of an Idea. Harvard University Press, Cambridge, MA

Lyell C (1997) Principles of Geology. Penguin Books Ltd, London

Rudwick MJS (1997) Georges Cuvier, Fossil Bones, and Geological Catastrophes: New Translations and Interpretations of the Primary Texts. The University of Chicago Press, Chicago

Westfall RS (1980) Never at Rest: A Biography of Isaac Newton. Cambridge University Press, Cambridge

Chapter 4
The Battle of the Parts

I can feel pieces of my brain falling away like a wet cake.

Bernard Black

WHILE DARWIN, Durkheim and others struggled to define a collectivity made up of individual organisms, a group of nineteenth and early twentieth century scholars, inspired by Darwin's work, explored collectivity at a smaller scale, sowing the seeds of a full-fledged hierarchical evolutionary theory. This work was primarily done in Germany, by Ernst Haeckel, Wilhelm Roux, and August Weismann, and was inspired by their studies of embryology.

Ernst Haeckel (Figs. 4.1 and 4.2) is perhaps best known today for the idea (now seen as evocative metaphor rather than scientific fact) that ontogeny recapitulates phylogeny. But in addition to his studies of embryological development (ontogeny) and its relation to evolutionary history (phylogeny), Haeckel suggested a full hierarchical theory of organic structure in his 1866 work *Generelle Morphologie der Organismen*. In what he referred to as the "doctrine of organic individuality", he postulated that organic bodies can occur, and be assembled in, six structural levels, each of which can be considered as an individual. At the lowest (meaning the smallest, not the least important) level were *plastids*, comprising cells and organelles. The second level contained *organs*, including not just organs such as liver, lungs and spleen, but also tissues and organ systems. The third level contained symmetrical body parts, which Haeckel called *antimeres*. Fourth came *body segments*, and fifth were what he termed *persons*, or, in Gould's phrasing, "vernacular individuals". Lastly, colonies of persons, which Haeckel termed *corms*. Haeckel emphasized that each of these levels should be considered as a type of individual; the primacy given to the level of persons was simply an accident of our own perspective as persons ourselves. He wrote that

> ... these 'true' or absolute individuals are, in fact, only relative ... the individuality of humans and higher animals leads us to the erroneous conception that morphological individuals of the fifth order are the 'true' organic individuals. This concept has become so general, and has been so strongly fixed in both scientific and vernacular consciousness, that we must mark it as the major source of the numerous and varied interpretations and debates that prevail on the subject of organic individuality. (quoted by Gould 2002, p. 210)

© Springer Science+Business Media B.V. 2018
S. Bahar, *The Essential Tension*, The Frontiers Collection,
DOI 10.1007/978-94-024-1054-9_4

Fig. 4.1 Ernst Haeckel as
a young man (1860).
Public domain (PD-1923)

Fig. 4.2 Illustration of sea
anemones from Ernst
Haeckel's 1904
Kunstformen der Natur.
Public domain

Fig. 4.3 Wilhelm Roux.
Public domain

A decade and a half later, Wilhelm Roux (Fig. 4.3) published a book titled *Der Kampf der Thiele im Organismus (The Struggle of Parts in the Organism)*, which applied Darwinian natural selection to cells and, most particularly, to organs. In a letter urging George Romanes to review the book for *Nature*, Darwin wrote that, making allowance for his limited ability to read German, the book[1] appeared to be "the most important book on Evolution which has appeared for some time… Roux argues that there is a struggle going on within every organism between the organic molecules, the cells and the organs" (quoted by Gould 2002, p. 210).

In Roux's model, organs metaphorically jostled each other in a struggle to obtain sufficient nutrition from the whole organism. He argued that "the construction of a harmonious and well-designed organism emerges from a struggle among the parts competing for limited nutriment" (Gould 2002, p. 211). Roux was explicitly drawn to this model on the basis of his observations of variability between cells. "No hepatocyte", he wrote, "is perfectly similar in size or shape to another; yet they all fit together in an efficient organ" (quoted – and translated – by Heams 2012). The second idea that inspired Roux was the idea of the Malthusian mismatch between the reproduction rates of organic beings and resources. "Charles Darwin and Alfred Russel Wallace have," he wrote, "demonstrated that because of geometrical growth of organisms, there must be a… struggle between them, and that due to the continuous variations of all parts within the organism, one could assume that only the best organisms would survive" (quoted by Heams 2012). Thus Roux derived his model for intracorporeal struggle from two of the three key premises of Darwinian evolution (later codified, with the inclusion of *heritability* as the third leg of the tripod, by Lewontin in 1970): *variability* and *competition*, leading to differential survival. Roux only discussed competition between components at the same level (cells with cells, organs with organs). He did not consider cross-level conflict because it would

---

[1] …which has still not been published in English translation (Heams 2012).

be akin to, in his words, trying to find the "sum of differential equations of different orders" (quoted by Heams 2012).

There exists a fundamental difference between selection at the organismal level and Roux's battle of the parts. The results of conflict between cells of the same type are not heritable beyond the lifetime of the containing organism. Limiting the natural selection analogy even further, victories or defeats in struggles between *organs*, which do not reproduce at all, cannot be transferred to the offspring of the organism containing them, unless one reverts to the idea of inheritance of acquired characteristics. Lung versus liver is "a theory of functional adjustment in development" (Gould 2002, p. 211), not a theory of evolution. Moreover, as Vernon Kellogg wrote in 1907, "[t]his competition chiefly depends on the hazard of position... Not the best qualified by the best situated fibers have vanquished the others by robbing them of food and thus finally destroying them" (quoted by Gould 2002, p. 211). Interestingly, however (and perhaps due to his lack of fluency in German), Darwin himself interpreted Roux's theory differently, concluding that the best qualified cells did indeed have a selective edge. According to Heams (2012), Darwin wrote to Romanes that the theory's "basis is that every cell that best performs its function is, in consequence, at the same time best nourished, and best propagates its kind." The phrasing of Roux himself seems consistent with this latter interpretation: "the cells more prepared to multiply in given conditions will indeed multiply faster than the others" (quoted by Heams 2012), depending on the precise meaning given to the phrase *more prepared to multiply*.

Roux was aware of the difficulties in his theory, writing that "[t]he magnitude of the struggle of cells depends on the number of cell generations required for this struggle to produce an effect, and this obviously depends on the moment when, during the lifetime of the individual [organism], the new character appears" (quoted by Heams 2012). However, there is no evidence that Roux *intended* to produce a theory of *how selection at lower levels drove evolutionary processes across deep time*. Thus he is perhaps criticized for not failing to accomplish something that he never set out to achieve in the first place.

A review of Roux's work in Weismann's *The Evolutionary Theory* (1904[2]) provides an insight into the subtlety of Roux's thought, and its resonance with current scientific concerns, that does not appear in the more modern assessments by Gould and Heams. Weismann pointed out that in Roux's view an organ was regulated according to "the strength of the stimulus applied to it" (Weismann 1904, Vol. I, p. 245). This suggests a process reminiscent of the strengthening of synapses proportional to their degree of activation (the Hebbian principle of "fire together, wire together"). Weismann explicitly referred to the strengthening of "brain elements" in response to repeated stimuli, writing that he had always been struck by how practice enables a musician to perform

---

[2]The German edition first appeared in 1902. An English translation was published in 1903 by Edward Arnold (London), and this is the edition cited by Gould in *The Structure of Evolutionary Theory*. The 1904 copy in my own collection appears to be a reissue of the 1903 translation.

quite unconsciously, when... thinking intensely of other things. It is in this case not the memory alone, but the whole complicated mechanism of successive muscle-impulses, with all the details of fast and slow, loud and soft, that is engraved on the brain elements, just like a long series of reflex movements which set one another a-going. Though in this case we cannot demonstrate the material changes which have taken place in the nervous elements, there can be no doubt that changes have taken place, and that these consist in a strengthening of definite elements and parts of elements. The strengthening causes certain ganglion-cells to give a stronger impulse in a particular direction, and this impulse acquires increasing transmissive power, and so on. (Weismann 1904, Vol. I, p. 243)

This description appears immediately after a review of the contrast between Lamarckian inheritance of acquired characteristics and the "Darwin-Wallace principle" of natural selection. Essentially, Weismann presents the nervous system as an example of strengthening by repeated use, revealing a Lamarckian logical structure in Roux's battle of the parts. If we leave aside the typical lack of reproduction of cells in the nervous system, Roux's model also contains a Darwinian element as well: the pride of place earned by the cells strengthened by repeated stimuli are inherited by their daughter cells.

If the above examples still seem to leave Roux in an uncertain position regarding whether his battle of the parts is a true process of natural selection, consider the development of the immune system in individual animals. It is now well understood that antibodies are produced in greater numbers depending on exposure to the corresponding antigen, and that antibody diversification is driven by a natural selection-like process. This idea, originally proposed by Jerne in 1955, was developed by Burnet (1957) as *clonal selection theory*. Despite many modifications resulting from an increased understanding of the genetics of antibody-producing cells, the theory still stands today (Hodgkin et al. 2007; Neuberger 2008). From this perspective, Roux's intracorporeal struggle maps onto biological contexts discussed today in explicitly evolutionary terms, even though they have no direct impact on the host organism's germ line, beyond facilitating its survival. Clearly, then, by current standards, a process need not have an impact on evolution in deep geological time in order to be interpreted as a form of natural selection. In this light, some of the criticisms of Roux seem a bit misplaced.

Roux introduced the idea that competition at the sub-organismal level directly affects organismal structure. This implies an interaction between selection at the organismal and sub-organismal levels. In this context, the essential tension makes its first appearance in a sub-organismal evolutionary context. Roux wrote that

[a]s the struggle of parts yields purposefulness within an organism... so does the analogous struggle for existence among individuals yield purposefulness with respect to external conditions of existence...To many the direction of this book [*Der Kampf der Thiele*] may well seem very strange – for it holds that, in an animal, in which everything is so exquisitely ordered, in which all the different parts interlock with such excellence, and work together in such perfected coordination, that a struggle of parts occurs, so that in one place, where everything works together according to firm principles, a conflict among the individual parts exists. But how can an entity exist, whose parts are at variance? ... How shall the good and the stable arise from struggle and battle? ... All good can only arise from struggle. (quoted by Gould 2002, p. 214)

Roux's last sentence is more metaphor than mechanism, but the idea impressed other evolutionary biologists, and not just those of the German Romantic school.[3] Gould suggests that Darwin, reading Roux's book in the last year of his life, was particularly struck by the step it took toward explaining the "organs of extreme perfection" that Darwin himself had struggled over. Darwin's difficulty lay in showing how "organismal selection constructs each barbule on every feather", which seems a nearly insurmountable task, even given the vast expanse of evolutionary time. Roux, Gould suggests, "offered Darwin a sensible exit [by suggesting that] selection builds the capacity for a functional response that can directly shape each organism in minutely adaptive ways through growth" (Gould 2002, p. 213). As Roux put it,

> [t]hrough the capacity of the struggle of parts, a much higher perfection, the purposefulness of the functioning part down to the last molecule, can arise, and occur much more rapidly, than if it had to originate, by the Darwin-Wallace principle, through selection of variation in the struggle for existence between individuals. (quoted by Gould 2002, p. 213)

Further, Roux continued,

> [i]f life conditions change, functional adaptation can lead to adaptive changes in all impacted organs, and this simultaneous action in millions of part[s] makes it different from the phenomenon of selection, which can only develop a few simultaneous adaptive properties. (quoted by Heams 2012)

Roux's emphasis on the role of mechanical interactions in providing structural constraints was a clear source of inspiration for D'Arcy Thompson's famous *On Growth and Form*. A modern form of his ideas can be seen in studies of developmental plasticity.

August Weismann, now best known for introducing the theory of germinal selection, was the third proponent of both selection below the organism level and of hierarchical selection theory during this era (Figs. 4.4 and 4.5). His germinal selection theory, predating the rediscovery of Mendel's work, was based on an assumption of unequal apportionment of genetic material during cell division. In his theory, germ cells contained genetic material while somatic cells did not. This is now, of

---

[3] It is interesting to note that elsewhere Roux makes the explicit comparison between competing cells forming a stable entity and social structures. Addressing those who might doubt that an entity can exist when its parts are at variance, Roux asked whether it is "impossible for the State to exist when citizens compete with each other" (quoted by Heams 2012). Elsewhere, he wrote that "[w]hen parts struggle against each other to acquire an ever greater efficiency, the overall performance should also increase, in the same way that the efficiency of an army increases when officers compete and when the best among them are selected to train the novice soldiers" (quoted by Heams 2012). Heams notes that Roux's professor, Rudolf Virchow, wrote also of "cellular democracies" and "republics of cells" when describing multicellular organisms (Heams, p. 27). Heams calls this is a "restrictive" reading of Darwinism, and that "equating darwinism and violent fight is misleading". However, Roux is making a more complex point. In both these quotations, he is pointing out how conflict at one level leads to stability at the next higher level. The metaphor of human society – as we have seen from the discussion of crowd theory in Chap. 1 – simply provides an example of how conflict at level n can lead to stability at level $n + 1$.

Fig. 4.4 August Weismann.
Public domain

course, known to be incorrect, but still stands as reasonable first pass at explaining the formation of gametes and differential gene expression among cell types.

Weismann developed his theory of germinal selection in an attempt to defend strict Darwinian selectionism from attacks by the philosopher Herbert Spencer. Early in his career, Weismann was a strict adherent of Darwinian selection at the organismal level. However, Spencer raised an argument about the disappearance of vestigial limbs as a challenge to the primacy of natural selection in driving evolutionary change. For example, it would be reasonable to suggest that the ancestor of a whale might benefit from a decrease in the size of its hind limbs, since this would make the animal a more streamlined swimmer. But what would happen when the limbs grew so small that their further reduction was, so to speak, below the threshold of selection? A decrease in the size of a limb with a surface area equal to the tiniest fraction of the animal's body surface could not conceivably produce a life-or-death advantage to its possessor. Even more, what of the continued decrease in limb size when the limb was so vestigial as to be contained within the animal's body?[4]

In an 1893 article entitled "The Inadequacy of 'Natural Selection'", Herbert Spencer proposed that a principle of use and disuse would best explain the disappearance of vestigial organs. Weismann replied with a work entitled "Die Allmacht Der Naturzuechtung", translated as "The All-Sufficiency of Natural Selection". The choice of the term *Allmacht*, which can also be rendered as "omnipotence" or more directly, but less flowingly in English, as "all-might", was a direct rebuttal to Spencer's charge of inadequacy (Gould 2002, p. 198). Drawing on his idea of the

---

[4] Note that an alternate explanation, based on the possible selective benefit of the reallocation of resources, also fails at this point due to the minimal benefit that could result from reallocation away from such tiny appendages.

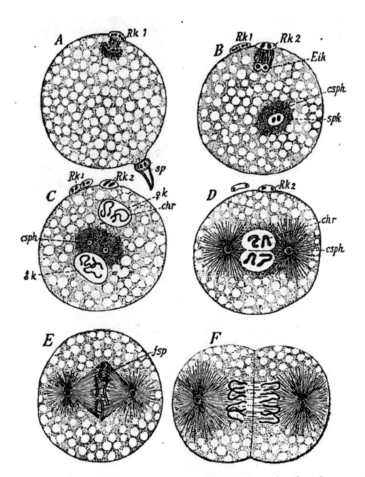

Fig. 75. Process of fertilization in *Ascaris megalocephala*, the thread-worm of the horse, adapted from Boveri and Van Beneden. *A*, ovum in process of the first directive division; *Rk1*, first polar body; *sp*, spermatozoon with two chromosomes in its nucleus, attaching itself to the ovum, and about to penetrate into it; a protrusion of the egg-protoplasm is meeting it. *B*, the second directive division has been completed; *Rk2*, the second polar body; *Eik*, the reduced nucleus of the ovum. The first polar body (*Rk1*) has divided into two daughter-cells, *spk*; the nucleus of the spermatozoon remains visible with its two centrospheres (*csph*). *C*, the sperm nucleus (♂ *k*) and the ovum nucleus (♀ *k*) have grown, each has two loop-like chromosomes; only the male nucleus has a centrosphere, which has already divided into two (*csph*). *D*, the two nuclei lie apposed between the poles of the nuclear spindle. *E*, the four chromosomes have split longitudinally; the spindle for the first division of the ovum (the segmentation spindle, *fsp*) has been formed. *F*, divergence of the daughter-chromosomes towards the two poles; division of the ovum into the first two cleavage cells or embryonic cells.

Fig. 4.5 A figure from Weismann's 1904 *The Evolution Theory* illustrating the fertilization of *Ascaris*. Note the level of detail in the drawing of the mitotic spindles and chromosomes. From author's collection

sequestration of germ cells, Weismann wrote that "Nature has carefully enclosed the germ-plasm of all germ-cells in a capsule, and it is only yielded up for the formation of daughter-cells, under the most complicated precautionary conditions." (quoted by Gould 2002, p. 201). If the germinal material is sequestered in only a few cells, then it is these cells and these cells alone that transmit evolutionary information to future generations. An immediate consequence of this is that inheritance of *acquired* characteristics is, in Gould's phrasing, "structurally impossible". From this starting point, Weismann struggled to construct arguments to explain the selective benefit of degeneration. In Gould's interpretation of Weismann's intellectual journey, "[d]egeneration acted as the lever that pried Weismann from his panselectionism, and led him through a chronological series of honorable changes that must be read, in one sense, as a retreat from a former pugnacious insistence on *Allmacht*, but that also represents a complexification and strengthening of his original views" (Gould 2002, p. 204). Ultimately, as we shall see, this led to a fully hierarchical theory of selection.

Weismann first suggested that degeneration, once a trait had fallen below the threshold of selection, could occur by progressive dilution of the genetic material; he termed this process *amphimixis* or *panmixia* (Weismann 1904, Vol. II, p. 206–226; Gould 2002, p. 205). This suggestion was somewhat plausible, given the fact that Mendel's experiments were unknown at the time. However, Spencer parried this argument by referring to Galton's principle of regression to the mean (Gould 2002, p. 206). There was no basis, Spencer asserted, for assuming that such dilution would result in loss of a trait rather than its enhancement, or, most likely, it net stability. Weismann's idea required a *directional* dilution, and there was no biological evidence to support this. Weismann accepted this reasoning, perhaps gratefully, since his dilution argument weakened the inherent *Allmacht* of selection.

Weismann next considered an alternate explanation for degeneration, based on selection *at a different level*. The work of Haeckel and Roux set a precedent for the hypothesis that selection could operate at the level of the germ-plasm. Weismann's next step was to argue that, since natural selection can produce morphological trends in evolution at the organismal level, directionality might occur in the selection of variation at the germinal level as well. This directionality, arising from germ-plasm-level selection, could possibly bias organismal evolution toward the degeneration of vestigial structures. In developing his germ-plasm theory, Weismann broke down Haeckel's subcellular level of plasmids into further categories (Weismann 1904, Vol. I, p. 349). The sub-microscopic molecular carriers of heredity he called *biophors*, which aggregated into *determinants*, which could be considered as somewhat equivalent to genes in modern terms. Determinants, which Weismann assumed to map uniquely onto phenotypic traits, were aggregated into *ids*, which could be considered analogous to histones. Multiple groups of ids aggregated into *idants*, analogous to chromosomes. In Weismann's theory, determinants competed for nutrients, and their ability to do so was decoupled from the fitness value of the phenotypic traits they encoded. Thus fitness at the determinant level was separate from fitness at other levels. Because the competition between determinants took place in the germ cells, its results could effect long-term evolutionary change, unlike Roux's battle of the parts among organs or somatic cells. In essence, Weismann preserved the *Allmacht* of selection by allowing it to act at multiple levels. As he wrote in 1896,

> Powerful determinants in the germ cell will absorb nutriment more rapidly than weaker determinants. The latter, accordingly, will grow more slowly and will produce weaker descendants than the former ... Since every determinant battles stoutly with its neighbors for food, that is, takes to itself as much as it can, consonantly with its power of assimilation and proportionately to the nutriment supply, therefore the unimpoverished neighbors of this minus [i.e., lesser] determinant will deprive it of its nutriment more rapidly than was the case with its more robust ancestors. (quoted by Gould 2002, p. 215)

Weismann's expansion of natural selection was not universally well received. In an archly dismissive 1896 review in *Science*, E. G. Conklin wrote dryly that

> [t]he insufficiency of natural selection to explain all the phenomena of phyletic transformation, Weismann attributes to the fact that this principle has been unduly limited in its field of operation; it has heretofore been regarded as applicable only to *persons*; it should be considered as applicable to every organic unit, whether visible or invisible, down to the hypothetical biophores...In brief, natural selection is still omnipotent if only it be regarded as omnipresent.

Conklin also chided Weismann for assuming the *allmacht* of natural selection as a foregone conclusion, and suggested that, to the extent that his theory of germinal selection explained how determinants, idants and biophors became stronger, rather than leaving more progeny, his theory lacked a mechanism of selection per se, and was thus suspiciously close to a Lamarckian model of strengthening by use and disuse (recall that the same criticism was leveled at some aspects of Roux's battle of the parts). Conklin finally attacked Weismann with the argument that he was proposing a model that was no more than an imaginary construct, complaining that "not a particle of evidence is adduced in proof of a single proposition named... In all seriousness, it seems to be that to class such a purely figurative and imaginary 'struggle' along with Darwin's principle, as Weismann does, is to wholly disregard the importance of evidence."

To an extent, Conklin's criticisms were justified – Weismann's theory was largely no more than a hypothetical construct. It is worth noting, however, that experimental work was being done on nuclear and cytoplasmic material at the time, and thus his concepts of ids, determinants and the rest were not entirely fanciful. Weismann, particularly early in his career, was no mean experimentalist himself (Churchill 2010), and was certainly well familiar with the work of Oscar Hertwig, who in 1875 had observed in his studies of sea urchins that fertilization involved fusion of the cell nuclei of the sperm and the egg. In the same decade, other scientists, exploiting recent advances in light microscopy and the discovery of aniline dye staining, observed that cell nuclei contained threadlike structures that were segregated from other cellular components during cell division. In 1884, Édouard van Beneden observed the reduction in the number of chromosomes in the parasitic nematode *Ascaris* during the formation of its germ cells; this observation was made during microscopic studies of the formation of polar bodies and their role in cell division.

The brilliant experiments of Theodor Boveri[5] and others were soon to clarify the role of chromosomes in inheritance, providing strong support for Weismann's theory of the germ-plasm (Madspacher 2008). Thus, while it was speculative, Weismann's theory was far from a fantasy.

In the mid 1890s, Weismann argued that germinal and organismal selection as working synergetically to account for the disappearance of vestigial organs (Gould 2002, pp. 218–219). A few years later, Weismann expanded his theory to encompass a fuller picture of hierarchical selection, and explicitly considered situations in which the direction of germinal selection could be decoupled from the direction of organismal selection.

> Now that we have made ourselves familiar with the idea of germinal selection we shall attempt to gain clearness as to what it can do, and how far the sphere of its influence extends, and, in particular, how whether it can effect lasting transformations of species *without the co-operation of personal [organismal] selection, and what kind of variations we may ascribe to it alone.* (Weismann 1904, Vol. II, p. 126)

Indeed, the problem of degeneration could be considered as an instance of decoupling of selective levels. An organ might be reduced in size due to organismal (or, in Weismann's terminology, "personal") selection alone. But once the organ had shrunk below the notice of this selective level, germinal selection completed the job. Thus the two levels worked in the same direction, but with complete independence, at least in the last stage, once personal selection had ceased to operate. "Useless organs," Weismann wrote in 1909, "are the only ones which are not helped to ascend again by personal selection, and therefore in their case alone can we form any idea of how the primary constituents behave, when they are subject solely to intragerminal forces" (quoted by Gould 2002, p. 222). In other words, the elimination of vestigial organs formed a control case in which only one level of selection was acting. In the independence of the two levels, Weismann wrote,

> lies the great importance of this play of forces [i.e., competition] within the germ-plasm, *that it gives rise to variations quite independently of the relations of the organism to the external world.* In many cases, of course, personal selection intervenes, but then it cannot directly effect the rising or falling of the individual determinants – these are processes quite outside of its influence. (quoted by Gould 2002, p. 221, my italics)

---

[5] Boveri's insights as well as his experiments were, in many ways, ahead of his time. In a passage quoted by Madspacher (2008), Boveri wrote in 1904 about the possibility of symbiosis between cells and subcellular components, in terms that echo the ideas about evolution, symbiosis and cell organelles proposed by Lynn Margulis. Boveri wrote that "[i]f we follow these structures [chromosomes] in their 'expressions of life' – how they branch out like rhizopodia during formation of the resting nucleus and contract again as it dissolves, how they propagate by division and from time to time copulate as pairs – then this indicates a level of 'expressions of life', [as] is ascribed to entire cells. The way the chromosomes form a unity with the protoplasm can be best described as an extremely close symbiosis. I think it is worth discussing the question of whether this might not be more than a metaphor. It might be possible that what we call a cell, and for which our mind demands simpler preliminary stages, originated from a symbiosis of two kinds of simpler plasmatic structures, such that a number of smaller ones, the chromosomes, settled within a larger one, which we now call a cell body."

Weismann's exploration of *decoupling* between levels of selection soon led him to consider *explicit conflict* between levels. Considering classic examples of harmful orthogenetic trends like the impossibly large horns of Irish elks or the curved teeth of saber toothed tigers, he suggested that, in such cases,

> the variation-direction which had gained the mastery in all ids could no longer be sufficiently held in check by personal selection, because the variations in the contrary direction would be much too slight to attain to selective value… In the majority of cases the self-regulation which is afforded by personal selection will be enough to force back an organ which is in the act of increasing out of due proportion to within its proper limits. The bearers of such excessively increased determinants succumb in the struggle for existence, and the determinants are thus removed from the genealogical lineage of the species. (quoted by Gould 2002, pp. 220–222)

In these cases, then, the level of personal selection snuffed out a group of runaway selfish determinants.

Weismann's writings ultimately emphasized a full range of levels in the evolutionary hierarchy, up to and including species selection.

> If the germ-plasm be a system of determinants, then the same laws of struggle for existence in regard to food and multiplication must hold sway among its parts which hold sway between all systems of vital units – among the biophors which form the protoplasm of the cell-body, among the cells of tissue, among the tissues of an organ, among the organs themselves, as well as among the individuals of a species and between species which compete with one another. (quoted by Gould 2002, p. 223)

He likened this to the action of gravitation at multiple scales of physical systems, writing that while

> [t]he great prominence thus given to the idea of selection has been condemned as one-sided and exaggerated, but the physicist is quite as open to the same reproach when he thinks of gravity as operative not on our earth alone, but as dominating the whole cosmos, whether visible to us or not. If there is gravity at all it must prevail everywhere, that is, wherever material masses exist; thus not only are the vital units which we can perceive, such as individuals and cells, subject to selection, but those units the existence of which we can only deduce theoretically, because they are too minute for our microscopes, are subject to it likewise. (Weismann 1904, Vol. I, p. ix).

Note how this comparison essentially restates Conklin's point about omnipotence and omnipresence, but from a diametrically opposed emotional valence (if such a thing can be said to exist in scientific commentary).

For all his emphasis on evolutionary hierarchy, it is a supreme irony that Weismann's "doctrine" was distilled into something else entirely by the mid-twentieth century scientists who codified the modern synthesis of evolutionary theory. Weismann's specific germ-plasm theory was shown to be incorrect, of course, but his idea of the sequestration of germinal information in certain cells is certainly partly correct, since, while most cells contain DNA, only gametes contain the information to create a new genetic individual. This aspect of Weismann's doctrine has become the one to which his name is primarily attached, and his emphasis on the genetic uniqueness of the individual provided an impetus for a renewed focus on selection at the organismal level alone. As described by Leo Buss in his groundbreaking book *The Evolution of Individuality*, the reasoning ran as follows. "If variation arising in the course of ontogeny is not heritable [beyond a short-lived cell lineage that will die with its host organism], the dynamics of cell lineages within the somatic environment are a matter of little direct evolutionary interest" (Buss 1987, p. 3).

Ironically, Weismann had envisioned a quite different intellectual legacy, writing in the preface to *The Evolution Theory* that the "extension of the principle of selection to all grades of vital units is the characteristic feature of my theories; it is to this idea that these lectures lead, and it is this – in my own opinion – which gives this book its importance. This idea will endure even if everything else in the book should prove transient" (Weismann 1904, Vol. I, p. ix). Instead, the idea was temporarily eclipsed. The distorted organism-centered perspective Haeckel had warned against opened its jaws and swallowed biology whole.

As a result of the narrowing focus, Buss writes, "[t]he modern synthesis was an intellectual event largely unattended by embryologists" (Buss 1987, p. 3). And in the words of John Maynard Smith, "[a]fter the publication of Darwin's *Origin of Species*, but before the general acceptance of Weismann's views, problems of evolution and development were inextricably bound up with one another. One consequence of Weismann's concept of the separation of the germ line and soma was to make it possible to understand genetics, and hence evolution, without understanding development" (quoted by Buss 1987, p. 12).

It is undeniable that vast advances were made with this organism-centered, embryology-free approach. But this approach ran counter to the entire thrust of Weismann's work. Even worse, it obscured a fundamental issue that lies at the core of current studies of the evolution of multicellularity, multi-level selection theory, and the evolution of individuality. The germ line is sequestered early in development in *only some types of living organisms*. These organisms can be studied as "genetic individuals" – though the picture is complicated, of course, by the new concept of the microbiome, as well as the fact that genetic individuals such as you, I, my cat, and the grumpy bookseller who is shipping me a copy of Weismann's 1904 volumes even as I type this, possess a mitochondrial as well as a nuclear genome. Yet in many other types of organisms, such as *Volvox*, the interplay between germ and soma is far more subtle, and this interplay lies at the boundary between a collection of cells and an individual multicellular organism. Weismann's insistence that "the principle of selection does rule over all the categories of vital units" (Weismann, Vol. I, p. viii), while now largely forgotten, has in fact never been more critically important.

References

Burnet FM (1957) A modification of Jerne's theory of antibody production using the concept of clonal selection. Aust J Sci 20(3):67–69

Buss LW (1987) The Evolution of Individuality. Princeton University Press, Princeton

Churchill FB (2010) August Weismann embraces the protozoa. J Hist Biol 43:767–800

Conklin EG (1896) Weismann on germinal selection. Science 3(76):853–857

Gould SJ (2002) The Structure of Evolutionary Theory. The Belknap Press of Harvard University Press, Cambridge, MA

Heams T (2012) Selection within organisms in the nineteenth century: Wilhelm Roux's complex legacy. Prog Biophys Mol Biol 110:24–33

Hodgkin PD, Heath WR, Baxter AG (2007) The clonal selection theory: 50 years since the revolution. Nat Immunol 8(10):1019–1026

Jerne NK (1955) The natural-selection theory of antibody formation. Proc Natl Acad Sci U S A 41:849–857

Lewontin RC (1970) The units of selection. Ann Rev Ecol Syst 70:1–18

Madspacher F (2008) Theodor Boveri and the natural experiment. Curr Biol 18(7):R279–R286

Neuberger MS (2008) Antibody diversification by somatic mutation: from Burnet onward. Immunol Cell Biol 86:124–132

Thompson DW (1992) On Growth and Form. Dover Publications, New York

Weismann A (1904) The Evolution Theory (trans: Thomson JA, Thomson MR). Edward Arnold, London

Chapter 5
Synthesis?

Shut up. I'm trying to make crème brûlée.

Bernard Black

HYPOTHESES REGARDING selection at a level below the organismal, as we have just seen in the work of Haeckel, Roux and Weismann, and ideas of selection at higher levels, in sociological works such as those of Durkheim, as well as key passages of Darwin's own writings, set the stage for twentieth century scientists to rigorously explore evolutionary processes at multiple levels. This did not happen. Instead, for much of the twentieth century, evolutionary theory was almost exclusively focused on the organismal level. Selection at the group level was pushed to the margins of discussion and, when it was brought up, often dismissed with ridicule. Stephen Jay Gould recalled suggesting, in the question session after a conference talk in 1973, that the small size of some island-dwelling mammal species in the Pleistocene era might be due to the advantage of large population size, which would render the population more resistant to extinction than their larger-bodied sister species. "Are you", the speaker replied, "really satisfied with a group selectionist argument like that?" Gould waited, but that was it. No actual content-based argument was forthcoming (Gould 2002, p. 554).

How and why did the scientific community turn so emphatically away from the interaction, balance, and tension between levels in biological systems? The shift to a nearly exclusive focus on adaptation at the organismal level was codified during period known as the Modern Synthesis, after Julian Huxley's 1942 book *Evolution, the Modern Synthesis*.

Sociological reasons have been put forward to explain this shift. Smocovitis (1996) has emphasized that the idea of "synthesis" in the sciences in general was quite popular early in the last century, influenced by the philosophers of the Vienna Circle. Synthesis was seen as a measure of the intellectual maturity of a field and its proponents (Gould 2002, p. 503). The catastrophic history of mid-century Europe may also have driven biologists toward a focus on adaptation and "improvement"; tragically, of course, this idea of improvement was also linked to some of the most terrible horrors of the twentieth century. This toxic aspect penetrated the work of some of the architects of the Synthesis, as in the sections of R.A. Fisher's work

© Springer Science+Business Media B.V. 2018
S. Bahar, *The Essential Tension*, The Frontiers Collection,
DOI 10.1007/978-94-024-1054-9_5

dealing with eugenics. Perceived "improvement", however, could be the cure for, as well as the cause of, historical catastrophe. Smocovitis suggests that

> [m]ore strongly selectionist models [were] favored by biologists who patterned themselves after physicists at the same time that they pointed the way to the 'improvement' of humanity and thus painted a progressive and optimistic picture of the … world… Evolutionary models favoring random genetic drift, which enforced a stochastic view of evolution – and culture – would not be favored in a post-war frame of mind seeking to 'improve' the world. (Smocovitis 1996, p. 131)

Ironically, the imprint of human history itself may be stamped on the rejection of the contingencies of history in evolution. Smocovitis suggests that this "drive for the improvement of humans and the increasing [perceived] necessity for progressive evolution within a positivistic theory of knowledge" were factors that led evolutionary biologists like Theodosius Dobzhansky and Sewall Wright, who initially advocated a significant role for genetic drift, to back off from their initial arguments (Smocovitis 1996, p. 131).

Gould, with inimitable archness of tone, suggested another possible sociological explanation. He commented that, while "some complex mixture of empirical and sociological themes may explain the adaptationist hardening of the synthesis", it was important not to "neglect the additional impetus of a cultural analog to drift and founder effects in small populations" (Gould 2002, p. 543). In other words, a few leading scientists focused their attention on building a "synthetic" theory of evolution around adaptation at the organismal level, and their students and acolytes jumped on the bandwagon.

A last sociological explanation, also investigated by Smocovitis, lies in the drive on the part of biologists to shape their discipline into something more like physics or chemistry: unified, and mathematically-based. While this helped inspire important advances like the development of the mathematical theory of population genetics (by Fisher, Haldane and Wright), it made explanations based on stochastic processes or pluralistic arguments appear unscientific. The call to "axiomatize" biology was led by scholars like J. H. Woodger, who wrote a book entitled *Biological Principles* in 1929, in which he sharply criticized biological science for remaining at a "metaphysical" stage of development. This critique was a nod to the Vienna Circle and their antecedents in the Positivist School, such as Ernst Mach, who held that fields matured as they underwent a transition from theology to metaphysics, before finally evolving into to a "positive" mathematical structure. While some historians of science have debated the importance of Woodger's influence, Smocovitis offers evidence that he had a significant influence on biologists such as Haldane (Smocovitis 1996, pp. 100ff). This approach certainly made an impression on Fisher, who proposed a "fundamental theorem" of natural selection that, he suggested, was as significant as the second law of thermodynamics. The law stated that "the rate of increase in fitness of any organism at any time is equal[1] to its genetic variance in fitness at that time" (Gould 2002, p. 511).

---

[1] Technically, this should read "proportional", rather than equal, since the rate of change of a quantity and its variance will not have the same units.

In contrast to Fisher, some of Haldane's statements appear far more measured. He did not buy into the inferiority complex some biologists exhibited toward physics, but still saw the development of evolutionary biology in the context of a conflict between fields, writing that "a meeting-point between biology and physical science may at some point be found, there is no reason for doubting. But we may confidently predict that when a meeting-point is found, and one of the two sciences is swallowed up, that one will not be biology." (Smocovitis 1996, pp. 106–107). In other statements, however, Haldane advocated a more pragmatic approach, both about the perspective that science demands, and the difficulty of attaining it. He concluded his book *The Causes of Evolution* with the statement that "[t]he truly scientific attitude, which no scientist can constantly preserve, is a passionate attachment to reality, whether it be bright or dark, mysterious or intelligible."

Whatever the origins of the narrowed view of evolution that emerged as a dominant view in the mid-twentieth century, we shall see below that it played a catalytic role as a consensus in the Kuhnian sense. The modern synthesis defined a clear tradition against which "loci of trouble" such as the problems of collective dynamics in biological systems could stand out and demand attention as the twentieth century drew toward a close.

*** *** ***

The Modern Synthesis can be divided into two stages, each marking a decisive turning point in the development of evolutionary theory. In the first stage, the rediscovery of the work of Gregor Mendel was folded into Darwinian theory, as genetics was shown to provide a substrate of variation upon which natural selection could act. The start of the second phase can be traced to the late 1930s and early 1940s, with the publication of key works by Julian Huxley and Theodosius Dobzhansky. In this period, various core subfields of biology, such as paleontology, systematics, cytology, botany and morphology were brought into alignment with the new "synthetic" theory linking Mendel and Darwin. This codification of a body of knowledge sounds like nothing so much as the development of a Kuhnian "well-defined and deeply ingrained tradition". Yet it had the result, as we shall see, of limiting the perceived actions of natural selection to the organismal level and marginalizing non-adaptive explanations for observed phenomena.

The two stages of the Modern Synthesis have been termed, by Sewall Wright's biographer William Provine, as well as by Stephen Jay Gould, *restriction* and *hardening*. The restriction phase was driven by the work of R.A. Fisher, who first showed, in a 1918 article in the *Transactions of the Royal Society of Edinburgh*, that the variation underlying Darwinian natural selection could have a Mendelian basis. The particulate nature of Mendelian inheritance had initially been seen as antithetical to the continuous range of variations observed in biological populations, and thus it had not been clear initially whether Darwin and Mendel could be reconciled. Indeed, the rediscovery of Mendel's work led to a resurgence of support for "saltationism", a discontinuous evolutionary process advocated by Francis Galton and others. Fisher suggested that particulate inheritance was essential to prevent the degradation and "smearing out" of traits that could result from Darwin's model of blending inheritance. The full mathematical development of Fisher's theory, published in 1930 as *The Genetical Theory of Natural Selection,* inaugurated the field of population genetics. In the following year, Sewall Wright extended the new field in a paper entitled "Evolution in Mendelian Populations". Fisher and Wright effectively eliminated Lamarckian inheritance as a viable model, since the genetic basis of organismal traits ruled out the possibility of inheritance of acquired characteristics. This is what Weismann had sought to do three decades earlier, but, without the work of Mendel to build on, his model of the germ-plasm lacked the empirical basis to be convincing.

Fisher's work provided "restriction" in the sense of eliminating models of evolutionary change such as saltationism and Lamarckism, which no longer seemed consistent with empirical evidence. His work placed a strong focus on adaptation, but did not entirely rule out a role for random, non-adaptive change. This idea was emphasized and explored in great depth by Sewall Wright, and genetic drift was initially called the "Sewall Wright effect". Wright's emphasis on this aspect of evolutionary change ultimately led to his being marginalized within his field, as the restriction phase moved into a phase of hardening, and non-adaptive explanations were almost uniformly rejected.

The lingering pluralism in the work of Fisher and Wright is also apparent in J. B. S. Haldane's 1932 book, *The Causes of Evolution.* As Gould notes, there is pluralism even in Haldane's choice to use "causes" rather than "cause" in his title (Gould 2002, p. 514). After demolishing arguments for Lamarckian evolution by inheritance of acquired characteristics, Haldane squarely addressed the existence (and importance) of random characteristics without adaptive value. "There remain," he wrote, "a host of morphological characters which have no obvious value to their possessor...there is no doubt that innumerable characters show no sign of possessing selective value and, moreover, these are exactly the characters which enable a taxonomist to distinguish one species from another." (Haldane 1932/1966, pp. 113–114) Haldane also allowed the possibility of sudden, saltationist change, though he admitted that the processes behind such rapid evolution were unclear.

Another concept addressed by Haldane was orthogenesis, or evolution constrained to proceed in a particular direction, irrespective of selective advantage. He cited examples in which, contrary to Darwin's formulation, variations do not all occur equally in every direction. "The Scythians, according to Herodotus, lived largely on the milk of mares, and if mares had varied in the same way [as cows and goats] there can be little doubt that man would have selected mares with a high milk production, as he has selected she-goats. But we no more breed milch mares than racing bulls." (Haldane 1932/1966, p. 140)

Haldane considered instances of opposing selective pressures, as in the sexual selection pressure for male peacocks to grow larger and more beautiful tails that provide no advantage in the peacock's own individual survival (Haldane 1932/1966, p. 120), and may even be detrimental, hampering the bird's ability to fly. He argued against the fallacy "that natural selection will always make an organism fitter in its struggle with the environment" (Haldane 1932/1966, p. 119), citing the detrimental

Fig. 5.1 Two pioneers of
evolution. The young
Julian Huxley, being held
by his grandfather, Thomas
Huxley. Public domain

effects of evolutionary arms races between members of the same species. The combination of traits needed for high fitness (number of surviving offspring) itself was the product of a delicate balance. Haldane explained this using the example of broody hens.

> In the wild state a broody hen is likely to live a shorter life than a non-broody one, as she is more likely to be caught by a predatory enemy while sitting. But the non-broody hen will not rear a family, so genes determining this character will be eliminated in nature. With regard to maternal instincts of this type selection will presumably strike a balance. While a mother that abandoned her eggs or young in the face of the slightest danger would be ill-represented in posterity, one who, like the average bird, does so under a sufficiently intense stimulus will live to rear another family, which a too-devoted parent would not. (Haldane 1932/1966, p. 207)

Haldane considered the puzzle of social insects, in terms that anticipate the kin selection theories of the 1960s. "In the case of social insects…the workers and young queens are samples of the same set of genotypes, so any form of behavior in the former (however suicidal it may be) which is of advantage to the hive will promote the survival of the latter and thus tend to spread through the species." (Haldane 1932/1966, p. 207) As a result, "altruistic behavior is a kind of Darwinian fitness and may be expected to spread as the result of natural selection" (Haldane 1932/1966, p. 131). Haldane did not explicitly investigate, however, whose fitness was being measured, and whether selection was working on the insect or the colony.

Pluralism is also evident Julian Huxley's *Evolution, the Modern Synthesis* (1942), which gave the scientific programme its name (Fig. 5.1). The second chapter of his book is even titled "The Multiformity of Evolution."

> Just as there is no one method of the origin of species, so there is no one type of variation. Different evolutionary agencies differ in intensity and sometimes in kind in different sorts

of organisms, partly owing to differences in the environment, partly to differences in way of life, partly to differences in genetic machinery. No single formula can be universally applicable; but the different aspects of evolution must be studied afresh in every group of animals and plants. We are approaching the time when evolution must be studied not only broadly and deductively, not only intensively and analytically, but as a comparative subject. (Huxley 1942, p. 46)

There is no arguing, however, that adaptation was the central pillar of Huxley's synthesis. He described it as "omnipresent", though he allowed that some adaptive traits could conflict with one another. Adaptation, he said, "cannot but be universal among organisms, and every organism cannot be other than a bundle of adaptations, more or less detailed and efficient, coordinated in greater or lesser degree. On the other hand, adaptations subserving different functions may be mutually destructive" (Huxley 1942, pp. 420–421). Elsewhere he noted that "genes may have their expression altered by modifiers so as entirely to change their selective effect" (Huxley 1942, p. 60). His view of adaptation, and its genetic basis, was quite subtle and multi-layered, even writing a decade before Watson and Crick, and many decades before the birth of epigenetics.

Huxley dealt explicitly with various possible sources of non-adaptive evolutionary trends. He noted that there were limitations on the range of possible variations and, like Haldane, suggested that these limitations might drive orthogenesis, which

depends on a restriction of the type and quantity of genetic variation. When dominant it prescribes the direction of evolution: when subsidiary it merely limits its possibilities...but we are not yet able to be sure in most cases whether a limitation of variation as actually found in a group is due to a limitation in the supply of mutations or to selection, or to other causes. (Huxley 1942, p. 524)

However, Huxley dismissed orthogenesis as a rare phenomenon, noted the lack of any known mechanism behind it, and described the explanation of any phenomenon in terms of orthogenesis as "provisional". He noted that

even if the existence of orthogenesis ... be confirmed, it appears to be a rare and exceptional phenomenon, and ... we have no inkling of any mechanism by which it may be brought about. It is a description, not an explanation. Indeed it runs counter to fundamental selectionist principles. (Huxley 1942, p. 509)

He allowed some room for the possibility of genetic drift, though he felt that it was unlikely to play a major role in shaping the differences between organisms, at least at higher taxonomic levels.

It may be presumed, on somewhat indirect evidence, that 'useless' non-adaptive differences due to isolation of small groups may be enlarged by the addition of further differences of the same sort to give generic distinction [i.e., differences at the level of genera], though it seems probable that differences of family or higher rank are always or almost always adaptive in nature. (Huxley 1942, p. 44)

Nonetheless, Huxley remarked over and over on the existence of drift ("the Sewall Wright effect"), calling it "perhaps the most important of recent taxonomic discoveries...deduced mathematically from neo-mendelian premises and...empirically confirmed both in general and in detail" (Huxley 1942, p. 260). While noting that it is likely to dominate only in small populations, he cited numerous of examples of drift, and repeatedly stressed its role. Discussing polymorphisms in freshwater fish, for example, he noted their diversity ("well over 120 patterns...mostly dependent on the recombination of 15 gene-pairs"). "Some of the excessive variation", he concluded, "...is apparently due to the fixation of 'accidental' characters by drift" (Huxley 1942, pp. 100–101). Elsewhere, noting again the prevalence of drift in small, isolated populations, he discussed the superimposition of adaptive and non-adaptive effects. "When isolation is relatively complete and when, in addition, the isolated populations are small, non-adaptive is superimposed upon adaptive divergence, often to a marked degree, chiefly owing to what we have called the Sewall Wright effect, or drift" (Huxley 1942, p. 155). Elsewhere, he praised Wright's work as a key achievement of the axiomatization of evolution.

> The proof given by Wright, that non-adaptive differentiation will occur in small populations owing to 'drift', or the chance fixation of some new mutation or recombination, is one of the most important results of mathematical analysis applied to the facts of neo-mendelism. *It gives accident as well as adaptation a place in evolution*, and at one stroke explains many facts which puzzled earlier selectionists, notably the much greater degree of divergence shown by island than mainland forms, by forms in isolated lakes rather than in continuous river-systems. Turesson (1927) uses the term 'seclusion types' for such forms in plants. Recently Kramer and Mertens ... have provided a quantitative demonstration of the principle in their work on Adriatic lizards (*Lacerta sicula*). (Huxley 1942, pp. 199–200, my italics)

Huxley then proceeded to provide a detailed synopsis of Kramer and Mertens's 1938 paper. He listed drift examples ranging from Hawaiian land snails to the famous Galápagos finches. In a subtle parsing of the overlap between isolation, chance and selection, he noted Muller's 1940 observation that "isolation *per se* is a cause of differentiation", and then clarified how this relates to the Sewall Wright effect, selection, and historical contingency.

> This is due to the nature of the evolutionary process, which proceeds by the presentation of numerous small steps, and by the subsequent incorporation of some of them in the constitution by selection, or in some cases by Sewall Wright's 'drift'. The improbability of the mutative steps being identical in two isolated groups, even if they be pursuing parallel evolution [i.e., experiencing similar selective drives], is enormously high, so that reproductive incompatibilities will in the long run automatically arise between them. If the direction of selection differs for the two groups, visible divergence will also automatically result, even in the absence of divergence due to drift. (Huxley 1942, p. 259)

In another comment on drift, Huxley wrote that although "the most frequent mode of geographical differentiation is broadly adaptive, there are many cases in which apparently non-adaptive differentiation has occurred, *either predominantly or superposed on a general adaptive divergence, or as a correlate of invisible physiological adaptation*" (Huxley 1942, p. 242, my italics). He does not mention the possibility of these non-adaptive changes being later co-opted to serve other purposes, but if this next logical step had been made, he could have introduced the concept that Gould and Vrba were to label, decades later, *spandrels* or *exaptations*.

While Huxley clearly allowed for the possibility of other evolutionary processes such as drift, adaptation was beyond any doubt the central focus of his conception of the modern synthesis. This is an important part of our story because adaptation is most clearly conceptualized at the level of the individual organism. It is therefore no accident that the focus of the modern synthesis on adaptation went hand in hand with a nearly exclusive study of evolution at the organismal level. Even the discussions of drift cited above all refer to traits at the organismal level.

The increasing focus on adaptation shifted researchers' views away from selection, drift, or any other process at other levels. As a consequence, it shifted the focus away from possible interactions between levels, and hence away from the core dynamic tensions that form the central theme of this book. For all Huxley's references to non-adaptive processes, he gives short shrift to problems of selection at levels higher than the organismal. In one of his later chapters, he briefly discussed intergroup vs. intragroup (or interspecific vs. intraspecific) selection, in the context of the evolution of altruism in social insects and human societies. He called the notion that natural selection and adaptation can "be for the good of the species as a whole, for the good of the evolving type pursuing a long-range trend, for the good of the group undergoing adaptive radiation" simply a fallacy. This might be possible "if there were a way in which selection could be restricted to effects on the species as a species. But as a matter of fact selection acts via individuals" (Huxley 1942, p. 483). Huxley cited Haldane's work to demonstrate that only societies like social insects, which have separate reproductive castes, can evolve altruistic behaviors that benefit the group at the cost of the lives of individual organisms.

Huxley used this restrictive language in reference specifically to the evolution of altruism. Group selection leading to the evolution of sociality, however, was something he described as unquestionably present in nature even though he devoted only a few pages of his book to it. He held that "[i]n one sense, almost all selection is intraspecific, in that it operates by favouring certain types within the species at the expense of other types. The only exceptions would be when species spread or become extinct as wholes" (Huxley 1942, p. 478). Yet he recognized that selection can act at multiple levels, and can even act at one level with an adaptive effect at another. For example, "[s]election for speed in an ungulate will *operate* intraspecifically in a broad sense, but it is *directed* interspecifically in being concerned with escape from predators" (Huxley 1942, p. 478, his italics). The phenomenon he is describing is essentially the concept of MLS1, multi-level selection of type 1, which we will discuss in much greater depth in a later chapter. He contrasted this with sexual selection that has no direct benefit to average species fitness, writing that selection for "plumage in male birds is *directed* intraspecifically, in being concerned with the advantage of one male over another in reproduction. It would thus be more correct to speak of selection concerned with intra- or interspecific *adaptation*" (Huxley 1942, p. 479, my italics). He emphasized the variable relative strengths of inter- and intraspecific selection, noting that they "will often overlap and be combined; but the intensity of one or the other component may vary greatly" (Huxley 1942, p. 479). Citing Wright's work on group selection, he noted that selection between groups "with a functional basis" could be called social selection, "since it will encourage the gregarious instinct and social organization of all kinds" (Huxley 1942, p. 479). Referencing Allee's studies of the selective benefits of large population

Fig. 5.2 Theodosius
Dobzhansky in 1928.
Embryo Project
Encyclopedia (1928).
ISSN: 1940-5030 http://
embryo.asu.edu/
handle/10776/2937.
Licensed as Creative
Commons Attribution-
NonCommercial-Share
Alike 3.0 Unported

size (the Allee effect), he wrote that once "aggregations of a certain size enjoy various physiological advantages over single individuals…selection will encourage behavior making for aggregation *and the aggregation itself will become a target for selection*" (Huxley 1942, pp. 479–480, my italics). This type of selection, acting on a population-level property that is not simply the average of some property of the individual members of the group, is classified as MLS2, multi-level selection of type 2. Quoting a 1940 statement by Allee, Huxley wrote that "sociality is seen to be a phenomenon whose potentialities are as inherent in living protoplasm as are the potentialities of destructive competition" (Huxley 1942, p. 480).

*** *** ***

The second phase of development of the modern synthesis – the "hardening" – witnessed the rising predominance of a nearly exclusive focus on adaptation. The hardening phase began with the linking of various areas of biological science to the Mendelian Darwinism codified during the first phase. This included works by Dobzhansky (Fig. 5.2) on genetics (*Genetics and the Origin of Species*, in 1937), George Gaylord Simpson on paleontology (*Tempo and Mode in Evolution*, in 1944) and Ernst Mayr (Fig. 5.3) on systematics (*Systematics and the Origin of Species*, in 1942). Other biologists folded their own fields into the modern synthesis during this period as well – Rensch working in morphology, White in cytology, and Stebbins in botany. However, by the time the second editions of these major works were being prepared in the early 1950s (Dobzhansky and Simpson) and the early 1960s (Mayr), all three authors had "moved from pluralism to strict adaptationism – and along a remarkably similar path" (Gould 2002, p. 522). The later work of Sewall Wright followed a similar trajectory.

Fig. 5.3 Ernst Mayr. From the Archives of the Museum of Comparative Zoology, Ernst Mayr Library, Harvard University

In the first edition of Dobzhansky's *Genetics and the Origin of Species*, he laid out a programme that allowed for pluralism and macroevolution. He considered the problem of discontinuity between species to be the fundamental problem of evolutionary biology (Gould 2002, p. 232). He urged a shift away from an exclusive focus on population genetics, even though *Drosophila* genetics was his own particular field of study. He wrote that

> [t]he origin of hereditary variations is…only a part of the mechanism of evolution…
> Mutations and chromosomal changes are constantly arising at a finite rate, presumably in
> all organisms. But in nature we do not find a single greatly variable population of living
> things which becomes more and more variable as time goes on; instead, the organic world
> is segregated into more than a million separate species, each of which possesses its own
> limited supply of variability which it does not share with the others…The origin of spe-
> cies…constitutes a problem which is logically distinct from that of hereditary variation.
> (quoted by Gould, p. 2002, p. 533)

Elsewhere, Dobzhansky addressed population-level traits and the degree to which the survival of individual organisms was subsumed beneath the group-level need for high variability. "Evolutionary plasticity can only be purchased at the ruthlessly dear price of continuously sacrificing individuals to death from unfavorable mutations" (quoted by Borello, 2010, p. 4).

Between the first (1937) and third (1951) editions of *Genetics and the Origin of Species*, Dobzhansky deleted two chapters that focused on non-adaptive aspects of evolutionary change; a small amount of the material in these chapters was sprinkled throughout the updated text, but most of it was excised entirely (Gould 2002, p. 526). He mentioned the role of drift in small populations, but with far less emphasis than Huxley, who, as we have seen, repeatedly referred to the work of Sewall

Wright. Dobzhansky agreed that "much evidence secured by different biological disciplines…attests the existence of phenomena which can most plausibly be accounted for by genetic drift" (Dobzhansky 1951, p. 176). As an example of such a phenomenon, he cited the distribution of the desert plant *Linanthus parryae* in the Mojave Desert, noting that "as to [its] microgeographic variation…, the most reasonable hypothesis is that it is caused by genetic drift" (Dobzhansky 1951, p. 169). His overall conclusion about drift, however, was the rather halfhearted concession that it might have some effect in small populations, but further studies were needed.

> It is not possible at present to reach definitive conclusions regarding the role played by genetic drift in evolutionary processes. If genetically effective population sizes in many species are small, at least at some periods of their evolutionary histories, the genetic drift would have to be recognized as an important factor. If, on the other hand, the population sizes are usually so large that they may be regarded for practical purposes as infinite, the genetic drift will remain only an interesting theoretical possibility. Some polemics, more acrimonious than enlightening, have arisen in biological literature concerning this problem. The only conclusion to be drawn from these polemics is that the available observational and experimental evidence is altogether insufficient, and that more work in this field is urgently necessary. (Dobzhansky 1951, pp. 164–165)

Perhaps most telling, however, are the statements in Dobzhansky's 1951 first chapter, "Organic Diversity", regarding adaptive landscapes. As Gould points out (2002, pp. 527–528), Dobzhansky took the idea of an adaptive landscape, originally introduced by Sewall Wright as a conceptual model for the co-existence of multiple possible states (adaptive peaks) of a population, with no state exhibiting a global fitness maximum, and recast it as a frozen landscape of optimal niches.

> The enormous diversity of organisms may be envisaged as correlated with the immense variety of environments and of ecological niches which exist on earth. But the variety of ecological niches is not only immense, it is also discontinuous. One species of insect may feed on, for example, oak leaves, and another species on pine needles; an insect that would require food intermediate between oak and pine would probably starve to death. Hence, the living world is not a formless mass of randomly combining genes and traits, but a great array of families of related gene combinations, which are clustered on a large but finite number of adaptive peaks. Each living species can be thought of as occupying one of the available peaks in the field of gene combinations. The adaptive valleys are deserted and empty… The hierarchic nature of the biological classification reflects the objectively ascertainable discontinuity of adaptive niches, in other words the discontinuity of ways and means by which organisms that inhabit the world derive their livelihood from the environment. (Dobzhansky, 1951, pp. 9–10)

This interpretation denies a place for historical contingency, a concept that is as closely linked to drift as the concept of adaptation is linked to organismal selection. It ignores entirely the fact that "[g]enealogy, not current adaptation, provides the primary source for clumped distribution in morphological space" (Gould 2002, p. 528).[2]

---

[2] This has been borne out by recent studies of computational evolutionary models, including some from my own research group, which show that organisms can cluster even on a *neutral* morphospace. Interestingly, the process by which this occurs parallels the mathematical process of a branching and coalescing random walk.

George Gaylord Simpson published *Tempo and Mode in Evolution* in 1944. Nine years later, he expanded it into a new work entitled *The Major Features of Evolution*. Like Dobzhansky, Simpson's work was initially pluralistic, and became much more narrowly focused on adaptation as the exclusive driving mechanism of evolutionary change. In *Tempo and Mode*, he explored the problem of macroevolutionary trends, with particular emphasis on the divergence of body plans within the fossil record. Are these discontinuities driven by the same processes that drive smaller-scale, continuous adaptations in individual populations, or are they driven by forces that are fundamentally different in kind? "This is related," he wrote, "to the old but still vital problem of micro-evolution as opposed to macro-evolution…If the two proved to be basically different, innumerable studies of micro-evolution would become relatively unimportant and would have a minor value to the study of evolution as a whole" (quoted by Gould 2002, p. 529).

In 1944, Simpson proposed the idea of quantum evolution in order to explain macro-evolutionary trends. He suggested that genetic drift might push a small population into an "inadaptive phase", off a local adaptive peak. The population would then be forced to change or die – to develop major new adaptations to enable it to thrive in its new circumstances, or to become extinct. For this reason, Simpson called it a "quantum" phenomenon: it was all or nothing. Simpson's idea allowed for the same underlying mechanism to drive macroevolutionary trends and small-scale adaptive trends. That mechanism was genetic variation. And yet there was a fundamental difference: natural selection and adaptation were present in the micro-evolutionary process, but *not* in the macroevolutionary one. He called his quantum evolution idea "perhaps the most important outcome of [his] investigation, but also the most controversial and hypothetical" (quoted by Gould 2002, p. 529). Elsewhere in *Tempo and Mode*, he emphasized that the quantum evolution idea dethroned selection from its role as the sole driver of evolutionary change.

> The aspects of tempo and mode that have now been discussed give little support to the extreme dictum that all evolution is primarily adaptive. Selection is a truly creative force and not solely negative in action.[3] It is one of the crucial determinants of evolution, although under special circumstances it may be ineffective, and the rise of characters indifferent or even opposed to selection is explicable and does not contradict this usually decisive influence. (quoted by Gould 2002, p. 530)

By 1953, Simpson's view of adaptation had become much more dogmatic, and his definition of quantum evolution had completely changed. He now argued that any advantageous evolutionary trend must have a selective advantage at its inception, thus ruling out the possibility for positive benefits to "hitchhike" on other adaptive features. He now declared that "[g]enetic drift is certainly not involved in all or in most origins of higher categories, even of very high categories such as classes or phyla" (Gould 2002, p. 530). Most strikingly, he now completely redefined quantum evolution as a sort of accelerated selection which drove the

---

[3] This is a reference to the old complaint that natural selection is a destroyer, not a creator, since it does not itself create variability in organisms. The false conclusion from this is that natural selection cannot drive evolutionary innovation.

"continuous maintenance of adaptation", and "not a different sort of evolution from phyletic evolution, or even a distinctly different element of the total phylogenetic pattern. It is a special, more or less extreme and limiting case of phyletic evolution... Indeed the relatively rapid change in such a shift is more rigidly adaptive than are slower phases of phyletic change" (quoted by Gould 2002, p. 531).

A similar shift in perspective from pluralism to exclusive adaptationism can be traced in the works of Ernst Mayr, *Systematics and the Origin of Species* (1942) and *Animal Species and Evolution* (1963). As a systematist, Mayr was intimately concerned with the process of speciation itself, which is certainly a problem that extends beyond adaptation at the level of the single organism. In 1942, Mayr displayed an interest in speciation as distinct from the problem of microevolutionary adaptation. The Modern Synthesis did mean that it was "feasible to interpret the findings and generalizations of the macroevolutionists on the basis of known genetic facts", but various factors such as "selection, random gene loss, and similar factors, together with isolation, make it possible to explain species formation on the basis of mutability, without any recourse to Lamarckian forces" (quoted by Gould 2002, p. 534). This ruled out the inheritance of acquired characteristics (as part of the sweeping-up operation of the "restriction" phase of the Synthesis), but did not define the speciation as reducible to selection and adaptation. Mayr clarified that factors "that promote or impede divergence...may be subdivided further into adaptive (selection) and non-adaptive factors." Among phenomena driven by non-adaptive factors, he gave the example of polymorphism within species, citing

> considerable indirect evidence that most of the characters that are involved in polymorphism are completely neutral, as far as survival value is concerned. There is, for example, no reason to believe that the presence or absence of a band on a snail shell would be a noticeable selective advantage or disadvantage. (quoted by Gould 2002, p. 534)

Mayr further emphasized the potentially nonadaptive nature of inter- and intraspecies geographical variation.

> It should not be assumed that all the differences between populations and species are purely adaptational and that they owe their existence to their superior selective qualities... Many combinations of color patterns, spots and bands, as well as extra bristles and wing veins, are probably largely accidental. This is particularly true in regions with many stationary, small, and well-isolated populations, such as we find commonly in tropical and insular species... We must stress the point that not all geographic variation is adaptive. (quoted by Gould 2002, p. 535)

What a difference two decades make. By 1963's *Animal Species and Evolution*, non-adaptive aspects of evolution had been relegated entirely to the sidelines. Mayr's focus was now exclusively on adaptation. "One conclusion emerges", he wrote, "more strongly than any other: every local population is very precisely adjusted in its phenotype to the exacting requirements of the local environment. This adjustment is the result of a selection of genes producing an optimal phenotype" (quoted by Gould 2002, p. 537). He now considered local populations to be virtually the exclusive product of "a continuing selection process". He did concede that not all species were absolutely optimized in all their characters, writing that, while "the genotype of each local population has been selected for the production of a well-adapted phenotype", it did not necessarily "follow from this conclusion, however, that every detail of the phenotype is maximally adaptive." Yet this lack of optimization was not a result of non-adaptive forces. Mayr no longer considered polymorphisms to be "probably largely accidental". He allowed that they might not be of selective value (variable spots on the back of a ladybug do not "necessarily mean that the extra spots are essential for survival"). But he no longer considered them to be the result of drift. Rather, their appearance "means that the genotype that has evolved in this area *as the result of selection* develops additional spots on the elytra" (quoted by Gould, p. 537, my italics). Thus non-adaptive characters arise via linkage with characters under selection (this can happen in nature, of course, but not to the *exclusion* of drift). And perhaps they really were adaptive after all. Mayr concluded this passage by remarking that "close analysis often reveals unsuspected qualities even in minute details of the phenotype."

In 1942, Mayr felt he "must stress the point that not all geographic variation is adaptive". In 1963, he stressed that it was. "The geographic variation of species," he wrote, "is the inevitable consequence of the geographic variation of the environment. A species must adapt itself in different parts of its range to the demands of the local environment. Every local population is under continuous selection pressure for maximal fitness in the particular area where it occurs" (quoted by Gould 2002, p. 538).

By 1963 Mayr had come to the conclusion that

> neutral genes are improbable for physiological reasons....It seems unrealistic to me to assume that the nature of the particular chemical (enzyme or other product) should be without any effect whatsoever on the fitness of the ultimate phenotype. A gene may be selectively neutral when placed on a particular genetic background in a particular temporary physical and biotic environment. However, genetic background as well as environment change continually in natural populations and I consider it therefore exceedingly unlikely that any gene will remain selectively neutral for any length of time. (quoted by Gould 2002, p. 538)

In fact, the idea that a genetic variation might be initially selectively neutral, but could later may take on selective value, is an important concept in current pluralistic evolutionary theory. We will revisit this idea in later chapters in the context of spandrels and exaptations.

During the two decades between the publication of Mayr's books, some traits that were originally thought to be neutral were found to have a selective advantage. This strengthened Mayr's growing assumption that neutral traits were so rare as to be virtually impossible. "Virtually every case quoted in the past as caused by genetic drift...has recently been reinterpreted in terms of selection pressures," he wrote in 1963 (quoted by Gould 2002, p. 538). As examples, he cited the recently discovered role in copulation of the tarsal combs in some *Drosophila* species and the role of patterns on snail shells previously viewed as "cryptic" in helping the snails evade predation. From these observations, however, he concluded that "[s]elective neutrality can be excluded almost automatically wherever polymorphism or character clines are found in natural populations" (quoted by Gould 2002, p. 538). If the

"almost" seems like a last fragile concession to pluralism, it did not outweigh
Mayr's certainty of *the impossibility of disproving* that a trait had selective value.
"One can never assert with confidence that a given structure does not have selective
significance," Mayr concluded (quoted by Gould 2002, p. 538). One could object
that it is theoretically possible to show, with statistical significance, that possessors
of all variants of some hypothetical trait have equal fitness, defined as number of
surviving offspring. Even leaving this objection aside, Mayr's statement certainly
does not constitute a proof that all structures *do* have selective significance. As
Gould comments, Mayr's work had rendered "the conceptual space of evolutionary
inquiry…so reconfigured that hardly any room (or even language) remain[ed] for
considering, or even formulating, a potential way to consider answers outside an
adaptationist framework" (Gould 2002, p. 539).

*** *** ***

Perhaps the most complex intellectual trajectory taken during this period was
that of Sewall Wright. Unlike many of his contemporaries, his critical contribution
to the Modern Synthesis was initially made not in a book, but in an article entitled
"Evolution in Mendelian Populations", published in *Genetics* in May 1931. Here,
he complemented the work of Fisher by solidifying a mathematical approach to
population genetics. Like Fisher, Wright had to assign some distribution of genetic
variability in the null case where natural selection does not act. This led to a focus
on the effects of genetic drift which, as we have seen above, was referred to for
many years as the "Sewall Wright effect". Wright repeatedly asserted that the action
of drift alone could be responsible for a significant amount of observed biological
diversity. "It appears," he wrote, "that the actual differences among natural geographical races and subspecies are to a large extent of the nonadaptive sort expected
from random drifting apart" (Wright 1931, p. 127). He cited the distribution of
human blood types as an example. He referred to "nonadaptive branching following
isolation as the usual mode of origin of subspecies, species, perhaps even genera,
[with] adaptive branching giving rise occasionally to species which may originate
new families, orders, etc." (Wright 1931, p. 153).

In a shorter and less mathematically dense paper that he presented to a genetics
conference the following year (having been asked to provide a more "accessible"
summary of his views), Wright wrote "[t]hat evolution involves nonadaptive differentiation to a large extent at the subspecies and even the species level is indicated
by the kinds of differences by which such groups are actually distinguished by sys-
tematists. It is only at the subfamily and family levels that clearcut adaptive differences become the rule… The principal evolutionary mechanism in the origin of
species must then be an essentially nonadaptive one" (quoted by Provine 1986,
p. 290).

As Provine pointed out, Wright's interpretation of the role of nonadaptive mechanisms was strongly driven by the contemporary scientific literature. Provine's survey of systematics publications of the 1920s showed a scientific consensus that
closely related species exhibited more divergence in nonadaptive than in adaptive

traits (Provine 1986, p. 292). (At the same time, there was also a school of thought which considered most differences between species to be adaptive; this view eventually gained dominance as many traits formerly thought to be nonadaptive were shown to have selective value, and these results strongly influenced the shift in Mayr's thought, as described above.)

Among the major works in the systematics literature that influenced Wright were those of Robson and Richards.[4] In a 1926 *Nature* paper whose themes were echoed in Robson's 1928 book *The Species Problem*, they objected to adaptationist "storytelling" in terms that, in a quiet and genteel way, presage Gould and Lewontin's seminal 1979 paper *The Spandrels of San Marco and the Panglossian Paradigm: A Critique of the Adaptationist Programme*. "In the first place," they wrote, "it is practically impossible to show that a character is not of value to an organism without an exhaustive knowledge of the life history and physiology of the latter. On the other hand, the adaptive value of a structure must not be presumed in default of evidence to the contrary" (Richards and Robson 1926).

A decade later, in their book *The Variation of Animals in Nature*, Robson and Richards issued a critique that reads like a direct rebuttal of Mayr's unshakeable faith in the ubiquity of adaptation.

> A survey of the characters which differentiate species (and to a less extent genera) reveals that in the vast majority of cases the specific characters have no known adaptive significance… It may be conceded that in a number of instances structures apparently useless may in the future be found to play an important part in the life of the species; further, many "useless" characters may be correlated with less obvious features which are of real use, but, even allowing for this, the number of apparently useless specific characters is so large that any theory which merely *assumes* that they are indirectly adaptive is bound to be more a matter of predilection than of scientific reasoning. (quoted by Provine, 1986, p. 293)

As the adaptationist viewpoint became increasingly dominant, advocates for nonadaptive change were pushed aside by mid-twentieth century biology. Yet Wright's story is far more complex than this. In later years, in interviews with both Gould and Provine, Wright argued that he had not advocated a major role for drift in driving evolutionary diversity. In fact, in 1967, he wrote "I have never attributed any evolutionary significance to random drift except as a trigger that may release selection toward a higher selective peak through accidental crossing of a threshold" (quoted by Provine 1986, p. 289). As we have seen, quotations from his earlier

---

[4] Richards and Robson's (1926) *Nature* paper (actually, the second of two short papers they published in *Nature* that year) also contains an interesting discussion of the role of drift and historical contingency in driving nonadaptive change. "Many animals are subjected to severe fluctuation in numbers through epidemics, bad weather, etc., and this has two possible effects. The usual result would be that the small number of individuals left over after, for example, and epidemic, would not include many of the uncommon variations, and the number of different variants in the population would be continually limited. After a minimum there is of course room for many more individuals than are actually found, so that during the subsequent multiplication there may be little or no competition. If an uncommon variant survived by accident, or if an unusually low minimum did not leave a random sample of survivors, then in the course of multiplication the character of the population would be changed. At present this seems to be the most likely means by which an entirely unadapative character could spread." (Richards and Robson 1926)

papers appear to flatly contradict this statement. However, it is true that Wright had never conceived of drift as *merely* a driver of nonadaptive change.

Wright was the first to introduce the idea of an adaptive evolutionary landscape. In this model, how could a population on a local adaptive peak ever explore the rest of the terrain? With his *shifting balance theory*, Wright suggested that genetic drift could lead a population down from a local peak, allowing the population, now subject to influence of natural selection, to explore other adaptive peaks. In his 1932 paper he noted that because "species will be shuffled out of low peaks more easily than high ones, [the total population] should gradually find its way to the higher general regions of the field [adaptive landscape] as a whole" (quoted by Provine 1986, p. 285). Thus Wright conceived of drift as an essential helpmate to natural selection, a source of noise which facilitated exploration of the adaptive landscape (Fig. 5.4), as a small increase in temperature can allow a protein to explore its folding landscape. As Provine noted, this idea was not present in Wright's 1931 work, and thus there is no evidence that his initial ideas of drift were introduced *in order* to explain how species explore the adaptive landscape. Nonetheless, it was a key aspect of Wright's work from 1932 onward, and claims of a standalone role for nonadaptive evolutionary change were increasingly relegated to the shadows.

Wright himself was pushed into the shadows as well. On the surface, his marginalization by the evolutionary biology community seems at odds with his increasing focus on the role of drift in facilitating adaptation. Certainly, even though he essentially turned his back on nonadaptive change as a significant evolutionary process, his name was forever associated with "the Sewall Wright effect", and this may have been a factor in how colleagues responded to him. But Gould has suggested that a more complex problem was at play (2002, pp. 555–556). Wright's shifting balance model allowed populations (demes) within a species to be pushed off locally adaptive peaks into valleys, and from there they could be driven by natural selection to explore other, perhaps higher, peaks, thus possibly contributing to an overall increase in adaptation for the population as a whole. But this process of drift followed by selection could result in various demes existing on different peaks; depending on the level of geographical isolation, these demes could then compete with one another, perhaps furthering the process of speciation. Thus *Wright's theory included the potential for selection between groups*. He even initially planned to call the model the "two-level theory". Variation among organisms provides a substrate for interdemic sorting. And if the model is extrapolated to species, it becomes a full model of multi-level selection. Gould suggested that Wright was marginalized because his contemporaries were growing so increasingly committed to the idea of selection at the organismal level alone that they simply could not take a multi-level theory seriously, even if it had adaptation at its core.

Gould's explanation may have been too hasty, however. While some of Wright's contemporaries may indeed have been unable to "handle the truth" of multi-level selection, it is not entirely clear that Wright was truly considering a case that could be defined as multi-level selection. Ambiguities abound in the interpretation of his concept of interdemic competition. Wright described his shifting balance theory as having three separate phases: random drift, intrademic (or intragroup) selection,

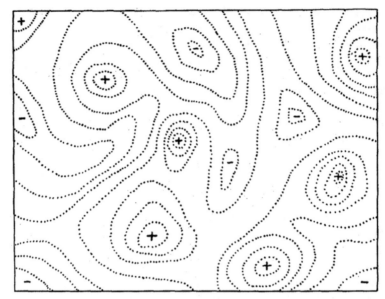

FIGURE 2.—Diagrammatic representation of the field of gene combinations in two dimensions instead of many thousands. Dotted lines represent contours with respect to adaptiveness.

Fig. 5.4 The first image of an adaptive landscape appeared in Wright's paper entitled "The roles of mutation, inbreeding, crossbreeding and selection in evolution", published in the Proceedings of the Sixth International Congress of Genetics (1932). Figure reproduced with the permission of the Genetics Society of America

and interdemic (or intergroup) selection. In his 1932 paper he described the spread of a population through the adaptive landscape.

> With many local races, each spreading over a considerable field and moving relatively rapidly in the more general field about the controlling peak, the chances are good that one at least will come under the influence of another peak. If [this is] a higher peak, this race will expand in numbers and by crossbreeding with the others will pull the whole species toward the new position. The average adaptiveness of the species thus advances under intergroup selection, an enormously more effective process than intragroup selection. (quoted by Provine 1986, p. 287)

If this is the essence of Wright's idea of interdemic selection, Provine notes that it has a fundamental flaw. It deals with neither selection between the demes, nor competition between them. It is describing, instead, "an interaction of individual selection (meaning intragroup or intrademic selection) with population structure and migration" (Provine 1986, p. 288). Thus this is not, in fact, a model of intergroup selection at all! However, Wright did also explicitly suggest competition between species, a case that cannot involve gene flow and cannot be reduced to intragroup selection. In the same 1932 paper he wrote that "effective intergroup competition leading to adaptive advance may be between species rather than races" (quoted by Provine 1986, p. 287). The multi-level aspect of Wright's work may have failed to persuade as much because of its ambiguity as because of the prevailing "organism only" view that formed the hardened core of the modern synthesis.

Ambiguities aside, it is fitting to end this survey of an increasingly narrow view of evolution with a taste of how the path would again widen into a rich field of pluralism. In Wright's adaptive landscape, variability among individuals provides a mechanism by which selection acts on populations. The particular organisms in a group give the group its distinctive characteristics, just as the particular genes in an organism give the organism its distinctive traits. If these characteristics are bound together in a reproductively isolated group, just as the traits are bound together in the phenotypic package of an organism, then the group can experience natural selection just as does the organism. The group is a less tightly bound package, but this comes as no surprise, since physical reality is not scale-free. Similar patterns of interaction may occur at multiple levels, but the inherent size scales of molecules, cells, organisms and populations necessitate that these interactions will have differences, as well as resemblances. Yet each level relies on the one below as a source of variation, and each level provides the variation for the one above.

In an interview late in his life, Sewall Wright told Stephen Jay Gould that he considered the major error of the modern synthesis to be its "exclusive focus on individual selection" (Gould 2002, p. 555).

References

Borello ME (2010) Evolutionary Restraints: The Contentious History of Group Selection. The University of Chicago Press, Chicago/London

Dobzhansky T (1951) Genetics and the Origin of Species. Columbia University Press, New York/London

Gould SJ (2002) The Structure of Evolutionary Theory. The Belknap Press of Harvard University Press, Cambridge MA/London

Gould SJ, Lewontin RC (1979) The spandrels of San Marco and the Panglossian paradigm: a critique of the adaptationist programme. Proc R Soc B 205:581–598

Haldane JBS (1966) The Causes of Evolution. Cornell University Press, Ithaca. (Reprint of 1932 edition)

Huxley J (1942) Evolution: The Modern Synthesis. Harper & Brothers, New York

Provine WB (1986) Sewall Wright and Evolutionary Biology. University of Chicago Press, Chicago/London

Richards OW, Robson GC (1926) The species problem and evolution. Nature 117(2941):382–384

Smocovitis VB (1996) Unifying Biology. Princeton University Press, Princeton

Wright S (1931) Evolution in Mendelian populations. Genetics 16:97–159

Wright S (1932) The roles of mutation, inbreeding, crossbreeding and selection in evolution. Proc Sixth Intl Congr Genetics 1:356–366

Chapter 6
Selfish Creatures, Huddled Together for Warmth

That's rubbish!! Start again!

Bernard Black

THOUGH THE major players in the Modern Synthesis were drifting toward an organism-only view, the problem of the interaction between individual animals in groups, colonies, populations and species did not go away. One of the pioneering investigators of animal interaction was the Chicago-based ecologist Warder Clyde Allee (Fig. 6.1). Allee's laboratory studied the positive benefits of interaction (or simply of co-existence) between individuals (see Courchamp et al. 2008 for an overview of Allee's work and influence). His studies, primarily on fish, emphasized the role of cooperation in promoting animal survival under conditions of stress or scarcity. In works such as *The Social Life of Animals* (1938), where he compiled the results of studies from many other laboratories in addition to his own, Allee demonstrated again and again that, while competition clearly occurs in nature, cooperation provides an important balancing tendency. He emphasized that, while under many conditions overcrowding could be deleterious to a population, under other circumstances *undercrowding* could be equally detrimental. This phenomenon is now known as the Allee effect, and appears regularly in nonlinear models of population dynamics, as well as in experimental studies ranging from the mating behavior of wild dogs to quorum sensing in the formation of bacterial biofilms.

The currently accepted definition of an Allee effect is "a *causal* positive relationship between (a component of) *individual* fitness and either population size or density" (Courchamp et al. 2008, p. 14, their italics). Allee effects thus relate to a situation where a group-level property affects individual-level fitness. It thus exists at the murky boundary between the concepts of fitness involved in multi-level selection of type 1 (MLS1) and multi-level selection of type 2 (MLS2). Fitness in the first case is defined as the average fitness of the individual members of the group, whereas in the second case it is a property that does not reduce to an average of the individual fitnesses (Okasha 2006). In fact, Allee effects could be interpreted as a sort of MLS1 in reverse, in which group properties affect individual fitness, rather than group fitness being defined by an average of individual properties.

One early and well-studied example of an Allee effect is the density-dependent reproduction rate of the flour beetle, *Tribolium confusum*. Living "in a flour beetle's

little world, a microcosm of thirty-two grams of flour" (Allee 1938, p. 104), populations at low density have difficulty surviving. But so do larger populations. A plot of
reproductive rate vs. initial population density had a peak at about four beetles per
32 g of flour. Describing studies performed by Thomas Park, Allee wrote that

> the results come from the interaction of two opposing tendencies. In the first place, adult
> beetles roam at random through their floury universe. They eat the flour, but they may also
> eat their own eggs as they encounter these on their travels. This habit of egg-eating tends to
> reduce the rate of population growth, the more so the denser the population. The second
> factor is the experimentally proven fact that up to a certain point copulation and successive
> re-copulation stimulate the female *Tribolium* beetles to lay more eggs, and eggs with a
> higher percentage of fertility. Thus, the more dense the beetle population, the more rapid its
> rate of increase. The interaction of these two opposing tendencies results in an intermediate
> optimal population in which more offspring are produced per adult animal than in either
> more or less dense populations. (Allee 1938, pp. 105–106)

Later in the same work, Allee modulated his characterization of these two tendencies as opposing. He described them as

> these two fundamental principles, the struggle for existence and the necessity for coopera-
> tion, both of which, consciously or unconsciously, penetrate all nature; and I shall say now
> that one may find that these two principles are not always in direct opposition to each other;
> that there is evidence that these basic forces have acted together to shape the course of
> evolution and even the evolution of social relations... (Allee 1938, p. 211)

Allee asked whether a minimal number of individuals was necessary to maintain a population. "Over and over again in the last half-dozen years," he wrote in 1938,

I have asked field naturalists, students of birds, wildlife managers, anyone and everyone who might have had experience in that direction, how many members of a given species could maintain themselves in a given situation. Always until this last summer I have found that, stripped of extra verbiage behind which they might hide their ignorance, the real answer was that they did not know. (Allee 1938, p. 107)

Allee found himself closing in on an answer when a visiting colleague from South Africa, J. Phillips, mentioned observations of minimal viable elephant populations in the Knysna and Addo Forests. From there, Allee began to collate studies of minimum populations in species ranging from tsetse flies to laughing gulls on Muskeget Island off Nantucket. Over and over he found that the minimal number was more than two. Clearly, populations were interacting in ways that were more complex than simply breeding. In some cases, positive population effects resulted from increased safety from predators or protection from the elements. That bobwhite quails, penguins, and other birds huddled together for warmth came as no surprise. A more complex example of temperature effects, however, was identified by Allee's research group in mice. Vetulani had found in 1931 that mice grown two to four in a cage reached a larger size by 20 weeks of age than those raised in isolation, and also grew larger than those raised under more crowded conditions, with five or more per cage. The stressful negative effects of crowding were easy to hypothesize. But why did the groups of two to four grow faster than the isolated mice? Loneliness? A more mechanistic interpretation came when E. Retzlaff, working in Allee's laboratory, was able to reproduce Vetulani's results at 16 °C but not at 29–30 °C. The interpretation seemed to be that isolated mice under these conditions had to expend energy on temperature regulation that could, if they had siblings to nest with, have been directed toward growth.

In other cases studied in Allee's own laboratory, enhanced group survival appeared to result from collective chemical effects. He found, for example, that groups of goldfish could precipitate a toxic solution of colloidal silver (and thus better survive exposure to its toxic effects), while isolated fish could not (Allee 1938, pp. 53–56).[1] In collaboration with J. Wilder, Allee exposed populations of *Planaria* to ultraviolet radiation, finding that those radiated in a group survived longer, even controlling for the possibility of one worm shielding another from the radiation under crowded conditions (Allee 1938, pp. 59–60). Here, the mechanism of the group protective effect was not immediately obvious. But in other cases, such as experiments performed by J. R. Fowler in 1931, *Daphnia* were found to survive better in groups when exposed to an excessively alkaline solution. "The reason here is simple," Allee explained. "The grouped animals give off more carbon dioxide,

---

[1] Note an apparent error in Allee's Table 1 (p. 54), where he states that the grouped fish survived for 182 min and the isolated fish survived for an average of 507 min, a difference in survival time with $p < 0.001$; this contradicts the statement on page 56 that "those in the groups of ten lived decidedly longer than their fellows exposed singly to the same amount of the same poison, and significantly so." Presumably the survival times are reversed in the table.

and this neutralizes the alkali. Long before the isolated individual can accomplish this, it is dead; in the group those on the outside may succumb, though if the number present is large enough even they may be able to live until the environment is brought under temporary control." (Allee 1938, p. 57)

Not all of Allee's experiments identified protective group effects against adverse conditions, however. Working with G. Evans, Allee found a faster rate of cell division in sea urchin eggs when crowded; they also observed a growth-inhibiting effect of undercrowding (Allee 1938, pp. 74–75). Citing experiments performed by F. Peebles in 1929, Allee speculated that these effects might be attributable to growth-enhancing and growth-inhibiting "extracts" made from sea-urchin eggs and larvae.

While Allee did not focus his work specifically on evolutionary biology, he noted, citing Sewall Wright, that a sufficient population size was necessary to maintain sufficient genetic diversity to adapt to changing conditions (Allee 1938, pp. 117ff). He was also keenly aware of his studies' implications for selection at the population level, writing that

> in the more poetic post-Darwinian days this struggle [for existence] was thought of as so intense and so personal that an improved fork in a bristle or a sharper claw or an oilier feather might turn the balance toward the favored animal. Now we find the struggle for existence mainly a matter of populations, measured in the long run only, and then by slight shifts in the ratio of births to deaths. (Allee 1938, p. 51)

*** *** ***

Another scientist who emphasized group-level dynamics – with, in contrast to Allee, a specific focus on evolution – during the mid-twentieth century was Vero Copner Wynne-Edwards (Fig. 6.2). An Oxford-trained ornithologist and ecologist, Wynne-Edwards was deeply struck by the pluralism of the early synthesis, particularly by the first (1937) edition of Dobzhansky's *Genetics and the Origin of Species*. He even concluded that a focus on the population level was a key finding of the Modern Synthesis, writing in 1948 that "the fundamental new idea is that populations, rather than individuals, are the basic units upon which evolutionary processes act" (Borello 2010, p. 4). Indeed, in the first chapter of his monumental treatise on *Animal Dispersion in relation to Social Behavior*, he described selection at the group and individual level together as fundamental tenets of "our Darwinian heritage", writing that

> selection operates largely or entirely at two levels, discriminating on the one hand in favour of *individuals* that are better adapted and consequently leave more surviving progeny than their fellows; and on the other hand between one *species* and another where their interests overlap and conflict, and where one proves more efficient in making a living than the other. (Wynne-Edwards 1962, p. 18)

Fig. 6.2 Wynne-Edwards
doing field work on Baffin
Island (Reproduced with
the kind permission of
Queen's University
Archives)

Wynne-Edwards began his studies at Oxford while Julian Huxley was directing the Department of Zoology there. When Huxley left for London, Charles Elton succeeded him, and Elton's views on animal ecology had a deep impact on Wynne-Edwards's later work. Elton emphasized, for example, the active role that animals may play in ensuring their own survival. Elton wrote that animals do not "sit about waiting for the environment mindlessly to select the fittest to survive, as plants must. They generally can and often do move from places in which they are not doing well into places in which they may do better" (quoted by Borello 2010, p. 44). This sort of thinking helped ignite Wynne-Edwards's lifelong interest in the role of animal behavior in driving evolutionary processes.

One of Wynne-Edwards's earliest works was a paper on "The Behaviour of Starlings in Winter", published in *British Birds* in 1929. Here, he explored concepts that were to become key themes in his later work. Previous studies had suggested that starling populations had risen dramatically in Britain in the wake of agricultural development, which had resulted in the increased availability of food. The evidence, Wynne-Edwards argued, suggested that food was not a limiting factor; he also pointed out that the timeframe of agricultural development and starling population rise did not match anyway. This suggestion that food is not a limiting factor in population growth directly counters the Malthusian idea of individual animals struggling against one another for survival. It startled Wynne-Edwards's contemporaries, but he was soon to collect more data in support of this surprising conclusion.

In 1930, Wynne-Edwards travelled to Canada to take a position at McGill University. On the trip across the Atlantic, he studied the distribution patterns of coastal, offshore and deep-water sea birds. He was deeply struck by the sparseness of these ecological communities, and this observation also played a fundamental role in the development of his ideas. As quoted in Mark Borello's *Evolutionary*

Restraints: the Contentious History of Group Selection, Wynne-Edwards painted an intense picture of the vast expanse of the North Atlantic.

> Nowhere on land, even in the Sahara, the prairies or the steppes of Asia, can such a vast expanse of monotony be found as on the great oceans…they present a uniformity of conditions unparalleled elsewhere on this earth. Yet it is hardly necessary to state that in spite of this prevailing sameness not all the birds primarily adapted to obtain their livelihood from the sea, even in a restricted area like the North Atlantic, belong to a single ecological community. Fulmars and cormorants, for example, might pass their whole lives without seeing one another, and could only do so at special times and places, for they belong to two communities as distinct as those of forest and fen, and their paths seldom cross. The factors which differentiate one community from another are not by any means understood, but present problems of no small interest. (Borello 2010, p. 48)

As Borello notes, this desolate description echoed the images evoked by Peter Kropotkin in his description of the ecology of Siberia and Manchuria, where life was brutal and scarce, and where, Kropotkin wrote, he "failed to find – although I was eagerly looking for it – that bitter struggle for the means of existence among animals belonging to the same species, which was considered by most Darwinists (though not always by Darwin himself) as the dominant characteristic of struggle for life, and the main factor of evolution" (quoted by Borello 2010, p. 49). Kropotkin, of course, went on to develop an early theory of group selection (of sorts) in his book *Mutual Aid*. In their fieldwork, both Kropotkin and Wynne-Edwards observed, rather than Nature red in tooth and claw, the solitary flower at the edge of the desert struggling for life against the drought. This last image comes, in fact, from Darwin's own words, and characterizes the breadth of his own definition of the struggle for existence (Darwin, *Origin of Species*, p. 90).

During his years in Canada, Wynne-Edwards continued his fieldwork, and he continued to find evidence contradicting the clichéd image of evolution as merely personal struggle between individuals, each trying to survive and produce as many offspring as possible. For example, he observed that at any given time only 33–40% of birds in breeding colonies of the sea bird *Fulmaris glacialis* were engaged in reproduction. Observations such as these would later lead to the development of his theory of how social interactions among animals had evolved to limit population growth in order not to overuse available resources.

In 1946, Wynne-Edwards returned to the United Kingdom, and began teaching at the University of Aberdeen. He continued to investigate the behavior of sea birds in sparse environments and, as an acknowledged expert in the field, was asked to review a new book by ornithologist David Lack (Fig. 6.3), entitled *Natural Regulation of Animal Numbers*. Lack argued that birds reproduce as fast as they can.

Lack's career had primed him to focus on competition between individuals in a crowded environment, just as Wynne-Edwards's personal trajectory had imprinted him with an image of sparsely distributed organisms struggling against a hostile climate. Lack did his fieldwork in the Galápagos, studying finches. Prior to his work, the standard view was that differences between species of finch were non-adaptive. His observations during the 1930s led him to conclude the opposite, and also to conclude that geographical isolation was essential for speciation (this is

Fig. 6.3 David Lack (*left*) and Chris Perrins in 1962, looking for chimney swift nests in Ithaca, New York, during a meeting of the International Ornithological Congress held at Cornell University (Photograph reproduced with the kind permission of the Lack family and Chris Perrins)

known as allopatric speciation, and is today considered to be the dominant, but not exclusive, mode of speciation).

Lack's studies helped push the Modern Synthesis deeper into its hardening, "adaptation-only" phase, by showing that species differences previously assumed to be nonadaptive, and likely due to drift, were adaptive after all. We have seen above how such examples were cited by Mayr. Indeed, Lack, who had studied in England with Julian Huxley, spent considerable time working with Mayr in the United States on his way back from the Galápagos. Mayr firmly held that speciation could only occur as a result of geographical isolation, and also advocated the principle of competitive exclusion, proposed by Gause, according to which two species cannot occupy precisely the same ecological niche. Lack also used this idea as the basis for concluding that bird numbers are limited by their food supply. In *The Natural Regulation of Animal Numbers*, he wrote that one

> reason for thinking that birds are limited in numbers by their food supply is that each species living in the same region depends on primarily different foods. If food were not limiting numbers, it is hard to see why such differentiation in feeding habits should have been evolved, but its evolution is essential to survival if food is limiting, since if two species compete for food, the chance of both being equally well adapted is negligible, so that one will eliminate the other. (Lack 1954, p. 148)

Note that Lack based his argument on the assumption that the principle of competitive exclusion is correct. In the following paragraph, he conceded that, when food is extremely abundant, exceptions can occur. He cited various species of European birds of prey feeding on voles during a "vole plague", and observed that four species of tit

> usually have different feeding stations, but when a particular food is temporarily superabundant, such as leaf-eating caterpillars in May or beechmast in autumn, all four often feed together…A similar situation was found in Darwin's finches (*Geospizinae*) in the Galápagos

Islands... In these and other cases, several species feed together only when a particular food
is very abundant, thus confirming the view that the normal differentiation in feeding sta-
tions has been evolved to avoid competition for food. (Lack 1954, p. 148)

It took Wynne-Edwards some time to write his review. "The trouble is," he wrote
to a colleague, "that I think Lack has failed to penetrate the first principles of the
subject, and in order to demonstrate this one must go very deep oneself" (quoted by
Borello 2010, p. 64). When he wrote it at last (in late 1955), he found much to criti-
cize, not least because he had indeed gone very deep himself, and had developed a
diametrically opposed model for the regulation of populations. Rather than populations reproducing at maximum capacity, Wynne-Edwards focused on data he and
others had collected regarding slowly-breeding species. These, he suggested, "have
evolved a series of interrelated adaptations, giving them a great measure of auto-
nomic control of their numbers." If these species "were adapted to impose their own
limit on the number and size of their breeding colonies (as an alternative to limiting
the minimum size of individual breeding territories)," he argued, "they could combine optimum feeding conditions with maximum numbers" (quoted by Borello
2010, p. 66). Note the different levels of selection at which the alternatives operate:
Wynne-Edwards suggests that a *species somehow imposes its own limit on the number and size of breeding colonies*, in contrast to *direct intraspecific competition
providing a limit on the size of individual breeding territories*. As for the "somehow", the means by which species imposed such limits, Wynne-Edwards proposed
social interaction itself. He suggested that members of population can signal population density to one another through what he termed *epideictic* displays. As a result
of such displays, the population can regulate its numbers, leading to a sort of
population-level homeostasis. As he wrote in a paper read at a contentious[2] meeting
of the British Ornithological Union in 1959,

> The hypothesis put forward here, therefore, suggests that animals have become adapted,
> with varying success, to control their own population densities, limiting them at the opti-
> mum level – this being the level that offers the best living to the largest number, consistent
> with safeguarding the food-supply from damage from so-called over fishing. It suggests
> that the result is achieved by interposing artificial, conventional goals as substitutes for
> direct competition for food. (quoted by Borello 2010, p. 70)

Wynne-Edwards repeatedly used the example of human over-fishing as a model and
metaphor for any species depleting its food supply due to over-use. He opened his
magnum opus, *Animal Dispersion in relation to Social Behaviour* (1962) with a
similar example. In extrapolating an argument from human activity to the natural
process of over-exploitation of natural resources, he was giving a nod to the opening

---

[2] See Borello (2010), p. 66ff., for a vivid description of the proceedings. Wynne-Edwards was in
the United States at the time of the meeting and thus had to ask one of his students, George
Dunnett, to read his paper for him. Dunnett's description of the meeting showed that some attend-
ees considered group selection as antithetical to Darwinian evolution, saying "if we are to believe
this, then there [is] no longer any possibility in believing in Darwinism and natural selection!"
(quoted by Borello 2010, p. 70) Lack, who attended the meeting, clearly held a similar view. One
of his arguments against Wynne-Edwards's theory was that "natural selection acts only on individuals" (Dunnett, quoted by Borello 2010, p. 71).

chapters of Darwin's *Origin*, which famously begins with a discussion of artificial, rather than natural, selection on pigeon populations.

Wynne-Edwards provided a massive compilation of data on the sparseness and patchiness of animal communities, as well as the evidence that many, if not most, species reproduce below the maximum rate of which they are capable. Examples included delayed maturation in birds, fishes, mammals and reptiles. He described his own observations of non-breeding adults in a breeding colony of fulmars, including histological evidence to confirm that the non-breeders had not simply reproduced in out-of-the-way, unobserved nesting sites. He discussed evidence of cannibalism and destruction of the young in many species. From this vast accumulation of data (*Animal Dispersion* has 23 chapters and 653 pages), Wynne-Edwards concluded that most habitats have some given carrying capacity, and that living organisms self-regulate in a sort of homeostasis, imposing control on their own numbers.

Wynne-Edwards provided an extensive survey of the various means by which social groups of animals communicate, with chapter titles such as "Social integration by the use of sound: land animals", "Social integration by underwater sound and low-frequency vibrations", and "Social integration by olfactory signals." He argued that each of these could serve as epideictic behavior and, in other cases, as "conventional" social behaviors onto which direct competition for food had been displaced. The term "conventional" is used here in the sense of social conventions, such as place in the social hierarchy (as in a pecking order). Wynne-Edwards argued that this was the evolutionary origin of all social behavior.

> Undisguised contest for food inevitably leads in the end to over-exploitation, so that a conventional goal for competition has to be evolved in its stead; and it is precisely in this – surprising though it may appear at first sight – that social organisation and the primitive seeds of all social behaviour have their origin. (Wynne-Edwards 1962, p. 14)

He emphasized that such conventions

> must, by their nature, always be properties of a concerted group, and can never be completely vested in or discharged by a lone individual in perpetual isolation; their observance has to be reinforced by the recognition and support of others who are bound by the same convention. In the absence of other parties they become meaningless. It is this concerted group that appears to constitute the primordial germ of the society... The social conventions themselves all ultimately spring from the need to develop substitute goals for mutual competition among rival members of the species – goals that are effective in preventing population-density from exceeding the optimum level. If we reduce this situation to its simplest terms it becomes possible...to define the elementary society as being *an organisation of individuals capable of providing conventional competition* among its members. (Wynne-Edwards 1962, p. 132, his italics)

Note that the group properties described here are of the MLS2 type: they depend on the interaction between individual members of the group, and are thus "emergent" properties that cannot exist without the group; they are far more complex than a simple MLS1 average of individual properties. Wynne-Edwards also emphasized the balance between competition and cooperation, poetically declaring that "this two-faced property of brotherhood tempered with rivalry is absolutely typical of social behavior; both are essential to providing the setting in which conventional competition can develop" (Wynne-Edwards 1962, p. 14). Elsewhere he wrote that social groups "combine the two apparently opposite qualities of cohesion, which draws the individuals together, and mutual rivalry, which tends to keep each individual at a distance from its neighbours; these might properly be called synagonistic and antagonistic tendencies" (Wynne-Edwards 1962, p. 133).

Wynne-Edwards's book received a mixed reception. While some lauded it as a landmark study, others, including Wynne-Edwards's Oxford mentor Charles Elton, critiqued it mercilessly. In a review in *Nature*, Elton described the writing style as that of "a bishop wearing blinkers", and the reasoning as "rather woolly" (Borello 2010, p. 86). Turning to substantive criticism, Elton noted that animal populations often oscillate wildly.[3] The homeostasis at the heart of Wynne-Edwards's theory, Elton argued, simply did not exist. Elton also complained that Wynne-Edwards provided no clear example of group selection actually at work. Another reviewer, F. W. Braestrup, provided bouquets of back-handed compliments, writing first "I am in perfect accordance with Wynne-Edwards" and following that up with "I think it may be said of the book, with a certain amount of truth, that most of what is sound is not new and most of what is new is not sound" (Borello 2010, p. 87). A more serious backlash, however, was not long to follow, and it effectively suppressed discussion of group selection for decades.

Both Lack and Wynne-Edwards set out to explain the regulation of animal populations in number and distribution. However, Lack held that the determining factor was a density-dependent *mortality* rate. This is natural selection acting at an individual level. Wynne-Edwards, in contrast, suggested density-dependent variation of *reproduction* rate. This is natural selection acting at the group level, since it depends on suppression of some individuals' immediate reproductive interests in exchange for a benefit to the group. The ire of all those with a distaste for group selection was soon directed toward Wynne-Edwards. Charles Sibley, an ornithologist, wrote to Lack that "[t]his matter needs to be exposed as the nonsense it is – and you're the one to do it!" (quoted by Borello 2010, p. 96). A few months later Sibley followed up his first letter, applauding Lack's decision to write a new book in response to Wynne-Edwards. "I do hope," he wrote,

> you will include an emphatic statement relative to the fact that group selection is impossible on genetic grounds simply because only individuals, not populations possess genes… One should concentrate on this basic fallacy and force the group selectionists to recognize that they must invent a totally new type of genetic system before their arguments have any basis. They tend to gloss over this pitfall and tend to go blithely on saying, in effect, 'Oh yes, but you don't quite understand' and so forth. They should not be permitted to leave this basic position until they have explained how the mechanism can possibly work with the type of genetic system evolved on this planet. All the rest is simply window dressing and nothing but a collection of interesting anecdotes misinterpreted on the basis of a false assumption right at the beginning. (Borello 2010, pp. 96–97)

---

[3] As Borello notes (2010, p. 87), the issue of the stability of natural populations was under considerable debate at the time.

Sibley, enjoying his vitriol, lost sight of a fallacy of his own. It is cells, not individual animals, that are the smallest units that possess genes (indeed, going smaller than that, one could consider genetic elements that can be transferred between cells, such as plasmids and transposons). Were Sibley's logic correct, natural selection at the organismal level could be considered be as fallacious as selection at the group level. (The argument that an individual organism has a single genome, present in each of its cells, is certainly valid for metazoans, but breaks down in the case of simpler organisms without germ-line sequestration.) Perhaps we will yet see the day when a splinter-group of physicists, in a similar spirit of false parsimony, will insist that correct physics can only be done at the level of the quark.

Sibley's critique (though it must be noted this was in a personal letter rather than in text intended for publication) also missed the fact that Wynne-Edwards had not, in fact, lost sight of genetics. While it was not a major focus of his work, Wynne-Edwards did discuss gene flow within populations. When there is much intermingling of nearby populations within the same species, he wrote, "there will be a constant and considerable interchange of genes between one population and another leading to a relative uniformity in the genetic make-up of populations scattered over wide areas; and conversely where it is weak a more effective reproductive isolation will facilitate the differentiation of local races" (Wynne-Edwards 1962, p. 463). Likewise, Allee devoted an extensive discussion to the role of minimal population size in maintaining sufficient genetic variation for a population to withstand environmental or other stresses. He discussed the role of population size in enabling gene fixation as a result of drift (Allee 1938, p. 121), and presented a subtle discussion of the need for a sufficient store of genetic variability in a population. Genes which prove "life-saving" in a small population

> may have been present in the species for a million years as a result of long past mutations, without having been of any value to the species in all that time. Now under changed conditions they may save it from extinction. It is important to note that organisms do not usually meet changed conditions by waiting for a new mutation[4]; frequently all members of a species would be dead long before the right change would occur. This means that since a species cannot produce adaptive changes when and where needed, in order to persist successfully it must possess at all times a store of concealed potential variability.[5] (Allee 1938, pp. 119–120)

In 1966, Lack published a book entitled *Population Studies of Birds*. He devoted much of the text, and one of the book's appendices, to rebutting Wynne-Edwards's arguments. He took issue, as others had done, with Wynne-Edwards's use of human overfishing as a model for the behavior of animal populations. He suggested that the idea of homeostasis in animal populations was far from new, having been previously discussed under the name of "self-balancing populations" by A. J. Nicholson in the 1930s. And, to seal the "what is correct is not new" portion of his argument, he

[4] In fact, many organisms do seem to change their mutation rates under environmental stress (see Chap. 14).

[5] This concept of "concealed potential variability" has much in common with Gould and Vrba's exaptations and spandrels, which we will discuss in depth in Chap. 15.

wrote that many "epideictic displays" had previously been described by other orni-
thologists, but "rightly, in my view, they have ascribed various functions to them
and not a single overriding one (epideictic). However, the general reader might not
be aware of this from Wynne-Edwards' book as he did not usually discuss the earlier
interpretations of the phenomena which he considered to be epideictic" (Lack 1966,
p. 311).

Lack's rebuttal of Wynne-Edwards is problematic in several aspects. He criticized Wynne-Edwards repeatedly for advocating control of reproduction rate via
group selection without sufficient evidence to support his claim. But when Lack
placed evidence for his own preferred view head to head with that of Wynne-
Edwards, he committed a similar omission. In a remarkable passage highlighted by
Borello, Lack wrote that

> [w]hen I wrote my earlier book of 1954, the existence of density-dependent mortality still
> rested largely on theoretical considerations, supplemented by data from laboratory popula-
> tions of insects which were, however, models rather than true experiments. The evidence
> from natural populations is not much stronger now, but nevertheless I believe that density-
> dependent mortality provides the best explanation of the balance between birth and death
> rates." (Lack 1966, pp. 7–8)

Elsewhere, after laying out inconclusive evidence on either side of an argument (on
the dependence of blackbird clutch size on environment and on age of the parents),
he selected his own argument because simply because he considered it "simpler
explanation" (Lack 1966, p. 122). And he considered it simpler because it was not
based on social interactions among the birds.

Lack's conflation of simple explanations with correct ones brings us back to the
hypothetical fallacy of revising all physics to be viewed from the quark-level because
quarks are small. The argument that group selection is wrong because it does not deal
with small things was taken to new, dizzying heights by George C. Williams, in his
critique of Wynne-Edwards, as we shall see in the next chapter. As rhetoric was
unleashed against the idea of group selection, much of it rested, sadly, on the incorrect
assumption – often explicitly stated – that *smaller* (individuals rather than populations, and later genes rather than individuals) always means *simpler* (already a stretch),
and *therefore inevitably* (here is the wild leap) simple and small things form the only
source for a logically defensible hypothesis. This error is a distortion of Occam's
razor. That logical principle states, in the vernacular, that the simplest explanation is
the best. This means that if you have a phenomenon that can be completely characterized by an algebraic equation with one independent variable, one dependent variable,
and three parameters, you do not need to describe it with a set of five equations in five
unknowns. That is not at all the same thing as the popular distortion of the rule: "only
look at small things". Conflating these notions leads to what Stephen Jay Gould, in his
inimitable style, calls "a disabling problem in logic". The problem is this: Occam's
razor "operates as a logical principle about the complexity of argument, not as an
empirical claim that nature must be maximally simple" (Gould 2002, p. 552).[6]

---

[6] Note the parallel of this slide from logical argument to empirical claim to that performed by Lyell
in his multiple interpretations of uniformitarianism (Chap. 3). Gibson (2000) made use of Occam's

Occam's original Latin states *non sunt multiplicanda entia praeter necessitatem*: things should not be multiplied more than necessary. This is not the same as stating that things don't multiply. Sometimes they do. And sometimes they interact with each other. And sometimes their interactions produce something entirely new.

References

Allee WC (1938) The Social Life of Animals. W. W. Norton & Company, Inc., New York

Borello ME (2010) Evolutionary Restraints: The Contentious History of Group Selection. The University of Chicago Press, Chicago/London

Courchamp F, Berec L, Gascoigne J (2008) Allee Effects in Ecology and Conservation. Oxford University Press, Oxford

Darwin C (2009) The Origin of Species by Means of Natural Selection. Modern Library, New York. This edition reprints the second edition of the *Origin*, from early 1860

Gibson G (2000) Evolution: Hox genes and the cellared wine principle. Curr Biol 10:R452–R455

Gould SJ (2002) The Structure of Evolutionary Theory. The Belknap Press of Harvard University Press, Cambridge, MA

Lack D (1954) The Natural Regulation of Animal Numbers. Clarendon Press, Oxford

Lack D (1966) Population Studies of Birds. Clarendon Press, Oxford

Okasha S (2006) Evolution and the Levels of Selection. Oxford University Press (Clarendon Press), Oxford/New York

Wynne-Edwards VC (1962) Animal Dispersion in Relation to Social Behaviour. Oliver and Boyd, Edinburgh/London

---

razor in a more logical way (to the extent that he privileged the null hypothesis over one hypothesis among many), in the service of an argument against knee-jerk adaptationism (see Chap. 15). He argued that "selection should only be invoked when the null hypothesis of neutrality cannot explain the data".

Chapter 7
The Vanishing Point Appears

Am I living in a beautiful vacuum?

R.E.M.

THE SHARPEST critique of Wynne-Edwards came from George C. Williams, in his 1966 book *Adaptation and Natural Selection: A Critique of Some Current Evolutionary Thought.* Williams was driven by a distaste for multi-level evolutionary thought that went back to his early years teaching at the University of Chicago in the mid-1950s.

> The triggering event may have been a lecture by A. E. Emerson, a renowned ecologist and termite specialist. The lecture dealt with what Emerson termed beneficial death, an idea that included August Weismann's theory that senescence was evolved to cull the old and impaired from populations so that fitter youthful individuals could take their place. My reaction was that if Emerson's presentation was acceptable biology, I would prefer another calling. (quoted by Borello 2010, p. 107)

Likewise, Williams's tone in a letter to David Lack indicates that he considered the idea of group selection rather ridiculous.

> You probably had some trouble with the wording of your discussion of Wynne-Edwards. The subject requires great care to avoid the appearance of sarcasm or ridicule. I know that when I got to that part about the epideictic function of the vertical movement of plankton [Wynne-Edwards's Chapter 16] I suddenly wondered if I had fallen for a really elaborate joke. (quoted by Borello 2010, p. 111)

Williams insisted that adaptations

> should be attributed to no higher a level of organization than is demanded by the evidence. In explaining adaptation, one should assume the adequacy of the simplest form of natural selection, that of alternative alleles in Mendelian populations, unless the evidence clearly shows that this theory does not suffice. (Williams 1966, pp. 4–5)

Conflating the principle of parsimony with the a focus on the small, Williams continued his argument in the following terms:

> Various levels of adaptive organization, from the subcellular to the biospheric, might conceivably be recognized, but the principle of parsimony demands that we recognize adaptation at the level necessitated by the facts and no higher. It is my position that adaptation need almost never be recognized at any level above that of a pair of parents and associated offspring. (Williams 1966, p. 19)

© Springer Science+Business Media B.V. 2018
S. Bahar, *The Essential Tension*, The Frontiers Collection,
DOI 10.1007/978-94-024-1054-9_7

Williams's approach creates an a priori bias against even considering the possibility of adaptation at higher levels (even though, of course, those adaptations could ultimately be traced back to alternative alleles in Mendelian populations). An important result of Williams's parsimony argument was ultimately to push back the "simplest" level from the individual to the gene. In a later passage, he reiterated his principle of parsimony, speaking of "genic selection" as "natural selection in its most austere form" (Williams 1966, pp. 123–124). What exactly was the "simplest" level on which parsimony dictates one should focus? Was it the individual or the gene? The focus shifted back and forth at various points in Williams's work. In the early pages of *Adaptation and Natural Selection*, he emphasized the role of natural selection in acting on "the genetic survival of individuals".

> With some minor qualifications to be discussed later, it can be said that there is no escape from the conclusion that natural selection, as portrayed in elementary texts and in most of the technical contributions of population geneticists, can only produce adaptations for the genetic survival of individuals. Many biologists have recognized adaptations of a higher than individual level of organization. A few workers have explicitly dealt with this inconsistency, and have urged that the usual picture of natural selection, based on alternative alleles in populations, is not enough. They postulate that selection at the level of alternative populations must also be an important source of adaptation, and that such selection must be recognized to account for adaptations that work for the benefit of groups instead of individuals. I will argue…that the recognition of mechanisms for group benefit is based on misinterpretation, and that the higher levels of selection are impotent and not an appreciable factor in the production and maintenance of adaptation. (Williams 1966, pp. 7–8)

There are several problematic points in this passage. Williams referred to higher-level adaptations as an "inconsistency", but did not explain why, thus inserting an inherent bias into the text. He also did not explain why, if Mendelian selection can result in alternative subgroups of alleles called individuals, it could not result in looser collective alternative subgroups called populations of individuals. In this omission, Williams not only negated the possibility of group selection as Wynne-Edwards (1962) envisioned it, but he also passed over the important work of the population geneticists, which deals with the frequency of gene distribution within populations.

Emphasis on genes is essential for understanding the mechanism of any evolutionary process. But Williams, in the passage above, slid toward the error Gould (2002) referred to as "mistaking bookkeeping for causality", confounding the locus of evolutionary bookkeeping with the locus of action of natural selection. Biological information will always, at least in our current biosphere, be encoded at the level of nucleotide sequences. But that does not mean that selection acts on individual alleles. It may. But there is no a priori reason why it must.

Williams proposed a new criterion that an entity must possess in order to be acted upon by natural selection: stability. This is a view that, as we will see below, heavily influenced Richard Dawkins. There can be no selection on somatic cells, Williams wrote, because

> [t]hey have limited life spans and (often) zero biotic potential. The same considerations apply to populations of somata. I also pointed out that genotypes have limited lives and fail to reproduce themselves (they are destroyed by meiosis and recombination), except where

clonal reproduction is possible. This is equally true of populations of genotypes. All of the genotypes of fruit-fly populations now living will have ceased to exist in a few weeks. Within a population, *only the gene is stable enough to be effectively selected.* Likewise in selection among populations, only populations of genes (gene pools) seem to qualify with respect to the necessary stability. (Williams 1966, p. 109, my italics)

The implication of Williams's example of the ephemerality of fruit fly genotypes is that *metazoan individuals cannot be subject to selection because of their lack of stability through evolutionary time.* Yet what is a genotype, after all, but a sampling from the gene pool, which Williams had already declared to have the "necessary stability" to be subject to selection? The validity of Williams's stability criterion itself can be called into question, as we will discuss below.

It is worth noting that, three decades later, Williams did somewhat moderate his tone. In the preface to the 1996 edition of *Adaptation and Natural Selection*, he claimed to have been misunderstood, and explained that while he had meant that group selection was not strong enough to produce what he termed "biotic" adaptation (defined below), "group selection can still have an important role in the evolution of Earth's biota" (Borello 2010, p. 110). But this ratcheting back could not undo the fact that his work had effectively placed the idea of group selection on ice for decades.

Even in the 1960s, however, Williams's tone regarding group selection was inconsistent. On page 109 of *Adaptation and Natural Selection*, he wrote of group selection that "there can be no sane doubt about the reality of the process. Rational criticism must center on the importance of the process and on its adequacy in explaining the phenomena attributed to it." To the extent that Williams undertook such criticism of the importance of group selection, he began his book with the assumption that true group selection must work at odds to selection at the individual or genic level. He noted the difference between adaptations of individuals[1] within a population, and adaptations of the population itself, referring to the former as "organic" adaptations, and to the latter as "biotic" adaptations. It was only these biotic adaptations, he wrote, that could be driven by group selection. The fact that cooperation in insect societies occurred only between genetically related individuals provided, he wrote, "cogent evidence of the unimportance of biotic adaptation" (Borello 2010, pp. 109–110).

In addition to insisting that selection above the level of the gene should be only an explanation of last resort, Williams considered adaptation itself to be a "special and onerous concept that should not be used unnecessarily" (Williams 1966, p. 4). Rather than assuming that adaptations grew up like mushrooms after a spring rain, Williams cautioned against the temptation to "recognize adaptation in any recognizable benefit arising from the activities of an organism. I believe that this is an insufficient basis for postulating adaptation and that it has led to some serious errors. A benefit can be the result of chance instead of design" (Williams 1966, p. 12). As an example, he asked the reader to imagine a fox walking through the snow toward a

---

[1] Note the contradiction with the passage quoted above regarding the instability of genotypes. That statement implies the inability of selection to act on individual organisms.

hen house. The fox has to wade through the snow, and is tired by the time it arrives to select its dinner. But if it follows the same path the next day, the snow will be tamped down a bit. Over repeated shopping expeditions, assuming there is no further snowfall, the fox will have constructed a path through the snow. This, however, does not mean that the fox's feet were adapted for the tamping down of snow. "At any rate, the concept of design for snow removal would not explain anything in the fox's appendages that is not well or better explained by design for locomotion" (Williams 1966, p. 13). Williams follows up this little anti-just-so story by dryly remarking that "the brewing of beer is not the function of the glycolytic enzymes of yeast". The interesting thing is the use to which Williams puts this argument. He does not explore the role of drift in evolution, or discuss the role of historical contingency in shaping the fossil record, or consider the future use of properties (later to be named exaptations and spandrels) not originally shaped by the process of selection as direct adaptations. Instead, Williams used the argument as a hammer against group selection. Many group adaptations, he suggested, were not adaptations at all, and therefore there is little, if anything, to be explained by an appeal to group selection. Subsequent chapters of his book, he promised, would "be primarily a defense of the thesis that group-related adaptations do not, in fact, exist." (Williams 1966, p. 93).

One the situations that might falsely lead to the conclusion of group-related adaptations, Williams argued, was the case where individual adaptations had fortuitously beneficial effects on a group of organisms. In more recent terminology, this corresponds to an MLS1 effect, where group-level fitness is defined as the average fitness of individuals within the group; a biotic adaptation would be an MLS2 effect.

> Benefits to groups can arise as statistical summations of the effects of individual adaptations. When a deer successfully escapes from a bear by running away, we can attribute its success to a long ancestral period of selection of fleetness. Its fleetness is responsible for its having a *low probability* of death from bear attack. The same factor repeated again and again in the herd means not only that it is a herd of fleet deer, but also that it is a fleet herd. The group therefore has a *low rate* of mortality from bear attack. When every individual in the herd flees from a bear the result is effective protection of the herd. (Williams 1966, p. 16, his italics)

But it would be as much a mistake, Williams argued, to consider this average fleetness as a group adaptation as to consider the fox's feet as snow-tamping devices. What would a group adaptation look like? After making the assumption that adaptation at the group level would necessarily convey greater collective fitness than a survival-related property averaged over the individuals, Williams asked the reader to

> imagine that mortality rates from predation by bears on a herd of deer would be still lower if each individual, instead of merely running for its life when it saw a bear, would play a special role in an organized program of bear avoidance. There might be individuals with especially well-developed senses that could serve as sentinels. Especially fleet individuals could lure bears away from the rest, and so on. Such individual specialization in a collective function would justify recognizing the herd as an adaptively organized entity. Unlike individual fleetness, such group-related adaptation would require something more than the natural selection of alternative alleles as an explanation. (Williams 1966, p. 17)

This passage raises two immediate questions. First, why would such a group-related adaptation *require a different explanation than the natural selection of alternative alleles*? Those alleles would be selected from the population's gene pool, after all. The process would entail natural selection of alternative alleles in a more statistical sense than the differential survival of two individual animals, but the genome would change nonetheless. The difference would be the size of the holes of natural selection's sieve – group-sized holes rather than individual-sized or cell-sized holes. But the sifting would change the frequencies of certain alleles. How else – leaving aside epigenetic and environmental effects – would the genome of a solitary bee species come to differ from the genome of a eusocial bee species?

The idea of eusocial insects brings us to a the second question raised by Williams's metaphor of sentinel deer. Eusocial insect societies practice cooperative brood care and show division of labor into breeding and non-breeding castes. Don't they do precisely what Williams described? And aren't, therefore, their social structures examples of group selection?

Not so fast. Williams defined a group as "something other than a family and... composed of individuals that need not be closely related" (Williams 1966, p. 93). This definition[2] became particularly important in his argument that eusocial insect societies cannot be considered as examples of group selection for the very reason that their constituents are related! In Chap. 7 of *Adaptation and Natural Selection*, called "Social Adaptations", Williams begins by noting that "[b]ehavioral or physiological mechanisms that operate between an individual and its own offspring are normally benign and cooperative, but interactions between unrelated individuals normally take the form of open antagonism, or, at best, a tolerant neutrality" (Williams 1966, p. 193). This assumption, combined with his definition of a group, effectively rules out cooperative interactions within a group, literally by definition

Let us follow Williams's argument. He first posed the question of how natural selection could possibly favor genes "that cause their bearers to expend resources to benefit their genetic competitors" (Williams 1966, p. 194). He then proceeded to consider the example of eusocial insects. Following the kin selection argument of W. D. Hamilton, Williams argued that genes that promote altruistic behavior among close relatives (and, being haplodiploid, many social insects are more closely genetically related than, say, birds in the same nest) "would be favorably selected because the aid provided would usually go to other individuals with the same gene" (Williams 1966, pp. 197–198). However, since most eusocial insect colonies are presumably[3]

---

[2] Note that Williams's definition retains some flexibility in its use of the phrase "need not be closely related" rather than "are not related".

[3] Williams did allow that "closeness of relationship between individuals of a colony can sometimes be seriously questioned. Even though multiple queens are normally supposed to be sisters, they would inevitably be genetically different and produce genetically different offspring. Genotypic diversity within such sister-queen colonies would be significantly greater than in the population as a whole. If it could be shown that there are thoroughly unified insect societies that normally contain several unrelated reproductives, they could only be explained as biotic adaptations resulting from effective group selection. The kinship of the reproductives would be a difficult proposition to prove one way or the other, but it is an extremely important point" (Williams 1966, pp. 200–201).

closely related, Williams's definition of a "group" effectively ruled out this phenomenon as a case of group selection.

Moving on to the behavior of other species, Williams conceded that "examples of cooperation and self-sacrifice ... [are] sometimes observed among individuals that are not closely related" (Williams 1966, p. 203). He argued, however, that any such behavior was a result of a misplaced reproductive instinct: animals may protect others in their herd because it is not to their advantage to bother turning off the protective behavior they show toward their young, especially in species with long breeding seasons. Is it possible that such "misplaced" instances of caring behavior might provide some benefit to the herd? Williams considered this possibility, but ruled it out as a possible biotic adaptation. After describing the warning signs that prey species such as rabbits or deer will give in order to warn their young of a predator's approach, and perhaps also to distract the predator, Williams wrote:

> As a result, rates of predation on deer and rabbit populations, even out of the breeding seasons, are probably somewhat reduced by the warning signals that these animals display when they take flight. This circumstance means that the vicinity of conspecific individuals has value as protection against predators, *and it undoubtedly contributes to selection pressures in favor of gregariousness in such species.* These developments, however, involve no biotic adaptation. They merely represent individual adjustments to opportunities presented by their ecological environments. (Williams 1966, p. 206, my italics)

But if this "undoubtedly" contributes toward selection pressure in favor of gregariousness, why is it not a biotic adaptation, and how can it simultaneously represent "merely individual adjustment"?

Focusing on the work of Allee, and foreshadowing the difference between adaptations and exaptations, Williams emphasized that the benefit provided by a trait should not be confused with its function. He noted that when a mouse huddled with others for warmth, it was doing so for its own benefit, not that of the group. Thus any benefits from this behavior to the overall mouse population, such as those described by Allee, are not biotic adaptations (in the sense that they, in modern terms, are MLS1 rather than MLS2 effects).

> There is no more reason to assume that a herd is designed for the retention of warmth than to assume that it is designed for transmitting diseases. The huddling behavior of a mouse in cold weather is designed to minimize its own heat loss, not that of the group. In seeking warmth from its neighbors it contributes heat to the group and thereby makes the collective warmth a stronger stimulus in evoking the same response from other individuals. (Williams 1966, pp. 211–212)

Williams criticized Allee's experiments on marine flatworms, which showed that the presence of multiple individuals facilitated survival in hypotonic water. Williams made the important point that Allee's experiments, indeed, did not allow a distinction

---

However, describing the communal behavior of a decidedly diploid and non-insect species, the California woodpecker, he wrote that he "*would predict* ... the societies of the California woodpecker, of the social insects and of all other such organized groups, will be found to be based almost entirely on family relationship" (Williams 1966, p. 202, my italics). Again, given his definition, their interactions could not be described as group selection. It is hard to avoid the conclusion that Williams had decided the case before even entering the courtroom.

between what we would call MLS1 and MLS2 effects. To make such a distinction, Williams pointed out that one would need to show that social cohesion increased with hypotonicity, or that the

> secretory machinery was activated by the deleterious change; that the substance secreted not only provided protection against hypotonicity, but was an extraordinarily effective substance for this protection. One or two more links in such a chain of circumstances would provide the necessary evidence of functional design and leave no doubt that protection from hypotonicity was a function of aggregation, and not merely an effect. (Williams 1966, p. 210)

He made similar arguments regarding the protective benefits of schooling in fish.

These points are important and well taken. "The statistical summation of adaptive individual reactions, which I believe to underlie all group action, need not be harmful. On the contrary, it may often be beneficial, perhaps more often than not" (Williams 1966, p. 211). In other words, MLS1 effects can be beneficial, but this does not prove that they are properties of the group *as a group*. However, notice the phrase at the center of this quotation, "*which I believe to underlie all group action.*" This is taking a position and defending it, not amassing evidence and drawing a conclusion from it.

Williams repeatedly considered examples that could be interpreted as biotic adaptations, and dismissed them without providing a rigorous argument. The dominance hierarchy shown by wolves, for example, "is not a functional organization. It is the statistical consequence of a compromise made by each individual in its competition for food, mates, and other resources. Each compromise is adaptive, but not the statistical summation" (Williams 1966, p. 218). He cited Allee on pecking order and Wynne-Edwards on dominance hierarchies, but did not refute them. He mentioned that "a few observations…suggest a functional organization in nonreproductive herds of mammals" (Williams 1966, p. 218) citing the example of adult musk oxen stationing themselves on the edge of a threatened herd "in what appears to be an attempt to defend the weaker members… This seems to be a functional division of labor and evidence for biotic adaptation, but there are other possible explanations. It may be that the defending bulls are showing misplaced reproductive behavior" (Williams 1966, pp. 218–219). But suggesting alternate explanation B does not disprove hypothesis A.

As a last example, Williams considered the case of apparently protective behavior of rams with respect to ewes and lambs, only to dismiss it because the rams were too far away from the ewes.

> Herds of bighorn sheep may segregate, with ewes and lambs staying close to the escape routes that lead to their high rocky havens, and the rams moving out on more level and more dangerous ground. That this is not a functional division of labor, with the rams there to protect the ewes and the lambs, is apparent from the distance between the two groups. The rams may move, as a group, to points several miles from their more timid relatives… If the rams happened to be close to the ewes and lambs when an enemy appeared and if they acted belligerently, they would give the appearance of the stronger attempting to defend the weaker individuals. (Williams 1966, p. 219)

But if they did, then, like the musk oxen, they would be presumably simply demonstrating misplaced parental instincts. Williams concluded by remarking that the possibility of functional organization at the group level does "warrant careful attention" by researchers studying gregarious animals. "Detailed and objective studies of wild populations," he wrote, "such as those by Altman, Hall, Lack, and Richdale, should provide important evidence on this point. That such studies have not yet furnished clear indication of the functional organization of large groups is already a matter of great significance" (Williams 1966, p. 220). Lack of evidence (or the existence of countervailing hypotheses) is evidence of lack.

In Williams's next chapter, "Other Supposedly Group-Level Adaptations", he reiterated his assumption that cooperation of genetically identical cells, such as within a single tissue, should not be considered as group-level selection. Since these cells share a genome, he considered selection on them to be selection acting at the gene level only, and therefore by definition not a case subject to group selection.

> We do not expect to find genetically different individuals cooperating in a single somatic system. I would explain this in the same way I would explain the general absence of functional social organization among genetically diverse individuals: only between-group selection could produce such organization, and this force is impotent in a world dominated by genic selection and random evolutionary processes. (Williams 1966, p. 221)

As an example, Williams noted the immune response to association between genetically different tissues in vertebrates, remarking on "an increasing tendency to avoid fusion…as one ascends the scale of histological specialization" (Williams 1966, p. 222). Plants are more tolerant of the introduction of foreign tissue, he noted, though grafting, of course, does not occur in nature. There exist genetically diverse root systems, but it is not clear what effect they have on the fitness of the constituent plants, and to what extent the captured roots contribute their genes "actively" to offspring or are "compelled to do so by the dominant individual."

As we will discuss in more detail in Chap. 10 individual cells of the slime mold *Dictyostelium discoideum* can aggregate into a communal individual composed of stalk cells that take on a somatic role, and a bulblike structure of cells that will contribute to the next generation. After reviewing the evidence then available that *Dictyostelium* cells might be genetically diverse, Williams concluded that aggregating cells were most likely composed of only a few clonal types and that a stalk cell "would, in assuming a somatic role, be favoring the reproduction of a group of cells that would usually contain a large portion of individuals genetically identical to itself" (Williams 1966, p. 224). He allowed for the possibility of selection even in a genetically inhomogeneous system, but stated that it would be "less effective" than a homogeneous system, and argued that "the behavior of the amoebae could still be interpreted as a purely organic adaptation. If the proportion of genetically identical cells is ordinarily small, biotic adaptation would be indicated" (Williams 1966, pp. 224–225). He did not speculate on the relative proportion of genetically identical cells that is necessary for a transition from organic to biotic adaptation. In the years since Williams wrote, the genetics of *Dictyostelium* have been investigated in great detail, and it is now known that *Dicty* (as it is affectionately known by many in the scientific community) is far more complicated than simply a group of clonal amoebae. In fact, R. H. Kessin described the *Dictyostelium* genome as "littered" with transposable elements.

Next, Williams turned to a property often suggested as a group-level adaptation – and the one that had initially sparked Williams's distaste for this style of thought – senescence. Is senescence, Williams asked, a biotic adaptation? First, he noted that it is general decay, not some sort of programmed cell death (though that terminology was not current at the time Williams wrote). Moreover, in wild populations, most organisms do not live long enough to die of old age. Williams noted the point made by Peter Medawar that variations in an organism's fitness after its reproductive period is complete will have no effect on how that individual reproduces, and therefore will not be "seen" by natural selection. As a result, there is no selective pressure against senescence. (Of course, this also raises the question of why some species have a finite reproductive period.) Williams argued that senescence may be exacerbated because "selection may…favor genes that produce slight increases in fitness in youth, even if they produce markedly deleterious effects later on" (Williams 1966, p. 226). He noted that it was important to "to distinguish the goal from the sacrifice", i.e., to separate function from effect. Leaves on lower tree branches no longer receive sufficient sunlight to be efficient photosynthetic machines. But this is not an adaptation. Rather, is a side effect of the redeployment of nutrients toward the higher branches (Williams 1966, p. 228).

After discussing a range of further examples, including toxins and bee stings, Williams turned to a more theoretical discussion, noting the difficulty of developing a clear criterion for fitness at the group level. "In the absence of objective or generally accepted criteria of population fitness, it has seemed pointless to attempt an evaluation of whether a supposed adaptation would contribute to the well-being of the group" (Williams 1966, p. 232). He noted colleagues who referred to group fitness in decidedly anthropomorphic terms. With criteria so slippery and so tainted by individual human bias, why even bother?

Williams turned next turned to the problem of population regulation. His critique of Wynne-Edwards was quite sharp and cogent, and was an important factor in the sharp decline of interest in group selection. (As Borello notes, another contributing factor was Wynne-Edwards's failure to adequately and thoroughly respond to Williams's critique.) Wynne-Edwards had likened population fluctuations around a stable environmental carrying capacity to the fluctuations of an organism's temperature around a homeostatically maintained value. But is population regulation a result of group selection, as Wynne-Edwards claimed, or was there a "simpler" explanation? In fact, Williams argued, population regulation is "a purely physical necessity. It is physically impossible for a population to exceed what its current environment is capable of supporting. The failure of a physical impossibility to occur is not something that we need attribute to evolved adaptations" (Williams 1966, p. 236). Population regulation, he wrote, occurs at the individual level. If resources are too scarce, an organism can adjust its fecundity to conserve its own resources and survive to the next breeding season, or it can "make gametes and starve to death". Any effect on a group-level property like population size is a statistical by-product of individual selection. Moreover, since these effects will set in as the population reaches its carrying capacity, what would be the advantage of developing a complicated set of epideictic displays to maintain populations at a homeostatic level slightly below the carrying capacity? There was certainly no evidence, Williams further argued, that such displays regulate population fluctuations rather than, or in addition to, population size, or that such regulation would be beneficial to a population.[4]

Williams next turned his attack toward the interpretation of territoriality. Wynne-Edwards had suggested that spacing between nests was a population-limiting factor, and that it was a result of group selection. Williams argued that it was more likely an organic adaptation, to optimize the feeding area for a bird and its nest, for example, with population "density regulation as merely an incidental statistical byproduct." He noted that sexual conflict in bird populations would typically affect male birds, rather than the number of breeding females or young, and thus would not have an immediate effect on the population size of the next generation.[5] This too, then, he argued, is likely to be a result of individual rather than group adaptation. Any number of hypothetically epideictic behaviors could be explained in this way, as individual adaptations with a secondary effect on population size. Williams suggested that the role of mass death in lemming populations may simply be a "psychosis" brought on by overcrowding, rather than a group-level adaptation.

What about symbiotic, mutualistic interspecies collaborations – algae and fungi living together in lichens, or the intestinal parasites of termites? Can these complexes act as single units of selection and adaptation? Williams sidestepped a direct answer. "This is certainly true in a sense," he wrote.

> Neither a termite nor its intestinal symbionts can become extinct without the other sharing its fate. Likewise the evolution of each would have been very different had the other not been there. The important question, however, is whether the selection of alternative alleles can simply and adequately explain the origin and maintenance of such relationships. *I believe that such an explanation is possible and plausible in every instance.* (Williams 1966, p. 246, my italics)

Allee had suggested that "ecosystems and, perhaps, the whole biota of the earth" could be considered as adaptive units (Williams 1966, p. 247). Williams mocked this suggestion by likening it to the idea that carrots were designed to be eaten by rabbits.

> Similarly the structure and behavior of a rabbit are more readily interpreted as means for escaping from predators than for supplying them with food. An ecosystem, as a machine, is highly inefficient for just this reason, the impediments raised by each trophic level to the passage of energy to the next higher level. It would seem absurd to belabor such an argument, but this is the critical evidence on the validity of the organization of the community

---

[4] In this context, note recent work reviewed by Scheffer et al. (2009, 2012) and by Jeff Gore's group at MIT (Chen et al. 2014; Dai et al. 2012, 2013) on fluctuations preceding population collapse. There may indeed be "value" in regulating population fluctuations (though this does not mean, of course, that such regulation actually has evolved as a group-level trait).

[5] It would affect the amount of genetic diversity in the population, however.

as a concept in any way analogous to the organization of an organism. (Williams 1966, p. 248)

In discussing organisms, however, Williams did not hesitate to emphasize chance and historical contingency, with all the imperfection they imply. Following his logic, we can conclude that imperfections in organisms are signs of history, while imperfections at higher levels are signs that the level does not exist as an entity at all.

Williams concluded his chapter returning to the question of population stability. He cited the work of oceanographer Maxwell Dunbar, who studied fluctuations in the size of Arctic populations and suggested that small population sizes had evolved because decreased reproductive rates would stabilize the population. Williams made the interesting suggestion that large population fluctuations might simply result from the small number of species in the Arctic environment, leading to a greater interdependence among species, and hence to increased mutual sensitivity. He reasonably criticized Dunbar for not explaining how group-selected limits on fecundity could achieve stabilization of the population size, but then concluded that any curtailment of fecundity "beyond what would be expected of adaptations designed to maximize each individual's currency of offspring would have to be explained by something other than genic selection" (Williams 1966, pp. 249–250).

The last chapter of Williams's book sketched out a research programme for the future. He made the important point that populations themselves are an environment to which individuals must adapt, but argued that this way of looking at populations is opposed to, and obscured by, "the tendency to think of a population as something adapted" (Williams 1966, p. 252). He then stated his overarching conclusions about what a species is, and, more importantly, what it is not. A species, he wrote,

> ...is a group of one or more populations that have irrevocably separated from other populations as a result of the development of intrinsic barriers to genetic recombination. The species is therefore a key taxonomic and evolutionary concept *but has no special significance for the study of adaptation. It is not an adapted unit and there are no mechanisms that function for the survival of the species.* The only adaptations that clearly exist express themselves in genetically defined individuals and have only one ultimate goal, the maximal perpetuation of the genes responsible for the visible adaptive mechanisms, a goal equated to Hamilton's 'inclusive fitness'. The significance of an individual is equal to the extent to which it realizes this goal. In other words its significance lies entirely in its contribution to one aspect of the vital statistics of the population. (Williams 1966, p. 252, my italics)

After laying out these parameters, Williams proposed to fight the "abundance of misinformation" that has prevented scientists from immediately recognizing the validity of his view by developing a new field of science, which called "teleonomy". The "first concern" of this field would be to answer the question, with regard to any biological phenomenon, "what is its function?" This would reveal, he suggested, the "function" of many properties previously misinterpreted as group-level adaptations to be merely a mirage. Many scientists have been unconsciously biased by an anthropomorphic view of nature, and an unconscious desire to impose a human concept of morality on nature, Williams argued. This led to the misconception of adaptation at the group level. What is the reason for which offspring are produced?

Is it, as is often stated, for the perpetuation of the species? Or are they produced, as I have maintained, to maximize the representation of the parental genes in the next and subsequent generations?...Each part of the animal is organized for some function tributary to the ultimate survival of its own genes. (Williams 1966, pp. 253–256)

The science of teleonomy should take as its programme, Williams suggested, to first identify the function of any putative adaptation, and then to attempt to develop an explanation for it based on the natural selection of alternate alleles. Failing this, he suggested with undisguised contempt, "a teleonomist may explore other possibilities, such as group selection or even mystical causes if he is so inclined" (Williams 1966, p. 258).

This research programme might seem to have drifted rather far from Williams's original conception of adaptation as an "onerous" concept. Williams was careful to note

that an effect can be called a function only when chance can be ruled out as a possible mechanism. In an individual organism an effect should be assumed to be the result of physical laws only, or perhaps the fortuitous effect of some unrelated adaptation, unless there is clear evidence that it is produced by mechanisms designed to produce it. (Williams 1966, p. 261)

However, this important caution was added almost as an afterthought to the intellectual programme of teleonomy, and readers encounter the question "what is its function?" long before they are cautioned to first ask "does it have a function?" The post hoc admonition seemed hardly sufficient to prevent pan-adaptationist bias in practitioners of the new science of teleonomy.

Williams concluded that his approach was not necessarily the truth, but certainly "the light and the way" (Williams 1966, p. 273). He suggested that it may take on a Kuhnian "normal science" role, a backdrop against which fruitful errors might stand out.

It is only by the rigorous application of a theory, however, that its imperfections can be recognized and rectified. We must take the theory of natural selection in its simplest and most austere form, the differential survival of alternative alleles, and use it in an uncompromising fashion whenever a problem of adaptation arises. When the best such explanation is complex and not very plausible, the way is paved for a better theory. (Williams 1966, p. 270)

However, the way was to grow much narrower before it began again to widen.

*** *** ***

Oxford evolutionary biologist Richard Dawkins was deeply influenced by the work of Williams and Hamilton, and extended their ideas in his two brilliantly argued and provocative works, *The Selfish Gene* (1976) and *The Extended Phenotype* (1982). Dawkins took the gene-centered view propounded by Williams, and mathematicized by Hamilton's work on kin selection, to its ultimate conclusion: natural selection acts on genes *only*, and everything at a higher taxonomic level, from the cell to the organism to the species, is a "vehicle" through which genes compete for their own survival.

Dawkins's view is particularly unique for its emphasis on *replication*. This was already noted by Williams as an important factor defining the central role of genes in evolution, a role defined by their stability from one generation to the next. But Dawkins built his entire theory around the idea of replication. To do so, he relied on criteria for evolvability significantly different from the "variation, heritability, and differential fitness" criteria defined by Richard Lewontin.

In a 1970 paper entitled "The Units of Selection", Lewontin defined the "logical skeleton" of Darwin's original argument as consisting of three core principles.

> 1. Different individuals in a population have different morphologies, physiologies, and behaviors (**phenotypic variation**). 2. Different phenotypes have different rates of survival and reproduction in different environments (**differential fitness**). 3. There is a correlation between parents and offspring in the contribution of each to future generations (**fitness is heritable**) (Lewontin 1970, my bold face).

These criteria are quite general and, as Lewontin discussed in his paper, are not limited to any particular level of biological organization.

Dawkins defined three quite different core principles: **fidelity**, **longevity** and **fecundity**. Let us follow Dawkins's argument in the early chapters of *The Selfish Gene* to trace the development of these criteria and the use to which he puts them.

Dawkins began with the postulate that "survival of the fittest" is a special case of a more general principle, "survival of the stable" (Dawkins 1976, p. 12). Going back to the earliest ideas of chemical evolution, he wrote that "[t]he earliest form of natural selection was simply a selection of stable forms and a rejection of unstable ones. There is no mystery about this. It had to happen by definition" (Dawkins 1976, p. 13). The next step in the origin of life likely arose when a molecule was formed that had the capacity for autocatalysis,[6] and ultimately the ability self-replicate. Dawkins termed these molecules *replicators*, and imagined that, with them,

> a new kind of 'stability' came into the world. Previously it is probable that no particular kind of complex molecule was very abundant in the [proverbial primordial] soup, because each was dependent on building blocks happening to fall by luck into a particular stable configuration. As soon as the replicator was born it must have spread its copies rapidly through the seas, until the smaller building block molecules become a scarce resource, and other larger molecules were formed more and more rarely. (Dawkins 1976, p. 16)

Dawkins allowed that, since "mistakes will happen", copying was not always perfect, but he maintained that accuracy, or copying fidelity, was more important than variation. This is where the core underpinnings of his conceptual framework diverge radically from those of Lewontin. Here, for example, is how Dawkins envisioned the relation between two populations of molecules with different mutation rates.

> If molecules of type X and type Y last the same length of time and replicate at the same rate, but X makes mistakes on average every tenth replication while Y makes a mistake only every hundredth replication, Y will obviously become more numerous. The X contingent in

---

[6] The ability of RNA to do this, discovered by Tom Cech in the self-splicing activity of RNA in *Tetrahymena thermophila,* led to the hypothesis of the "RNA world", in which genetic information was initially encoded in RNA rather than DNA.

the population loses not only the errant 'children' themselves, but also all their descendants, actual or potential. (Dawkins 1976, p. 17)

Dawkins clearly envisioned something quite different from a lineage branching and expanding through time within a space of possibilities. The point was *not to evolve, but to stay the same.* He said as much, noting that the reader

> may find something slightly paradoxical about the last point. Can we reconcile the idea that copying errors are an essential prerequisite for evolution to occur, with the statement that natural selection favors high copying-fidelity[7]? The answer is that although evolution may seem, in some vague sense, a 'good thing', especially since we are the product of it, nothing actually 'wants' to evolve. Evolution is something that happens, willy-nilly, in spite of all the efforts of the replicators (and nowadays of the genes) to prevent it happening. (Dawkins 1976, pp. 17–18)

As pointed out by Griesemer (2000), emphasis on fidelity possesses a fundamental conceptual problem: the processes that maintain fidelity are themselves the *products* of evolution. Okasha wrote that fidelity

> characterizes the evolutionary process in terms of features that are themselves the product of evolution. The longevity and copying fidelity of replicators (such as genes) and the cohesiveness of interactors (such as organisms) are highly *evolved* properties, themselves the product of many rounds of cumulative selection. The earliest replicators must have had extremely *poor* copying fidelity…, and the earliest multicelled organisms must have been highly *non-cohesive* entities, owing to the competition between their constituent cell-lineages… If we wish to understand how copying fidelity and cohesiveness evolved in the first place, we cannot build these notions into the very concepts used to describe natural selection. (Okasha 2006, p. 16)

Having established fidelity as a core criterion, Dawkins added the more obvious ones of fecundity and longevity. The more times a replicator can replicate, the more it can dominate the population; the longer it lives, the more times it can replicate. Each of these properties, he emphasized, are a type of stability. (They could also be interpreted as properties that increase fitness, defined as number of offspring.) It was only at this point that Dawkins introduced the idea of competition, as an ancillary property to his core three. In contrast, competition played a Lewontin's three core principles from the start, as a possible outcome of differential fitness. Differential fitness, Lewontin emphasized, could result in direct competition between organisms, but could also be realized via the struggle of organisms with the environment. Competition was thus a subset of Lewontin's second criterion, while it did not even merit a place on the podium in Dawkins's scheme. But competition in Dawkins's primordial soup was nonetheless important.

> There was a struggle for existence among replicator varieties. They did not know that they were struggling, or worry about it; the struggle was conducted without any hard feelings, indeed without feelings of any kind. But they were struggling, in the sense that any mis-copying that resulted in a new higher level of stability, or a new way of reducing the stability of rivals, was automatically preserved and multiplied. The process of improvement was

---

[7] I would argue, and I believe so would many other scientists, that natural selection does nothing of the kind. It favors fitness, it favors offspring who survive, even if they are a bit different from their parents. Indeed, natural selection sometimes favors an increase in mutation rate, as will be discussed below in Chap. 14.

cumulative. Ways of increasing stability and of decreasing rivals' stability became more elaborate and more efficient. Some of them may have even 'discovered' how to break up molecules of rival varieties chemically, and to use the building blocks so released for making their own copies. These proto-carnivores simultaneously obtained food and removed competing rivals. Other replicators perhaps discovered how to protect themselves, either chemically, or by building a physical wall of protein around themselves. This may have been how the first living cells appeared. Replicators began not merely to exist, but to construct for themselves containers, vehicles for their continued existence. The replicators that survived were the ones that build *survival machines* for themselves to live in. The first survival machines probably consisted of nothing more than a protective coat. But making a living got steadily harder as new rivals arose with better and more effective survival machines. Survival machines got bigger and more elaborate, and the process was cumulative and progressive. (Dawkins 1976, p. 19, his italics)

Note the essential role of variability in this progression of events. I would argue that this exposes a fundamental inconsistency in Dawkins's insistence on fidelity as a core principle. However, this insistence is essential for the next step of his argument. Since fidelity is key, it is the replicators, not the messy 'vehicles', that change from generation to generation, that matter. The story of life is theirs, not ours.

But do not look for them floating loose in the sea; they gave up that cavalier freedom long ago. Now they swarm in huge colonies, safe inside gigantic lumbering robots, sealed off from the outside world, communicating with it by torturous indirect routes, manipulating it by remote control. They are in you and they are in me; they created us, body and mind, and their preservation is the ultimate rationale for our existence…replicators have built a vast array of machines to exploit… A monkey is a machine that preserves genes up trees, a fish is a machine that preserves genes in the water; there is even a small worm[8] that preserves genes in German beer mats. (Dawkins 1976, pp. 20–21)

Genetically speaking, Dawkins concluded, we are "like clouds in the sky or dust-storms in the desert" (Dawkins 1976, p. 34).

Dawkins noted that genes are naturally "gregarious". But before one could suggest that he is ignoring the group selection inherent in the ganging up of genes,[9] he was quick to emphasize that the aggregations of genes are extremely temporary.[10] Using the analogy of a continually shuffled deck of cards, he reminded the reader that the crossing-over inherent in the process of meiosis, the generation of gamete cells for sexual reproduction, leads to such extensive shuffling between chromosomes that genes can hardly be said to carry permanent associates with them through evolutionary time.

---

[8] *Panagrellus redivivus*. They also live in book-binding glue.

[9] Okasha (2006) makes precisely this argument, referencing the work of Szathmáry and Demeter. "[W]hat Dawkins misses is that [the 'ganging up' of genes] in effect invokes group selection. From the selective point of view, replicating molecules combining themselves into compartments is strictly analogous to individual organisms combining themselves into colonies or groups (Szathmáry and Demeter 1987). But Dawkins is an implacable *opponent* of group selection, insisting on the impotence of selection for group advantage as an evolutionary mechanism. Clearly, Dawkins does not appreciate that evolutionary transitions necessarily involve selection at multiple levels." (Okasha 2006, p. 222)

[10] This echoes the statements made by Williams regarding the ephemerality of genotypes. However, aggregations need not be permanent to be subject to group selection; see the discussion of MLS1 in Chap. 13 below, and David Sloan Wilson's concept of trait groups (Wilson 1975).

When he initially introduced this argument in *The Selfish Gene*, Dawkins failed to mention that not all species undergo sexual reproduction, and indeed that bacterial fission, which typically does not involve crossing-over, arose well before sexual reproduction (Narra and Ochman 2006; Goodenough and Heitman 2014). Sexual reproduction has been suggested to provides a fitness advantage at the species level, for the very reason that it increases the variation within a population, and thus the ability of the species to explore the evolutionary space of possibilities.

"Some people," Dawkins wrote,

> regard the species as the unit of natural selection, others the population or group within the species, and yet others the individual. I said that I preferred to think of the gene as the fundamental unit of natural selection, and therefore the fundamental unit of self-interest. What I have done now is to *define* the gene in such a way that I cannot really help being right! (Dawkins 1976, p. 33, italics and exclamation point his)

This statement is quite remarkable, not only for the strategic reveal in the last sentence, but also because of the assumption implicit in the first two. Dawkins does not even allow the possibility that a scientist could pick more than one level to be "the" unit of selection. This is the Highlander approach to evolution: *there can be only one*!

There are indeed cases where DNA can behave selfishly, just as a cell lineage in a metazoan body behaves selfishly when it becomes cancerous. One example of this is *meiotic drive*, in which genes manipulate the production of gametes in order to increase their own frequency within the population (Lindholm et al. 2016; the term is used for such effects even when they do not occur during meiosis). Examples include sex-linked genes that act to distort the ratio of offspring. One dramatic case of meiotic drive is the t haplotype in mice, which arises from variants of genes on chromosome 17. While promoting its own survival, this collection of genes produces outcomes such as male sterility and embryonic death: a clear conflict between levels of selection (Silver 1993).

The evolutionary interests of genetic material can also conflict with those of the containing organism during lateral transmission of genetic elements such as plasmids (Werren 2011). A recent study suggests that stem cells use an RNA-based "immune" system to defend themselves against mobile genetic elements (Haase 2016). It has been speculated that selfish genetic elements played a role in eukaryotic evolution (Hurst and Werren 2001; Werren 2011) and other evolutionary transitions (Ågren 2014; Koonin 2016).

Dawkins extended his ideas of replicators and vehicles by applying the ideas of Williams and Hamilton to problems of competition, cooperation, and altruism. Dawkins discussed evolutionary game theory from the perspective of competing genes. Discussing various types of collective animal behavior, he interpreted the actions of individual animals as driven by the self-preservation of their genes. Most notably, Dawkins presented Hamilton's kin selection argument for collective behavior in eusocial insects from the gene's eye view.

Hamilton's theory of inclusive fitness (also called kin selection) offers an explanation of the phenomenon of altruism, most notable in social insects. Inclusive fitness holds that altruism can evolve because genes correlated with altruistic behavior can spread not only if the animal that carries them survives, but if any of that animal's relatives survive as well, since the relatives will have a high probability of carrying the same gene. Thus an organism will behave in a way that maximizes its "inclusive fitness", a measure that includes not only the organism's fitness, but also that of its relatives. This idea is well known in the form of Hamilton's rule, $r > c/b$, where r is the relatedness of two individuals, one of whom behaves altruistically toward the other, b is the reproductive benefit gained by the recipient of the altruistic act, and c is the cost to the altruistic individual of performing the act. The relatedness is typically defined as the percentage of genes shared by the two individuals, which also gives the likelihood that the altruistic gene will be present in both genomes. According to Hamilton's rule, the frequency of a gene for the altruistic act will increase in the population if this inequality holds. In other words, the altruistic behavior will be favored if relatedness exceeds the cost-benefit ratio. Hamilton's rule was epitomized by J. B. S. Haldane's statement that he would "lay down his life for two brothers or eight cousins".

Behaviors have been observed in mammals that support Hamilton's rule, such as squirrels adopting orphaned young to whom they are related, but shunning unrelated orphans (Gorrell et al. 2010). The most dramatic examples of kin selection are found in social insects. Most insects of the order *Hymenoptera*, which includes bees and wasps, are haplodiploid. The queen, after having been fertilized on her "maiden flight", can regulate the fertilization of her eggs once she returns to the nest; eggs that are fertilized become female workers (with a diploid genome, two copies of each chromosome, one from each parent), and eggs that are not fertilized become male drones (with a haploid genome, one copy of each chromosome, from their mother only). This means that the males are all genetically identical to each other, and to the queen. The sisters are thus less closely related to their mother than their brothers are. But, more importantly, *they are more closely related to each other than to their mother*. The sisters are essentially identical twins on their father's side; being haploid, the males will produce sperm that are all identical. In contrast, sisters have a 50% chance of sharing genes that come from their mother, since any one of her genes may have been shuffled with a paternal gene during crossing-over. This means that sisters have a relatedness ratio of ¾, rather than the typical ½ one would expect in a non-haplodiploid population.

From the gene's eye view, worker sterility is not some sort of altruistic sacrifice "for the group". Rather, since a worker is more closely related to her sisters than she would be to her offspring, it is to the benefit of her genes to help preserve her sisters and "farm her own mother as an efficient sister-making machine" (Dawkins 1976, p. 175) rather than to be a queen herself. Dawkins noted that eusociality (sterility in a worker caste) typically occurs in haplodiploid *Hymenoptera* species, where it has likely arisen multiple times *independently* through evolutionary history. (Eusociality does occur in diploid insects such as termites, where both males and females are sterile workers, and has been observed in mammals such as the naked mole rat.) Dawkins cited the work of Trivers and Hare (1976), who showed that some *Hymenoptera* species do indeed try to bias their mothers to produce sisters rather than brothers, to whom they would be less closely related, and discussed the factors that might enable the workers to "win" this manipulative struggle with their mother, to whose genetic advantage it would be to produce an equal number of sons and daughters.

Dawkins did not disagree entirely with the organism-centric view that natural selection acts on bodies. He wrote that by

> any sensible view of the matter Darwinian selection does not work on genes directly. DNA is cocooned in protein, swaddled in membranes, shielded from the world and invisible to natural selection. If selection tried to choose DNA molecules directly it would hardly find any criteria by which to do so. All genes look alike, just as all recording tapes look alike. The important differences between genes emerge only in their *effects*. (Dawkins 1976, p. 235, his italics)[11]

Where Dawkins diverges most radically from many other evolutionary theorists, however, is in his interpretation of what evolution is *about*. Dawkins is ultimately unconcerned with the evolution of entities that are acted upon by natural selection, be they genes, organisms, or species. These biological entities were defined by Hull as *interactors*, who emphasized their role as "cohesive wholes" interacting with the environment in such a way "that this interaction causes replication to be differential" (Griesemer 2000). By reducing *interactors* to *vehicles*, Dawkins radically displaced the locus of the evolutionary process away from change itself. Dawkins likened his conceptual shift to that induced by staring at a Necker cube until its image appears to flip itself inside out. To Dawkins, this was a radical and illuminating shift; to Gould, it was no more than "mistaking bookkeeping for causality".

In the last chapter of *The Selfish Gene* and in his next book, *The Extended Phenotype* (1982), Dawkins introduced a new idea that radically broadened the reach of the gene. This idea was presented both in its own right and as part of a larger argument to demote the importance of the organism. With the extended phenotype, Dawkins took the idea of a gene's *effects* to a startling but perhaps inevitable limit. If genes interact with the world, competing for a chance to replicate, via the organisms they inhabit, must they not also interact via the *behavior* of those organisms, their interactions, their created artifacts, their manipulations of the external world? A gene's extended phenotype is its effect, not on the body it inhabits and has helped to create, but *on the world*.

Dawkins suggested two broad categories of extended phenotypes: gene effects on the external world, and gene effects on other bodies. Examples of the first include the elaborate nests of weaver birds (Fig. 7.1), the little stone houses built by the larvae of caddisflies (Fig. 7.2)[12], and beaver dams (Fig. 7.3). The second category can be divided into direct and indirect effects. Direct effects arise in parasitism,

---

[11] This statement is somewhat belied, or at least muddied, by the adoption, early in *The Selfish Gene*, of a definition of the gene in terms of its role as a unit of natural selection. "The definition I want to use," Dawkins wrote, "comes from G. C. Williams. A gene is defined as any portion of chromosomal material that potentially lasts for enough generations to serve as a unit of natural selection" (Dawkins 1976, p. 28).

[12] Caddisflies will construct nests out of whatever material is available in their environment, whether small stones or plant matter, as shown in (Fig. 7.2). The artist Hubert Duprat "collaborates" with caddisflies by providing them with jewels and precious metals with which to build their houses.

Fig. 7.1 The extended phenotype: a Baya Weaver bird (*Ploceus philippinus*) in Kolkata, West Bengal, India. Photo by J. M. Garg, https://commons.wikimedia.org/wiki/File:Baya_weaver_at_nest_I_IMG_5101.jpg. (Creative Commons Attribution-Share Alike 3.0 Unported license)

Fig. 7.2 The extended phenotype: a caddisfly emerges from a nest built of plant material. Photograph by MyForest - Own work, CC BY-SA 3.0, https://commons.wikimedia.org/w/index.php?curid=11679443

when the parasite's genes have a specific effect on the body of the host. As an example, Dawkins described flatworms (flukes) that parasitize certain species of snail, inducing them to produce unusually thick shells. One might initially assume that thicker shells would benefit the snails. In fact, a thicker-than-necessary shell is a costly waste of resources better devoted, from the point of view of the snail and its

Fig. 7.3 The extended phenotype: a beaver dam near Olden, Jämtland, Sweden. Photograph by Lars Falkdalen Lindahl (https://commons.wikimedia.org/wiki/File:Beaver_dam_Jämtland.JPG, licensed under licensed under the Creative Commons Attribution-Share Alike 4.0 International, 3.0 Unported, 2.5 Generic, 2.0 Generic and 1.0 Generic license)

genes, to reproduction. The fluke thus induces the snail to enact a behavior that increases its longevity, but not its fecundity, because while "snail genes stand to gain from the snail's reproduction, fluke genes don't" (Dawkins 1976, p. 241). Nonetheless, "the change in a snail shell is a fluke adaptation…come about by Darwinian selection of fluke genes" (Dawkins 1976, p. 242).

Another example of parasite and host genes at odds is the protozoan *Nosema*, which infects flour beetle larvae, producing a substance similar to the beetle's own juvenile hormone, prolonging the larval stage. The giant larvae, twice the size of normal adults, cannot mature (and thus cannot reproduce) but make excellent hosts. Dawkins also described the parasitic castration of certain crab species by *Sacculina*, a parasite related to the barnacle.

> It drives an elaborate root system deep into the tissues of the unfortunate crab, and sucks nourishment from its body. It is probably no accident that among the first organs that it attacks are the crab's testicles or ovaries; it spares the organs that the crab needs to survive – as opposed to reproduce – till later… Like a fattened bullock, the castrated crab diverts energy and resources away from reproduction and into its own body. (Dawkins 1976, pp. 242–243)

Returning to the snail and fluke to draw his selfish gene conclusions, Dawkins remarked that the genes of the fluke and those of the snail can both be viewed as parasites in the snail body.

> Both gain from being surrounded by the same protective shell, though they diverge from one another in the precise thickness of shell that they 'prefer'. This divergence arises, fundamentally, from the fact that their method of leaving this snail's body and entering another one is different. (Dawkins 1976, p. 243)

This raises an important distinction between types of genetic parasitism. In the three examples above, parasite genes depend on host survival but not on host reproduction, and thus can wantonly squander the reproductive chances of their host. But in other cases, such as when the parasite is transmitted through the host's eggs, the reproductive interests of both sets of parasitic genes align. Bacteria parasitic on the haplodiploid ambrosia beetle are transmitted through the host's eggs, and actually provide a necessary stimulus to unfertilized eggs that will develop into males. Likewise, an alga transmitted via the egg of a hydra species provides its host with oxygen, rather than depleting the hydra's resources (as do other parasitic algae that live on hydra but are not transmitted this way). In such cases, Dawkins suggested that a parasite is likely to ultimately merge into the host body, since its genes "share the same destiny" as those of the host. One could ask what selective processes would drive such an evolutionary transition,[13] but Dawkins did not.

The cases above can be interpreted from a "gene's eye view", but that was not Dawkins's only purpose in setting out these examples. Extending the reach of the gene, he was simultaneously loosening the boundaries of the organism. The coup de grâce came with the second type of gene effects on bodies: the indirect ones. Examples include an ant that induces other ants to kill their own queen and raise the invader's young as their own, and the almost drug-like trance induced in small birds by the wide red gape of the cuckoo's mouth. In these cases, the parasite genes are acting on the *behavior* of the host.

Having made short shrift of organismal integrity, Dawkins spent the last few pages of *The Selfish Gene* (as well as the last chapter of *The Extended Phenotype*, "Rediscovering the Organism") at pains to reconstruct it. Why, he asked, did genes "gang up in cells? Why did cells gang up in many-celled bodies? And why did bodies adopt…a 'bottlenecked' life cycle?" (Dawkins 1976, p. 257). In answering the first of these questions, Dawkins came perilously close to a group selection argument, or at least to suggesting something suspiciously reminiscent of the division of labor. "Why did those ancient replicators give up the cavalier freedom of the primeval soup and take to swarming in huge colonies?" (Dawkins 1976, p. 257). The answer, he suggested, lies in the fact that genes code for individual proteins, which cannot mediate complex biochemical pathways in isolation.

---

[13] Such a transition might resemble that likely undergone by mitochondria and other organelles during the evolution of eukaryotic cells, as suggested by Lynn Margulis (whom Dawkins does not mention).

A whole set of enzymes is necessary, one to catalyse the transformation of the raw material into the first intermediate, another to catalyse the transformation of the first intermediate into the second, and so on. Each one of these enzymes is made by one gene. If a sequence of six enzymes is needed for a particular synthetic pathway, all six genes for making them must be present. (Dawkins 1976, p. 257)

But if they are each necessary, and thus more likely to survive when replicated in this group of six than alone, their common product would seem to fit the criteria of an MLS2 property. Surely the six genes coding for the six enzymes could be viewed as a group subject to a common selective pressure. However, Dawkins admonished the reader that

[t]empting as it is, it is positively wrong to speak of the genes for the six enzymes…as being selected 'as a group'. Each one is selected as a separate selfish gene, but it flourishes only in the presence of the right set of other genes. (Dawkins 1976, p. 258)

Dawkins dropped the argument at this point, failing to explain why it was so "positively wrong" to consider the genes as being selected as a group. Based on his previous arguments, one can infer that he might have been thinking of the crossing-over that occurs during meiosis as a means for separating the individual selfish genes from one another, so that they cannot move through evolutionary history together, but must find their fitness randomly when thrown together by accident (in other words, sorting rather than selection). But what of the vast number of species that do not undergo meiosis? And what of linkage disequilibrium, the non-random association between alleles at different loci? Dawkins did not address these issues, but instead moved right along to why cells "gang together" into "lumbering robots".

Even if organisms can be said to have some sort of collective integrity separate from their role as vehicles, they are doomed to comparative insignificance under Dawkins's scheme from the start, because they are not replicators. As for reasons why cells form multicellular organisms, Dawkins suggested overall size as an advantage under some circumstances, as well as the ability to produce specialized organs (while still serving the same genetic goals, since all the cells in a multicellular organism can be expected to be clones[14]).

Why the bottlenecked life cycle? Why return to a single cell to start a new generation? Dawkins answered these questions with the metaphor of two plant species, 'bottle-wrack' and 'splurge-weed'. Splurge-weed reproduces by simply breaking off pieces of itself every so often, while bottle-wrack reproduces by sporulation. The regularly repeating life cycle allows bottle-wrack to "return to the drawing board" in each generation, and provides a fixed sequence and calendar for embryological development. It also allows for genetic unity within a single plant.

In splurge-weed, cell lineages are broad-fronted… It is therefore quite possible that two cells in a daughter will be more distant relatives of one another than either is to cells in the parent plant… Bottle-wrack differs sharply from splurge-weed here. All cells in a daughter plant are descended from a single spore cell, so all cells in a given plant are closer cousins (or whatever) of one another than of any cell in another plant… In bottle-wrack, the indi-

---

[14] This is not the case in aggregative multicellular organisms such as *Dictyostelium*, however (see Chap. 10).

vidual plant will be a unit with a genetic identity, will deserve the name individual... Selection will therefore judge rival plants, not rival cells as in splurge-weed. (Dawkins 1976, pp. 262–263)

These individual plants, however, are still subservient to "the fundamental unit, the prime mover of all life, [which] is the replicator" (Dawkins 1976, p. 264). But Dawkins allowed group selection to slip, just for a moment, and "strictly for those with a professional interest". One can find "an analogy here," he wrote

with the argument over group selection. We can think of an individual organism as a 'group' of cells. A form of group selection can be made to work, provided some means can be found for increasing the ratio of between-group variation to within-group variation. Bottle-wrack's reproductive habit has exactly the effect of increasing this ratio; splurge-weed's habit has just the opposite effect. (Dawkins 1976, p. 263)

Curiously, this argument occurs a mere five pages after the assertion that it was "positively wrong" to consider genes encoding the members of a complex enzymatic pathway as being selected as a group.

*** *** ***

In one of the later chapters of *The Selfish Gene*, Dawkins noted the similarities between genetic and cultural transmission. Dawkins suggested that ideas themselves, or scraps of ideas, essentially transmit themselves from mind to mind, taking on a life, of sorts, of their own, and competing for brain space with other scraps of thought. He suggested the term *meme* to describe them, and the concept has become a good example of itself, growing in usage and popularity as it bounces from mind to mind. In fact, it can even be said to have speciated, becoming not only a unit of cultural transmission but also a unit of internet social currency, involving Grumpy Cat and a snarky caption.[15]

Like genes, memes possess the properties of fidelity, longevity and fecundity in varying degrees, and can form "co-adapted stable set[s] of mutually assisting memes", such as the idea of hellfire co-adapting with the idea of faith (Dawkins 1976, p. 197).[16] The idea of cultural evolution had been suggested before but, as with his ideas of genes, Dawkins emphasized that

---

[15] At a recent advertising festival at Cannes, Dawkins pointed out that internet memes do not mutate by chance, but produced by human attempts at creativity. (In 2016, we learned that they can be mutated for political purposes as well.)

[16] In another example, Dawkins rather acidly remarks, in terms that will not surprise readers of his more recent works, that "the meme for blind faith secures its own perpetuation by the simple unconscious expedient of discouraging rational inquiry" (Dawkins 1976, p. 198). Furthermore, in his review ("Caricature of Darwinism") of *The Selfish Gene* in the March 17, 1977 issue of *Nature*, Lewontin remarked, with even lower pH, that Dawkins has failed to consider another, far more plausible explanation for the perpetuation of the idea of hellfire, namely that it is not self-perpetuating at all, but simply is "perpetuated by some people because it gives them power over other people" (Lewontin 1977).

a cultural trait may have evolved the way that it has, simply because it is *advantageous to itself...* All that is necessary is that the brain should be capable of imitation: memes will then evolve [to] exploit [that] capability to the full. (Dawkins 1976, p. 200)

The exploitative aspect of the meme concept has fervently captured the public imagination: imagine, our brains colonized by a writhing mass of ideas! This idea was further developed by Susan Blackmore in *The Meme Machine* (1999). In mesmerizing but horrifying terms, Blackmore probed every aspect of culture, reducing everything from listening to jazz to preferring pasta for dinner to a meme. Memes, she wrote,

> can gain an advantage by becoming associated with a person's self concept. It does not matter how they do this – whether by raising strong emotions, by being especially compatible with memes already in place, or by providing a sense of power or attractiveness – they will fare better than other memes. These successful memes will more often be passed on, we will all come across them and so we, too, will get infected with self-enhancing memes. In this way our selfplexes are all strengthened. (Blackmore 1999, p. 232)

Have we floated up to such a fine-grained level of selection that *everything* has dissolved into a dust-storm in the desert? Is the idea of a collectivity completely checkmated by ephemeral, selfish aggregations? Or, combining the meme concept with the increasingly documented effects of interactive media on the human brain, are we heading toward a pitiful Kurzweilian future where no one will remember, let alone have the attention span to contemplate, Darwin's footsteps on the Sandwalk? In fact, the story of the collectivity, and of the essential tensions that enable its existence, is far from over. The next part of the story will take us through a tour along the frontier of the most recent scientific studies of collective behavior and experimental evolution. But before we begin, let me throw out one more quote from Blackmore, particularly ironic given the fact that she inherited the meme of the meme from Richard Dawkins. Noting that "group selection is favoured by mechanisms that reduce the differences in biological fitness *within* the group and increase the differences *between* groups, thus concentrating selection at the group level", Blackmore suggested that "[m]emes may provide just this kind of mechanism." Citing Boyd and Richerson's suggestion that group selection is favored when behavioral variation is culturally acquired, she concluded that "memes can have precisely the effect of decreasing within-group differences and increasing between-group differences" (Blackmore 1999, p. 198).

So, however, may cell adhesion molecules. Let's take a look.

References

Ågren JA (2014) Evolutionary transitions in individuality: insights from transposable elements. Trends Ecol Evol 29(2):90–96

Blackmore S (1999) The Meme Machine. Oxford University Press, Oxford

Borello ME (2010) Evolutionary Restraints: The Contentious History of Group Selection. The University of Chicago Press, Chicago/London

Chen A, Sanchez A, Dai L, Gore J (2014) Dynamics of a producer-freeloader ecosystem on the brink of collapse. Nat Commun 5:3713

Dai L, Vorselen D, Korolev KS, Gore J (2012) Generic indicators for loss of resilience before a tipping point leading to population collapse. Science 336(6085):1175–1177

Dai L, Korolev KS, Gore J (2013) Slower recovery in space before collapse of connected populations. Nature 496(7445):355–358

Dawkins R (1976) The Selfish Gene. Oxford University Press, Oxford. (1989 edition)

Dawkins R (1982) The Extended Phenotype. Oxford University Press, Oxford

Goodenough U, Heitman J (2014) Origins of eukaryotic sexual reproduction. Cold Spring Harb Perspect Biol 6(3):a016154

Gorrell JC, McAdam AG, Coltman DW, Humphries MM, Boutin S (2010) Adopting kin enhances inclusive fitness in asocial red squirrels. Nat Commun 1:22

Gould SJ (2002) The Structure of Evolutionary Theory. The Belknap Press of Harvard University Press, Cambridge, MA/London

Griesemer J (2000) The units of evolutionary transition. Selection 1(1–3):67–80

Haase AD (2016) A small RNA-based immune system defends germ cells against mobile genetic elements. Stem Cells Int 7595791:2016

Hurst GD, Werren JH (2001) The role of selfish genetic elements in eukaryotic evolution. Nat Rev Genet 2(8):597–606

Koonin EV (2016) Viruses and mobile elements as drivers of evolutionary transitions. Philos Trans R Soc Lond B 371(1701):20150442

Lewontin RC (1970) The units of selection. Ann Rev Ecol Syst 70:1–18

Lewontin RC (1977) Caricature of Darwinism (review of The Selfish Gene). Nature 266:283–284

Lindholm AK, Dyer KA, Firman RC, Fishman L, Forstmeier W, Holman L, Johannesson H, Knief U, Kokko H, Larracuente AM, Manser A, Montchamp-Moreau C, Petrosyan VG, Pomiankowski A, Presgraves DC, Safronova LD, Sutter A, Unckless RL, Verspoor RL, Wedell N, Wilkinson GS, Price TA (2016) The ecology and evolutionary dynamics of meiotic drive. Trends Ecol Evol 31(4):315–326

Narra HP, Ochman H (2006) Of what use is sex to bacteria? Curr Biol 16(17):R705–R710

Okasha S (2006) Evolution and the Levels of Selection. Oxford University Press (Clarendon Press), Oxford/New York

Scheffer M, Bascompte J, Brock WA, Brovkin V, Carpenter SR, Dakos V, Held H, van Nes EH, Rietkerk M, Sugihara G (2009) Early-warning signals for critical transitions. Nature 461(7260):53–59

Scheffer M, Carpenter SR, Lenton TM, Bascompte J, Brock W, Dakos V, van de Koppel J, van de Leemput IA, Levin SA, van Nes EH, Pascual M, Vandermeer J (2012) Anticipating critical transitions. Science 338(6105):344–348

Silver LM (1993) The peculiar journey of a selfish chromosome: mouse t haplotypes and meiotic drive. Trends Genet 9(7):250–254

Szathmáry E, Demeter L (1987) Group selection of early replicators and the origin of life. J Theor Biol 128(4):463–486

Trivers RL, Hare H (1976) Haploidploidy and the evolution of the social insect. Science 191(4224):249–263

Werren JH (2011) Selfish genetic elements, genetic conflict, and evolutionary innovation. Proc Natl Acad Sci U S A 108(Suppl 2):10863–10870

Williams GC (1966) Adaptation and Natural Selection: A Critique of Some Current Evolutionary
    Thought. Princeton University Press, Princeton
Wilson DS (1975) A theory of group selection. Proc Natl Acad Sci U S A 72(1):143–146
Wynne-Edwards VC (1962) Animal Dispersion in Relation to Social Behaviour. Oliver and Boyd,
    Edinburgh/London

Part II
At the Frontier

Chapter 8
Flocking, Swarming, and Communicating

Lordy, I hope there are tapes.

James Comey

AS WE have seen, Durkheim and others attempted to explain how groups of humans come together as a collective. One of the most beautiful innovations of twentieth century physics was the ability to analyze such behavior with statistical methods and computational models. This has enabled researchers to quantify the specific aspects of a group of individuals that change as the individuals coalesce into a crowd. For the first time, a quantitative step could be taken toward answering the question *when does a group become a separate individual in its own right?*

Ironically, one of the first steps along this road was taken, not by a physicist at all, but by a biologist – and one who was using a computational argument to support his stance *against* group selection, no less. This was none other than W. D. Hamilton, whose work so inspired Richard Dawkins. In a 1971 paper entitled "Geometry for the Selfish Herd", Hamilton proposed a model by which marginal predation drove the evolution of gregariousness. This was designed to explicitly counter the group selection arguments of Allee and Wynne-Edwards. To set the stage for his model, Hamilton asked his readers to picture a circular lily pond.

> Imagine that the pond shelters a colony of frogs and a water-snake. The snake preys on the frogs but only does so at a certain time of day – up to this time it sleeps on the bottom of the pond. Shortly before the snake is due to wake up all the frogs climb out onto the rim of the pond. This is because the snake prefers to catch frogs in the water. If it can't find any, however, it rears its head out of the water and surveys the disconsolate line sitting on the rim – it is supposed that fear of terrestrial predators prevents the frogs from going back from the rim – the snake surveys this line and snatches *the nearest one.* (Hamilton 1971, his italics)

Hamilton followed this thought experiment with computer simulations of the frogs, and of other prey animals in a two-dimensional landscape. Hamilton built on earlier ideas presented by Galton and others, as well as mid-twentieth century mathematical models of optimal packing, to illustrate how "selfish avoidance of a predator can lead to aggregation" (Hamilton 1971). In one sense, this was a masterful demonstration of how individual behavior can lead to the emergence of a group, and therefore a debunking of Hamilton's group selection bugbear. There was no need to resort to any "good of the group" argument, or even to the misplaced parental instincts

© Springer Science+Business Media B.V. 2018
S. Bahar, *The Essential Tension*, The Frontiers Collection,
DOI 10.1007/978-94-024-1054-9_8

suggested by Williams in his rebuttal to Wynne-Edwards. And yet, whether or not it can be described as an individual in its own right (and thus susceptible to selective pressure), a group *did emerge* from Hamilton's model, as the frogs crowded together, each trying to avoid being nearest to the snake. In the years following Hamilton's paper, as statistical physicists and nonlinear dynamicists turned their interest toward biological physics, more precise means were developed for quantifying this sort of "group".

***   ***   ***

In the late 1980s, the problem of flocks, herds and schools caught the attention of people working in computer graphics. While an animator could painstakingly create a flock frame by frame, wouldn't it be far more realistic to simulate flock movement based on realistic principles of animal motion? In order to tackle this problem, Craig Reynolds realized that he would have to devise an algorithm that broke the movement of birds in a flock down to a few fundamental rules. These began, he suggested, with *interaction between* the birds. "This approach", he wrote,

> assumes a flock is simply the result of the interaction between the behaviors of individual birds. To simulate a flock we simulate the behavior of an individual bird (or at least that portion of the bird's behavior that allows it to participate in a flock) … If this simulated bird model has the correct flock-member behavior, all that should be required to create a simulated flock is to create some instances of the simulated bird model and allow them to interact. (Reynolds 1987)

He named the interactors in his model "boids" (from "bird-oids", not from a New Jersey accent) noting that he used this term generically "even when they represent other sorts of creatures such as schooling fish". Not only, Reynolds found, did the boid model reduce to interactions between individuals, but those actions were *purely local*. Reynolds explained the biological motivation for this requirement as follows.

> There is no evidence that the complexity of natural flocks is bounded in any way. Flocks do not become "full" or "over-loaded" as new birds join. When herring migrate toward their spawning grounds, they run in schools extending as long as 17 miles and containing millions of fish… Natural flocks seem to operate in exactly the same fashion over a huge range of flock populations. It does not seem that an individual bird can be paying much attention to each and every one of its flockmates. But in a huge flock spread over vast distances, an individual bird must have a localized and filtered perception of the rest of the flock. A bird might be aware of three categories: itself, its two or three nearest neighbors, and the rest of the flock… [T]he amount of thinking that a bird has to do in order to flock must be largely independent of the number of birds in the flock. Otherwise we would expect to see a sharp upper bound on the size of natural flocks when the individual birds became overloaded by the complexity of their navigational task. This has not been observed in nature. (Reynolds 1987)[1]

---

[1] Rather ruefully, Reynolds pointed out that the computer scientist is not so lucky: computational modeling of flock behavior does indeed get significantly more difficult as the flock size is increased.

In addition to these logical arguments, Reynolds cited the experimental studies by Brian Partridge (1982) showing that schooling fish are more influenced by their close neighbors than by distant fish.

With this motivation, Reynolds defined three behaviors that would lead to a flock exhibiting "non-colliding, polarized, aggregate" motion. First, *collision avoidance*; second, *velocity matching*, in which each boid would attempt to align with its near neighbors; and lastly, *flock centering*, whereby boids attempt to remain close to their nearest flockmates. (Note the similarity of this last rule to that used by Hamilton). Using these rules, Reynolds generated quite realistic simulations. He noted that his simulations might be more realistic for animal herds or fish schools than for actual bird flocks, since birds have excellent vision and a wide field of view (~300°) and thus would be less likely to be limited to perceiving only a few close neighbors than, say, fish schooling in muddy waters. Future work, he suggested, might include actual simulation of visual perception by the boids. This challenge was taken up by Garry Peterson, who introduced a "zooid" model in 1993, which incorporated field of view and directionality of perception (it was easier for a zooid to perceive those ahead of it rather than those following it, a realistic requirement lacking in Reynolds's boids). Zooid flocks split into polarized clusters, switched flock leadership from one zooid to another, and formed stable "carousels", similar to the circular movement patterns exhibited by ants.

Animators and computer graphics experts continued to develop more detailed and realistic models in the years that followed, but by this time statistical physicists were turning their attention to the problem of flocking, and it is to their efforts that we now turn our attention as well.

*** *** ***

Much earlier in this book, I used the analogy of spin state alignment as a metaphor for interactions between like-minded individuals in a crowd. In fact, the comparison goes far beyond mere analogy. The tremendous mid-twentieth century achievements in the study of phase transitions in magnetic materials paved the way for models of collective dynamics extending far beyond Hamilton's selfish herd or Reynolds's boids.

The statistical physics of phase transitions was developed[2] in large part to model the mutual alignment of spin states in a material placed in an external magnetic field. As a function of the applied field and the temperature, nearby spins will interact and force each other into alignment, resulting in a phase transition to a magnetized state. This is an example of a *continuous* or *second-order* phase transition, and it possesses a remarkable set of properties. During the transition, the value of an *order parameter*, such as the energy of the system or the net number of unpaired spins, undergoes a sharp, but not discontinuous, change as a control parameter (typically the external magnetic field strength or the temperature) is varied. In the control

---

[2] See Brush (1967) for a history of the early days of this field and the development of the iconic Ising model.

parameter range where this transition takes place, the order parameter exhibits large fluctuations, and the magnetized regions of the material exhibit *critical scaling*. This means that there are magnetized patches of all length scales, and if renormalization is applied to the system, the scale-free behavior will persist (see Yeomans 1992 for an elegant introduction to this field). In more technical terms, the correlation length of the system exhibits power-law scaling as a function of the difference between the current value of the control parameter and its critical point, the precise value of the control parameter where the transition takes place. Other measures of system behavior also exhibit power-law scaling, and, taken together, the critical exponents according to which these measures scale in the transition region define the *universality class* of the phase transition.

In addition to its role in the study of phase transitions, the property of being "scale-free" is particularly resonant to physicists because it is also observed in nonlinear systems such as fractals and chaotic attractors. The fractal fern, for example, generated by iteration of random choices between four affine transformations (see Barnsley 1993) contains an infinitely nested sequence of infinitely lacy ferns, persisting to the smallest imaginable spatial scale. Likewise, the layers of trajectories inside a chaotic attractor repeat to an infinitely fine level of detail as a result of the stretching and folding that generates chaotic behavior – stretching, the endless pulling apart of nearby trajectories caused by a positive Lyapunov exponent, and folding caused by the volume contraction inherent in the dissipative nature of the system (Strogatz 1994). The combination of stretching and folding provides another example of an essential tension, a set of orthogonal conceptual axes analogous to the competition and cooperation whose story we are tracing through the pages of this book.

Physicists were quick to notice the relevance of phase transitions and nonlinear dynamics for fundamental problems in biology. While some turned to look for critical scaling in heartbeat fluctuations, others, far more successfully, explored the inherent nonlinearity of the electrophysiological excitation of neurons and cardiac myocytes. Still others were inspired to study fractal growth phenomena in expanding bacterial colonies. And along with the study of growth came the study of *aggregation*.

The application of statistical physics tools to the study of collective animal behavior was pioneered by Tamás Vicsek, of Eötvös University in Budapest, and his colleagues. In a seminal *Physical Review Letters* paper in 1995 and a detailed follow-up study in *J. Phys. A* in 1997, they drew the analogy between the alignment of spin states and the alignment of animal velocities during flocking behavior. From this, they developed a computational model that ultimately led to an elegant theory of animal aggregation, capable of describing everything from the massive murmurations of starlings (Fig. 8.1) that cause power cuts in the United Kingdom to how the placement of columns in a theater can aid the escape of a panicked crowd during a fire (Helbing et al. 2000). Flocking has even been used to analyze and model data collected from human motion in mosh pits (Silverberg et al. 2013). Computer modeling of crowd behavior is also now widely used in practical situations such as crowd control and risk analysis. Many of these methods are reviewed in the 2014

Fig. 8.1 A murmuration of starlings flying near Springfield, Dumfries and Galloway, Great Britain, before settling in for a night's sleep in the trees. Photograph copyright Walter Baxter (http://www.geograph.org.uk/photo/1069366), reproduced under a Creative Commons Attribution ShareAlike 2.0 License (https://creativecommons.org/licenses/by-sa/2.0/)

book *Introduction to Crowd Science*, written by Keith Still of Manchester Metropolitan University, who has developed a range of sophisticated computational techniques to model crowd behavior under stress.

The assumptions made by Vicsek and his colleagues in their 1995 paper "Novel Type of Phase Transition in a System of Self-Driven Particles" are remarkably simple. "The only rule of the model," they wrote,

> is [that] *at each time step a given particle driven with a constant absolute velocity assumes the average direction of motion of the particles in its neighborhood of radius r with some random perturbation added...*our model is a transport related, nonequilibrium version of the ferromagnetic type of models, with the important difference that it is inherently dynamic: the elementary event is the motion of a particle between two time steps. Thus the analogy can be formulated as follows: The rule corresponding to the ferromagnetic interaction tending to align the spins in the same direction in the case of equilibrium models, is replaced by the rule of aligning the *direction of motion* of particles in our model of cooperative motion. The level of random perturbations we apply are in analogy with the temperature. (Vicsek et al. 1995, their italics)

The biological basis of the model is a flying group of birds trying to avoid collisions. Particles adjust their motion in each time step by the average velocity of their neighbors in region *r*, with some added variability, which provides the temperature-like

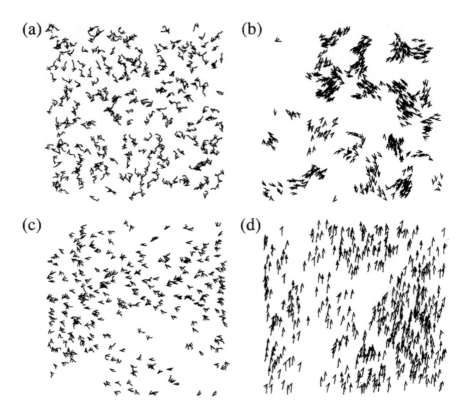

Fig. 8.2 Particle velocities (indicated by *arrows*) and recent particle trajectories (indicated by *short continuous curves*) in the Vicsek model. Flocking patterns are shown for different particle densities and levels of noise In panel **b**, for small density and noise, the particles form small groups moving together, but in random directions. In panel **d**, for higher density and small noise, the particles take on an ordered, collective motion. Reprinted with permission from T. Vicsek, A. Czirók, E. Ben-Jacob, I. Cohen, and O. Schochet, *Physical Review Letters*, Vol. 75, 1226–1229, 1995. Copyright 1995 by the American Physical Society

noise in the system. The assumptions in Vicsek's model are strongly reminiscent of Reynolds's model, though without the "flock centering" assumption.[3]

Using the average normalized velocity (or, in other cases, the average normalized momentum) of the particles as the order parameter, Vicsek and colleagues found that, as the noise parameter was decreased (or, in other simulations, as the particle density was increased), the simulated organisms became increasingly aligned, formed local clusters and then, as the parameters were changed further, formed a large flock moving in the same direction (Figs. 8.2 and 8.3). Moreover, the order

---

[3] Vicsek et al. appear not to have been aware of the boids model at the time they submitted their 1995 paper. Toner and Tu (1995) noted that D. Rokhsar alerted them to Reynolds's work. Following their citation, nearly all statistical physics studies of flocking cited Reynolds's boids model as an important precursor.

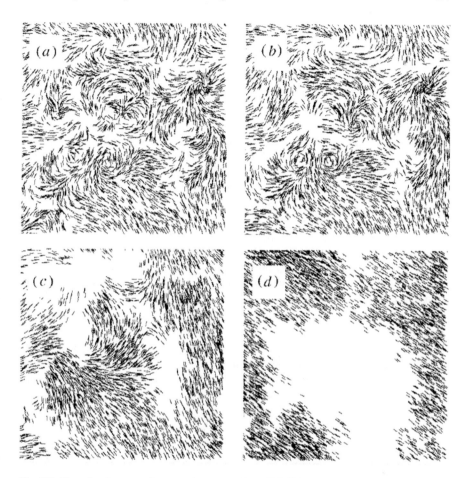

Fig. 8.3 Time development of a swarming system from Czirók et al., 1997 (Figure 9). The system initially exhibits unstable vortices (panels **a** and **b**) before the onset of long-range order (panel **d**). Reproduced with permission from A. Czirók, H. E. Stanley, and T. Vicsek, Spontaneously ordered motion of self-propelled particles, *J. Phys. A: Math. Gen.* 30: 1375–1385, 1997

parameter exhibited critical scaling in the transition region. Vicsek and colleagues identified the critical exponents quantifying the scaling behavior, and found them to be unique, placing their model in a novel universality class. They concluded their paper by noting that self-propelled particles, while rare in purely physical systems, abound in biology. Vicsek et al. called for the future use of this approach in models of molecular motors and traffic models, and the addition of a "hard core" to the particles, representing the "social distance" maintained – up to a point – by humans in crowds.

A virtual avalanche of studies followed, with both major theoretical developments and comparison to experimental data. For example, Simha and Ramaswamy (2002) introduced a model incorporating the "fluid" within which the particles swam and interacted. Only a few months after the publication of the Vicsek et al. model, Toner and Tu published a *Physical Review Letters* paper entitled "Long-Range Order in a Two-Dimensional Dynamical *XY* Model: How Birds Fly Together", in which they compared Vicsek's model to a model of spins in two dimensions; this model can essentially be considered as a two-dimensional analogue of the Ising model. The Ising model can, of course, exist in two spatial dimensions, but the variable of interest, the spin itself, can only align along a single axis (the direction of quantization). In the *XY* model, spins rotate in the plane of the lattice. Toner and Tu showed that, when all particles in Vicsek's model were assigned zero velocity, it reduced exactly to the *XY* model. Thus the long-range order in the Vicsek model resulted *from the fact that the system was dynamic*: population structure arose from the very fact that the particles were moving.

Recall that the initial Vicsek et al. model did not include Reynolds's (and indeed, Hamilton's) "flock centering". In other words, while the model created group *alignment*, it did not create *cohesion*. As Grégoire and Chaté pointed out in 2004, this meant that, as it approached the limit of zero density, the model would become unable to exhibit flocking at all. Grégoire et al. (2003) and Grégoire and Chaté (2004) modified the Vicsek et al. model to incorporate cohesion. They added a term representing interaction between the particles that was repulsive at short distances to represent a hard core, and then attractive (up to a limiting distance), similar to the form of a Lennard-Jones potential. Depending on parameters controlling the relative magnitudes of the interaction and alignment forces, the model exhibited both static clusters and also states in which clusters drifted and moved, sometimes separating into smaller sub-clusters connected by "filaments" within which the particles exhibited no alignment. The characteristics of the trajectory of the center of mass also varied as a function of the system parameters.

Grégoire et al. identified liquid- and solid-like states in their model. Particles diffused with respect to one another in the liquid-like state, but not in the solid-like state. Diffusion behaviors also depended on a particle's position within the flock. For example, particles within the leading edge of the flock exhibited a much faster increase in their mean square separation than did those in the core or the center region. This result suggested an interesting interpretation regarding possible mechanisms for the onset of the division of labor. Organisms such as *Dictyostelium,* which coalesce from individual amoeboid cells into an "organism" with somatic and reproductive cells, typically do so as they crawl along a surface under the influence of some chemotactic signal. It is possible that amoebae closer to the leading edge, which develop into somatic cells, may, as in the Grégoire model, experience different relative rates of diffusion than cells closer to the core or at the lagging edge, which develop into the spores. The difference in pressure and cell-cell interaction in these different regions might have stimulating effects on the expression of different genes, triggering spore-specific gene expression in the posterior region and stalk-specific gene expression in the anterior region. In this way, aggregation itself could lead to differential environments for the constituent members, resulting in differentiation into distinct cell types. We will return to this issue in Chap. 10, which focuses on *Dictyostelium.*

Anisotropy was also evident during the transition from one phase to another, rather than simply during the motion of an aggregate in a constant phase. During the transition from the solid to the liquid phase, Grégoire et al. noted that the particles toward the edges of the group tended to exhibit local order parameter values characteristic of the liquid phase, while particles closer to the center exhibited characteristics of the solid state (and showed increased spatial ordering).

Surprisingly, Grégoire et al.'s self-propelled particle models seemed to exhibit discontinuous (first-order) phase transitions, rather than the continuous transitions originally identified by Vicsek and colleagues. This implied that the models did not exhibit critical scaling at all, and could not be described by a set of critical exponents and placed within a universality class. The difference between one type of transition and another is actually quite important for the flocking agents themselves. As Aldana et al. explained,

> [t]he question as to whether the phase transition is first order or second order is not just of academic interest. Consider for instance a school of fish. For low values of the noise, the nature of the phase transition is irrelevant because the fish peacefully move in the same direction in a dynamical state far away from the phase transition. However, when the system is perturbed, for instance by a predator, the entire school of fish collectively responds to the attack: The school splits apart into smaller groups, which move in different directions and merge again later, confusing the predator. A possible hypothesis for this collective behavior to be possible is that, when attacked, the fish adjust their internal parameters to put the whole group close to the phase transition. If this phase transition is of second order, then the spatial correlation length diverges[,] making it possible for the entire school to respond collectively. However, this collective response would be very difficult to mount if the phase transition was first order, since in such a case the spatial correlation length typically remains finite. (Aldana et al. 2009)

Grégoire and colleagues performed simulations using the Vicsek model in its original form, without any interactions between the particles, with various values of the system size L, the linear dimension of the area in which the simulated particles moved. As the system size was increased, the transition from disorder to order as a function of added noise became increasingly sharp, i.e., increasingly discontinuous. This suggested that the originally identified continuous nature of the transition was an artifact of finite size effects rather than a fundamental characteristic of the system. In further support of the discontinuous nature of the transition (comparable to solid/liquid/gas transitions in physical materials), Grégoire et al. noted a bimodal distribution of order parameter values in their version of the model with cohesion: during the transition there were two non-overlapping metastable states, but no intermediate critical regime. A continuous phase transition would have exhibited an order parameter distribution with two overlapping peaks and a non-zero intermediate regime. Grégoire et al. showed that the order parameter distribution had a non-vanishing intermediate zone for smaller system sizes. But this region dropped to zero as the system size was increased, leaving two non-overlapping peaks, as expected for a discontinuous transition.

The disparity between the results of Vicsek and Grégoire generated considerable debate. After a flurry of publications, the issue appeared to be, at least partially, laid to rest, with the general consensus being that the Vicsek model likely *does* exhibit a continuous phase transition (Aldana et al. 2009). Baglietto et al. (2012) pointed out that, since the Vicsek model takes place in a finite region of simulated space, the correlation length of the system cannot exceed the system size L. As a result, comparison to infinitely large systems is not appropriate, and one must use statistical physics-based techniques such as finite-size scaling and data collapse to assess the behavior of the system, which Grégoire et al. (2003) and Grégoire and Chaté (2004) had not done. Using these techniques, Baglietto et al. obtained critical scaling exponents for the Vicsek model and showed that they satisfied a hyperscaling relationship consistent with a continuous transition.

A further exploration of the difference between continuous and discontinuous transitions in swarming models came from an analysis of the role of the added noise term by Aldana et al. (2007) and Pimentel et al. (2008). In the original Vicsek model, each agent updated its velocity based on an average of its neighbors' velocities, with some noise then added. This represented variability in the agent's decision-making process. Grégoire and Chaté (2004) had reproduced this model exactly, but had also developed another version of the model in which noise was added to the velocity of each neighboring agent *before* averaging. This would correspond to an agent experiencing variation in its ability to assess the behavior of a neighbor, rather than variation in its decision-making process. Pimentel et al. (2008) referred to the former type of noise as *intrinsic*, and to the latter type as *extrinsic*, since it could be caused by a blurry environment, rather than by an internal decision-making process.

Though they had not employed finite-size scaling or data collapse techniques, Grégoire and Chaté had identified discontinuous phase transitions in models with both noise types, as a function of the noise amplitude. At first glance, it was not at all surprising that the two types of noise would have similar effects. However, Aldana et al. (2007) argued that they should in fact be expected to produce quite different types of phase transitions. In response, Chaté et al. (2007) argued that the network models used by Aldana et al. to demonstrate this, however, had "no bearing" on Vicsek-like models, since they did not account for the movement of the self-propelled particles and "the local coupling between order and density, which is well known to be crucial for understanding collective properties of active particles". Chaté et al. therefore concluded that the Vicsek model, regardless of the type of noise, did indeed undergo a discontinuous transition.

The controversy continued as Pimentel et al. (2008) showed that, for *identical* parameter values, the model with *intrinsic* noise, originally suggested by Vicsek et al., exhibited a *continuous* transition, while the model with *extrinsic* noise, proposed by Grégoire and Chaté, exhibited a *discontinuous* transition. Pimentel et al. implemented two simplified versions of the Vicsek model, a "voter" model in which the alignment of opinions is implemented as the flipping of binary spin-like variables, and a "vectorial network" model which essentially provided a mean field version of the original Vicsek model. These two simplified models could be solved analytically; Pimentel et al. showed that both models, in the case of extrinsic noise, underwent a discontinuous transition, while with intrinsic noise they experienced a continuous one. They also considered cases in which the noise was partly intrinsic and partly extrinsic. Here, the voter model always showed a continuous transition, while the vectorial network model exhibited discontinuous transitions for some regions in parameter space and continuous transitions in other regions.

Subsequent models were developed to investigate heterogeneity among interacting agents. Vicsek and Grégoire had added noise terms to the velocity equation for each agent, representing error in the process by which it assessed the alignment of its neighbors. But what if the actual behavioral parameters such as velocity or body size (implemented via the size of the hard core in the Lennard-Jones-type interaction term) varied from organism to organism? Studies by Romey (1996), Couzin et al. (2002), and Hemelrijk and Kunz (2005) explored these and other possibilities, and found that inter-agent variability led to self-sorting, in which individuals with different characteristics took different positions within the flock.[4] For example, faster individuals were found farther from the center of the group, and individuals with larger error clustered at the back of the pack. "By changing the way they respond to others," wrote Couzin et al. "individuals can change the structure of the group to which they belong." Their study also extended agent-based flocking models to three dimensions. Here too, flocking resulted from local interactions. Couzin et al. also distinguished between various different types of collective behavior, drawing a distinction between a swarm (an aggregate with cohesion but a minimal polarization), a toroidal configuration (milling about an empty core) and "parallel groups" which exhibited significant velocity alignment. All three behavior types occur naturally in groups of organisms. Toroidal milling is observed in fish schooling in open water (Couzin et al. 2002). Vortex swarming behavior has also been observed experimentally and modeled computationally in the zooplankton *Daphnia* (Ordemann et al. 2003; Vollmer et al. 2006).

Strefler et al. (2008) identified additional complexities that arise upon passing from two dimensions to three (Fig. 8.4). In both cases, they found noise-induced transitions from translational motion to a toroidal formation. However, in three dimensions, the transition from translation to rotation occurred at a different noise value than the opposite transition, from rotation to translation, leading to a hysteresis curve and bistability. Couzin et al. (2002) also observed hysteresis and bistability, as a function of the control parameter defining the "zone of orientation", the region of neighbors with which an agent attempts to align (Fig. 8.5).

Transitions to ordered behavior can occur in one-dimensional agent-based models as well, and these accord well with empirical observation of the "marches" undertaken by locusts and crickets. Buhl et al. (2006) demonstrated the density-dependence of marching behavior in locusts in a laboratory environment. As the density of individuals was increased, random behavior underwent a transition to alignment in a coordinated marching group. The onset of ordered marching exhibited characteristics of a continuous phase transition. Buhl et al. showed that a one-dimensional version of the Vicsek model provided a good fit to the observed data. A later empirical study showed that cannibalism is a driving force in the emergence of

---

[4] Giardina (2008) provides an excellent review summarizing the results of these and many other contributions to the field.

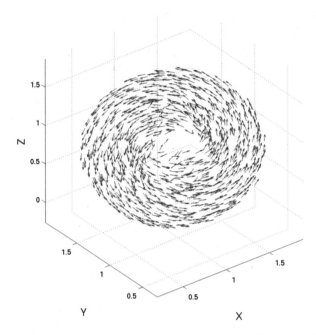

Fig. 8.4 Rotational motion emerges from a swarming model in three dimensions. Reprinted with permission from J. Strefler, U. Erdmann and L. Schimansky-Geier, *Phys. Rev. E* 78, 03127 (2008), Copyright 2008 by the American Physical Society

such marching formations. Aware of the potential threat from behind, each individual keeps moving forward (Bazazi et al. 2008). Similar behavior has been observed in some cricket species, where the propensity for cannibalism is exacerbated by deprivation of salt and protein (Simpson et al. 2006).

> This demonstrates that coordinated mass migration in animal groups may be driven by highly selfish and aggressive behavior. In the case of locusts, both tactile and visual stimuli from behind are necessary, and the major source of these in a group will come from other locusts, which are demonstrably cannibalistic. Migration is widely viewed as an adaptation to exploit spatiotemporally variable environments… However, cannibalism is perhaps one of the mechanisms that catalyzes the alignment of individuals and subsequently drives the directional mass movement of insects in migratory bands. This suggests a new perspective to our understanding of collective motion… At high population densities, individuals in migratory bands can benefit by reducing predation risk…but can find themselves serving as a source of potentially limiting food or water for cannibalistic conspecifics… Our results indicate that the defensive response to this risk, movement away from the attack, provides a general mechanism that results in marching bands being autocatalytic: Aggressive interactions stimulate motion in others, which increases encounter probabilities, and thus further aggressive acts. (Bazazi et al. 2008)

Other researchers explored "Euclidian" interaction models, where agents lived and interacted on a lattice grid. Taking such models to their continuum limit produced convection-diffusion models similar to hydrodynamic descriptions of fluid flow. Toner and Tu (1998; see also Toner et al. 2005) investigated the propagation of

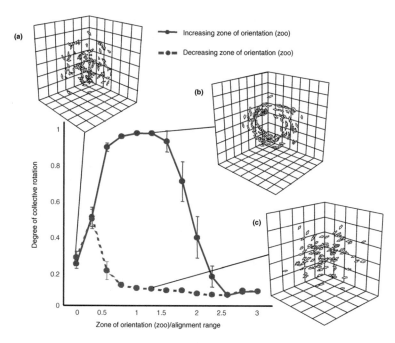

Fig. 8.5 Results from a computational model show changes in group shape from swarm (**a**) to torus (**b**) and to polarized alignment (**c**) as a result of changing local interactions. Note that the behaviors in (**b**) and (**c**) occur for the same zone of orientation, indicating bistability and hysteresis. Reprinted from *Trends Cog. Sci.* 13(1), I.D. Couzin, Collective cognition in animal groups, pp. 36–43, 2009, ©2008, and from *J. Theor. Biol.*, 218, I.D. Couzin et al., Collective memory and spatial sorting in animal groups, pp. 1–11, © 2002, with permission from Elsevier

density waves in a model of bird flocking described by equations reminiscent of the Navier-Stokes equations. A drawback of these Euclidian models, however, is that they are coarse-grained and, when taken to the continuum limit, are population-rather than individual-based. They are thus better at characterizing the behavior and structure of a flock once it has formed than at modeling the flock formation process itself.

Collective behavior of animal groups lends itself to the efficient transmission of information in the absence of any overall group leader. In a sense, it allows for a form of "crowd-sourcing". Recall that Couzin et al. (2002) had identified conditions under which flocks formed parallel groups with aligned velocities. This alignment, they argued, facilitated information transfer within the group.

> The tendency of individuals to align with one another within the parallel group types is important not only in minimizing collision between individuals and facilitating group movement, but also in allowing the group to transfer information. For example, if an individual were to turn sharply, as a response to avoiding a predator or an obstacle, the alignment tendency allows this turn to influence the orientation of neighbours (which need not directly detect the stimulus), facilitating a transfer of information (turning) over a range greater than the individual interaction radius. (Couzin et al. 2002)

They suggested the swarm configuration, with high cohesiveness but minimal alignment, would be less effective in producing information transmission since "the relatively high variance of individual orientation means that the change in direction of
group members, as a result of detecting a predator for example, is propagated less
efficiently". Couzin and colleagues also speculated that local directional polarization in a group undergoing toroidal motion might facilitate information transmission, while permitting the group to stay relatively stationary, which might allow the
group to rest or avoid certain predators.

In a computational study published in *Nature* in 2005, Couzin et al. tested their
predictions. In an agent-based model, they showed that information transfer within
groups can occur without any direct signaling mechanism beyond the propensity of
agents to align locally with their neighbors. This can occur even when only a few
agents possess "knowledge" of a desired direction of travel, and even when no
organism is explicitly aware of which of its flock-mates possess the information.
For larger groups, a smaller proportion of informed individuals was needed in order
to steer the group in a particular direction. In other words, a biased directionality
among a very few agents was sufficient to drive the entire group in a particular
direction, thus providing "effective leadership" in a group with no designated leader,
and where the agents were only aware of the motion of their near neighbors.

In a further and quite striking result, Couzin et al. (2005) showed that groups of
agents in which a few were biased in one direction, and a few in another, could reach
a consensus, pulling the entire flock in one direction or another. The consensus-
making decision, however, underwent a bifurcation as a function of the difference
of opinion between the two biased groups. If the difference of opinion was small,
the entire flock settled on a direction that was the average of the preferred direction
of the two factions. As the difference of opinion passed a threshold value, however,
the group experienced a symmetry-breaking behavior, and followed one or the other
of the two factions, depending on the trial. Shortly after the publication of this
paper, Biro et al. published a study (2006; see also Couzin 2009) showing that homing pigeons actually do exhibit precisely this decision-making behavior (Fig. 8.6).
Using pairs of pigeons trained to take different routes, Biro et al. found that when
the routes were similar, the pair of birds traveled a compromise course that essentially averaged the two routes. When the distance between the two routes passed a
threshold value, however, the pair of birds broke for one route or the other. A similar
consensus-making process was observed by Dyer et al. (2008) in groups of human
subjects (university students at the University of Leeds and the University of Hull,
as well as local school children).

Choice between different paths has also been explored in empirical and computational studies of a rather different kind: ant trails. Here, collective motion arises,
as in the flocks and swarms we have considered previously. However, the interaction
between agents is not entirely without signaling. Ants lay down a pheromone trail
as they forage, leaving behind a signal for their colleagues to follow. This process
allows selection of an optimal (shortest) path, driven by the fact that pheromone
accumulation occurs faster on shorter paths, and conversely has less time to evapo-
rate before the next ant comes along and deposits further pheromone (Deneubourg

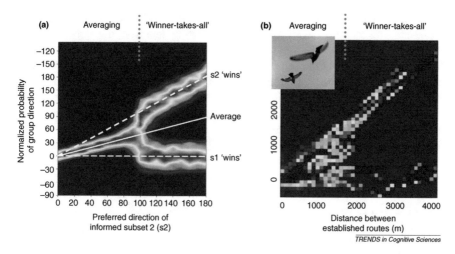

Fig. 8.6 (**a**) Simulations show that, when there are two substes of individuals with a difference of opinion regarding direction of motion, the group chooses a consensus direction below a critical difference in opinion. Above this critical difference, the group follows a winner-takes-all strategy. (**b**) Experiments by Biro et al. in homing pigeons show similar behavior, with consensus routes chosen in the case of a small difference in opinion and a winner-takes-all strategy obtaining for larger differences of opinion. Reprinted from *Trends Cog. Sci.* 13(1), I.D. Couzin, Collective cognition in animal groups, pp. 36-43, 2009, © 2008, with permission from Elsevier. Panel (**a**) reprinted by permission from Macmillan Publishers Ltd, *Nature* 433, Couzin et al., Effective leadership and decision-making in animal groups on the move, pp. 513–516, © 2005. Panel (**b**) reprinted from *Curr. Biol.* 16(21), Biro et al., From compromise to leadership in pigeon homing, pp. 2123–2128, © 2006, with permission from Elsevier

et al. 1990). Beekman et al. (2001) showed that colonies with a greater number of foraging ants engaged in organized foraging forays along pheromone trails, while colonies with fewer foragers performed disorganized, random individual foraging trips. As colony size (and therefore number of foragers) was increased, the change in behavior exhibited characteristics of a phase transition from a disordered to an ordered state. In experiments with ant colonies traveling in artificially constructed branching paths, Dussutour et al. (2004) found that, at low population density, the ants selected one of the two paths, while a higher density led to symmetrical traffic flow through the two branches. Ant-based models for path optimization have been used to develop new computational techniques (Dorigo et al. 1996). This approach has come to be described as "swarm intelligence", now a full-fledged field in its own right.

Comparison of computational models to real-world data is clearly of paramount importance, and empirical studies of ant foraging, locust and cricket marches, and consensus choice coincide well with the predictions of flocking models. But how realistic are these models? For example, how well do these models reflect the actual fine detail of a flock of starlings? To answer this question, one must confront an extremely challenging problem in data collection and analysis: how to deal with data from large numbers of very similar organisms moving in three dimensions.

Imaging the data from a single angle is completely insufficient to provide a full representation of the behavior. Even after collecting data from a variety of angles, the investigator must deal with another very difficult task: the "matching problem". Specifically, how does one correlate bird A in an image taken from angle θ_1 with the same bird A in an image taken from angle θ_2? One approach to this problem, inspired by data collected from starling flight, was presented in a software analysis package (called STARFLAG) by Cavagna et al. (2008a, b, c), which combined optimization, statistical mechanics and computer vision techniques. This approach was used in analysis of the flights of flocks of ~3000 starlings (Ballerini et al. 2008a, b), which "were found to be relatively thin and to slide parallel to the ground, a feature that is completely missed by any two-dimensional reduction of the group, as one performed by simple photographs or far away observers" (Giardina 2008). Birds were also observed to have a "repulsion zone" similar to that assumed by Couzin and others in their computational models; this zone was observed to be approximately the radius of a wingspan (Giardina 2008).

Another surprising result from these studies was the observation that "the angular distribution of neighbors around a focal individual revealed a strong anisotropy, where nearest neighbors are more likely to be found on the sides rather than in the direction of motion" (Giardina 2008). This was consistent with an earlier study of schooling in herring, cod, and saithe (Partridge et al. 1980). Equally striking was the finding that flocks were denser at the edge than at the core. Ballerini and colleagues also observed that, as flocks wheeled and rotated, the overall shape of the flock was maintained, and the birds' velocity vectors rotated around the main axis of the flock. Most fundamentally for future studies, however, was the finding that it is the *number of neighbors* rather than the *neighbors within a given distance* that each bird tracks as it makes its velocity adjustments. In other words, birds pay attention to their immediate *topological*, rather than their *metric,* environment (Ballerini et al. 2008a; Giardina 2008). Birds were found to follow the motion of their six or seven nearest neighbors. The researchers suggested that awareness of topological distance might be far more useful to birds than metric distance, and indeed "indispensible to maintain cohesion … in spite of large density fluctuations, which are frequent in flocking due to predator attacks" (Giardina 2008). For example, if a startled flock scatters as a predator approaches, and birds increase their relative distances beyond their metric limit, the flock's cohesion would be immediately lost, and the birds would lose the protection of flocking precisely when it was most needed. If the interactions were topological, however, the flock could quickly regain cohesion. In a two-dimensional computational simulation, Ballerini et al. (2008a) set up pairs of flocks that were identical in all respects save that one had metric interactions and the other had topological interactions. The flocks were then subjected to a simulated attack from a predator represented by a repulsive force at the center of the flock. Following the attack, the topologically connected flocks were much more likely to retain cohesion than the metric groups, which were prone to splitting off into subgroups and losing stragglers.

As a result of these studies by Ballerini et al., investigators have begun to focus more on models based on the topological distance between agents in a flock. For example, Bialek et al. (2012) developed a statistical mechanics model for flocking based on maximum entropy; consistent with experiment, this model predicts that the number of neighbors with which a bird attempts to align itself is independent of flock density, consistent with the conclusion that the interactions are topological rather than metric.

Like the original Vicsek model, the predator attack simulations performed by Ballerini et al. (2008a) did not include an explicit attraction term between the agents. Camperi et al. (2012) undertook a series of detailed simulations to expand the model to three dimensions, and to investigate the role of cohesion. Starling flocks exhibit weak cohesion between birds, with a radial pair correlation function more similar to interactions in a gas than a liquid or solid (Camperi et al. 2012). Camperi and colleagues designed a modified version of the model introduced by Grégoire and Chaté, scaled up to three dimensions, and with three possible types of interactions: *metric* (aligning and interacting with neighbors within a given radius), *topological* (aligning and interacting with a set number of nearest neighbors), and *topological interactions with angular resolution*. In this last case, neighbors were chosen topologically, but if more than one neighbor was found within a certain angular resolution, then only one within that resolution was used. This allows neighbors to be chosen "in a *balanced* way": were the angular resolution to be on the order of π, a bird would have, on average, only two neighbors, but they would be on opposite sides.

Using similar parameters for each type of interaction, Camperi et al. ran simulations in the presence of noise for a large number of generations. They found that the metric model broke into a number of sub-flocks (ranging from 6 to 16, depending on the run). In contrast, the topological models exhibited much greater cohesion. The simple topological model never broke into more than three sub-flocks, and then only in a very few of the 400 simulations performed for each model. The topological model with angular resolution, however, never broke into sub-flocks at all. Next, using parameter values for each model that provided optimal cohesiveness and the closest fit to actual data, Camperi et al. applied an external perturbation to the simulated flocks, in the form of a large obstacle along the axis of the flock's center of mass. Again, the metric model performed worst, with the flock ultimately breaking into a large number of subgroups. Of the two topological models, the balanced one performed best.

The initial simulations Camperi et al. performed had a large number of neighbors in the topological cases (22), and an interaction radius chosen in the metric case to give approximately the same number of neighbors. They noted, however, that as the number of neighbors decreased, the stability of even the balanced topological model decreased, and the flocks began to form filaments, with incipient sub-flocks connected by small bands of individuals. In order to determine the optimal number of neighbors such that the balanced topological model would cohere without breaking into filaments, Camperi et al. measured the fraction of particles belonging to filaments as a function of the number of neighbors. Depending on the parameters of the model, this occurred for between five and ten neighbors, consistent with the value of seven obtained from the actual starling data.

While sticking with a set number of "alignment buddies" may promote flock cohesion, the particular six or seven individual neighbors with which a bird aligns may shift over time (Cavagna et al. 2013). Individual birds have been shown to exchange neighbors as a result of stochastic fluctuations; birds at the edge of the flock exchanged positions with their neighbors less frequently than those near the center. Cavagna et al. suggested that this

> could be a consequence of the fact that birds compete with each other for a place in the interior of the flock. This struggle for the occupation of the same internal space would imply that when attempting to move inward, a border bird experiences a repulsion produced by its internal neighbours, pushing it outside again. The attempt to move inward is then reiterated, until by some fluctuation the bird successfully leaves the border. (Cavagna et al. 2013)

This process is consistent with the observation that starling flocks are denser at the edge than at the core.

In the context of the phase transition debate between Vicsek and Grégoire and Chaté discussed above, it is interesting to note that another study by Cavagna et al. (2010) identified scale-free correlations in starling flocks. Barberis and Albano (2014) implemented a Vicsek-like model with extrinsic noise but with topological rather than metric interactions between neighbors. In other words, the agents in their model attempted to match velocities with a set number of nearest neighbors, rather than with nearest neighbors within a set distance. Using the case of seven nearest neighbors, they analyzed the phase transition behavior as a function of extrinsic noise, essentially following the analysis of Baglietto and colleagues. They identified a continuous phase transition from a fluctuating to an ordered state as the noise was decreased. Strikingly, they identified critical exponents quite close to those obtained in the case of metric interactions: the two models belong to the same universality class. Barberis and Albano also found that the spatial distance among the seven nearest neighbors fluctuated significantly near the critical value of the noise, and suggested that natural selection for birds capable of responding to such fluctuations might have played a role in evolution of optimal flock sizes.

*** *** ***

Collective motion is a field where interactions become exponentially more complex at smaller scales. Interactions of ensembles of single cells occur at the molecular level, at a size scale where the medium itself can plays a significant role in mediating the interactions. The swarming of bacteria, then, is far more challenging to model and to interpret than the collective movement of a bird flock; collective movement at this level must also take hydrodynamic effects into account, rendering models of the Vicsek type insufficient to characterize the dynamics. However, ensembles of single cells do assemble to perform collective motion, and this can be characterized by some of the same general techniques as those described above. Spurred by the results on bird flocking and fish schooling, various groups of biophysicists, including Ray Goldstein, John Kessler, Lev Tsimring, Jeff Hasty and others, as well as Hugues Chaté and Tamás Vicsek, have performed both experiments and computational studies on the collective motion of single cells.

In 2006, Szabó et al. published their investigation of collective migration of keratocytes derived from goldfish scales. Cells were grown at various densities and their motion was imaged over a 24-h period using time-lapse videomicroscopy. Szabó et al. observed a "sharp transition from random motility to an ordered collective migration of dense islands of cells as the density was increased", as shown in Fig. 8.7. Similar to the original Vicsek et al. (1995) study, they characterized the system using the average normalized velocity as an order parameter, measuring the velocity of 20–30 cells per experiment from their frame-to-frame displacement. Szabó et al. found "that a kinetic phase transition takes place from a disordered into an ordered state as cell density exceeds a relatively well-defined critical value".

Collective groups of keratocytes cannot align their motion with that of their neighbors, however, in the conscious manner that birds and fish may be assumed to. Instead, migrating groups of cells interact via direct physical contact. In order to model this type of behavior, Szabó et al. had to develop a model that did not include any explicit velocity alignment term. Instead, they used pairwise intercellular forces that, similar to the Lennard-Jones-type potential used by Grégoire and Chaté, was repulsive for small distances, attractive for intermediate distances, and zero beyond a threshold radius. The model also included a noise term representing variability in each cell's ability to adjust its direction of motion in response to interactions with neighboring cells, and it reproduced the experimental results with good agreement. The computational model showed phase transition behavior as the noise magnitude was varied, with the density held constant (though a companion experimental observation was not achievable, as there was no mechanism by which to vary the noise experimentally). Analysis of critical exponents in the model gave values nearly identical to those obtained from the original Vicsek et al. (1995) model. Lastly, Szabó and colleagues showed experimentally and computationally that adding closed boundary conditions induced vortex-like circular motion. They suggested that their results might have applications for the collective migration of cells during wound healing or embryogenesis.

The collective behavior of tissue cells was studied extensively in the subsequent decade. Battersby (2015) provides a brief overview of recent work beginning with the unforgettable phrase "flesh really does crawl". And perhaps it should, indeed: Deisboeck and Couzin (2009) suggested that cancer cells might exhibit emergent collective migratory behavior. Chang et al. (2013) provided a detailed computational model for the interaction of tumor cells with the surrounding normal tissue, which also predicted collective migratory activity. Another computational model showed that a cell-cell interaction process known as contact inhibition of locomotion[5] was sufficient to induce ordered cell migration, suggesting that this might be

---

[5] Collective inhibition of locomotion is a process whereby cells, at least *in vitro*, change their direction of motion upon contact. It has some commonalities with the nematic collisions observed between microtubules (see below), and is impaired in malignant cells. See Mayor and Carmona-Fontaine (2010) for an overview and evidence for the phenomenon *in vivo*.

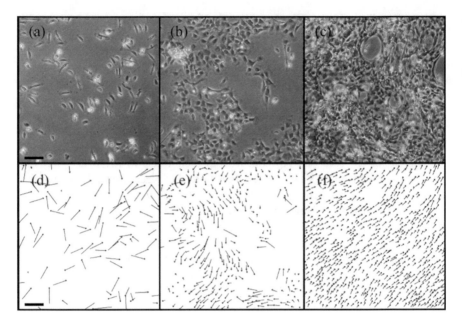

Fig. 8.7 Results from Szabó et al. show clustering in migrating keratocytes (*top row*) and the corresponding cell velocities (*bottom row*). Panels **a**, **b** and **c** show phase contrast images for cells at three different densities; scale bar shows 200 μm. Panels **d**, **e** and **f** show cell velocity vectors; scale bar corresponds to 50 μm/min. Reproduced with permission from B. Szabó, G. J. Szöllösi, B. Gönci, Zs. Jurányi, D. Selmeczi, and T. Vicsek, *Physical Review E* 74, 061908, 2006. Copyright 2006 by the American Physical Society

"a possible mechanism behind collective cell migration that is observed, for example, in neural crest cells during development, and in metastasizing cancer cells" (Coburn et al. 2013).

Molecular mechanisms of cell migration are currently being investigated by many laboratories; to cite one example among literally thousands of studies, a protein called merlin (the product of a tumor suppressor gene, neurofibromin 2) acts as a mechanochemical transducer to coordinate the movement of sheets of epithelial cells (Das et al. 2015). Future computational studies will surely include interaction terms inspired by such experimental studies.

Collective behavior has also been observed in single-celled organisms. Rod-like bacteria at high concentrations exhibit behavior reminiscent of liquid crystals (Dombrowski et al. 2004). In an experimental study of *E. coli*, for example, Volfson et al. (2008) observed "a dynamical transition from an isotropic disordered phase to a nematic[6] phase characterized by orientational alignment of rod-like cells". Interaction between bacteria is complicated by the fact that interactions change as a

---

[6] In a nematic phase, typically observed in systems such as liquid crystals, the constituents align parallel to each other, but do not form ordered layers. In bacteria, this has been referred to as a "bio-nematic" phase.

function of colony density; at lower concentrations, chemotactic interactions dominate, while at higher concentrations direct biomechanical interactions take over (Volfson et al. 2008). Working at high concentrations in order to focus on these latter interactions, Volfson and colleagues showed an increased degree of ordering in a growing bacterial colony as the bacteria reproduced and the density increased. They developed continuum and discrete element models of the system, including bacterial growth rate, velocity and pressure as well as an order parameter characterizing bacterial alignment, which agreed well with the experimental results. The authors found that the

> phenomenon is fundamentally different from the nematic transition in liquid crystals and polymers ... and vibrated granular rods..., where ordering is primarily driven by the combination of steric exclusion and fluctuations. The mechanism of the ordering transition reported here is related to cell growth and therefore should be ubiquitous in "living granular matter." (Volfson et al. 2008)

From this perspective, the ordering of birds and fish is perhaps closer to the liquid crystal case than is the bacterial case. Bacteria, too, can interact with one another via quorum sensing and can undergo a transition into a biofilm, a topic we will explore in the next chapter.

Wioland et al. (2013) studied the motion of *Bacillus subtilis* in an oil emulsion. The diameter of the droplet in which the bacteria were suspended served as a control parameter for the onset of ordered vortex-like motion. While they did not explicitly characterize phase transition behavior in the system, the investigators measured a sharp increase in a "vortex order parameter" as a function of the drop diameter; the increase has qualitative features similar to a continuous phase transition. Another study, by Chen et al. (2012) was also highly suggestive of critical scaling in bacterial colonies. Again working with *B. subtilis,* Chen and colleagues found long-range scale-invariant correlations in the fluctuations of bacterial velocity and orientation. They commented that these

> correlations may give some evolutionary advantages... With such correlations, a change in the state of an individual influences that of all others in the system; information of an external stimulus, such as a predator or food, can propagate quickly through the whole system and the system can respond coherently to maintain its integrity. (Chen et al. 2012)

A previous study from the same group, led by Harry Swinney at the University of Texas, Austin, also found that the sizes of bacterial clusters in a similar experimental setup followed truncated power law scaling, indicating the sort of scale-free distribution expected in the critical regime of a continuous phase transition, though truncated due to the finite system size (Zhang et al. 2010).

Extending flocking models to include the *shape* of the agents, Ginelli et al. (2010) demonstrated the onset of nematic order in simulations of self-propelled rod-like particles as a function of noise amplitude. The authors observed *nematic collisions* between their simulated particles, in which "particles incoming at a small angle align 'polarly' [in other words, they take the same direction after the collision], but those colliding almost head-on slide past each other", continuing along in opposite directions.

In a groundbreaking theoretical study, Baskaran and Marchetti (2009) developed a model for a suspension of active organisms that included hydrodynamic effects. The model included a variety of organism types such as "shakers" (active particles that are not self-propelled), "movers" (active and self-propelled), "pullers" (a "mover" subtype corresponding to contractile swimmers such as the algae *Chlamydomonas*, propelled by flagella) and "pushers" (a "mover" subtype propelled from the rear, such as *E. coli*). Baskaran and Marchetti generated a phase diagram for swimmer behavior, predicting the conditions under which a suspension of a particular type would become unstable.

The theoretical investigation of nematic collisions by Ginelli et al. is reminiscent of a set of studies performed on the interactions, not of bacteria, but of *single molecules*. In 2012, Sumino and colleagues observed the formation of vortices in an experimental system of microtubules moving on a surface of dynein molecules. The microtubules underwent nematic collisions, leaving the interaction in a common direction if they approached each other at an angle smaller than 90°, and in opposite directions if their initial angle of approach was greater than 90°. The microtubules formed vortices rotating either clockwise or counterclockwise (though individual molecules appeared to have a slight preference for counterclockwise rotation), and individual molecules were observed to occasionally "jump from one vortex to another and change their rotational direction" (Vicsek 2012). Sumino et al. developed a computational model for the microtubule interactions, and investigated the effect of the confined microtubule environment on vortex formation. As Vicsek wrote it in a "News and Views" piece in *Nature*, "it seems that owing to the persistence of curvature in the individual microtubules' motion, curved 'walls' of microtubules are spontaneously formed and enforce the formation of vortices". Sumino et al. speculated that the onset of ordered behavior in this system, which combined "smooth, reptation-like motion and sharp nematic alignment", might belong to a new universality class.

In another study of collective motion in cytoskeletal molecules, Schaller et al. (2010) recorded the movement of fluorescently labeled actin molecules on a substrate seeded with heavy meromyosin molecules. Here, as in the Sumino et al. study, collective behavior emerged at high densities, but the molecular interactions were polar (always aligning in the same direction after collision) rather than nematic.[7] Butt et al. (2010) measured actin alignment at densities comparable to those observed in the cell, and found that the ordered regions are on the order of the size of a cell (10–100 μm).[8] Similar results were obtained by Hussain et al. (2013), using shorter actin filaments moving along a surface seeded with heavy meromyosin. These studies were consistent with theoretical predictions (Kraikivski et al. 2006) of the onset of nematic ordering of cytoskeletal filaments interacting via molecular motors, except at high actin filament concentration and high myosin density. Here,

---

[7] This difference in interaction type does not prevent the collective behavior of the actin filaments from being described as "nematic", in the sense that they align but do not form ordered layers.

[8] The ordered regions observed by Schaller et al. (2010) and Sumino et al. (2012) were in the same size range (up to 400 μm).

the order parameter measured experimentally by Hussain et al. dipped sharply, a behavior quite atypical for the theoretically predicted critical phase transition.

The observation of the size scale of the ordered regions in the studies by Butt et al. (2010), Schaller et al. (2010), and Sumino et al. (2012) raises the important question of the biological relevance of ordering at the molecular scale. The ordered behavior of a bird flock or fish school may be interpretable in terms of predator avoidance or deterrence, and bacterial swarming may be related to antibiotic resistance (Lai et al. 2009; Butler et al. 2010; Benisty et al. 2015). But the biological usefulness of pattern formation in cytoskeletal molecules *in vitro* remains unclear. It is possible, of course, that this phenomenon does not occur naturally. Sumino and colleagues remark that "there is a striking analogy between our analysis of microtubule collisions and that performed in plant cell cortex [in the transport of fluid], although there is only treadmilling and no actual displacement of microtubules in the cortex." Brugués and Needleman (2014) showed that active liquid crystal models, informed by confocal microscopy of fluorescent-labeled tubulin in *Xenopus* oocytes, could predict the cytoskeletal arrangement of spindles during cell division. Likewise, Loose and Mitchison (2014) observed self-organization (likely via polymerization) of two cytoskeletal proteins, FtsZ and FtsA, which participate in the formation of the Z-ring, a structure involved in initiating bacterial cytokinesis. Using fluorescence microscopy, they observed the formation of molecular bundles and vortex structures. More studies revealing the biological relevance of molecular ordering within the cell are surely forthcoming.

*** *** ***

Collective behavior emerges from the interaction of ensembles of individual animals, cells, and molecules. In all these cases, long-range order is driven by local interactions, and the transition to an ordered state typically occurs as a continuous phase transition, with critical scaling in the transition regime. The examples given here provide a just brief taste of recent results in this very active field. We have seen, if nothing else, that groups of individual organisms can form collective structures characterized by order parameters, becoming "something in their own right" in a sense far more quantitative than the crowd descriptions given by Plato, Carlyle and others. But there has been little evidence of a "tension", or a balance of forces, here; we have simply seen individual local forces act from the bottom up to create a larger structure. To see how such larger structures behave as "things in themselves", we will need to involve a second layer of forces, pushing, as it were, from the outside. We will need to visit biological contexts where the process of aligning with one's neighbors does not only promote one's safety, but also leaves one trapped in a genetic bottleneck. We will need to take the perspective of the biological unit that most often confronts this tradeoff: the cell.

References

Aldana M, Dossetti V, Huepe C, Kenkre VM, Larralde H (2007) Phase transitions in systems of self-propelled agents and related network models. Phys Rev Lett 98:095702

Aldana M, Larralde H, Vazquez B (2009) On the emergence of collective order in swarming systems: a recent debate. Int J Mod Phys B 23:3661–3685

Baglietto G, Albano EV, Candia J (2012) Criticality and the onset of ordering in the standard Vicsek model. Interface Focus 2:708–714

Ballerini M, Cabibbo N, Candelier R, Cavagna A, Cisbani E, Giardina I, Lecomte V, Orlandi A, Parisi G, Procaccini A, Viale M, Zdravkovic V (2008a) Interaction ruling collective animal behavior depends on topological rather than metric distance: evidence from a field study. Proc Natl Acad Sci U S A 105:1232–1237

Ballerini M, Cabibbo N, Candelier R, Cavagna A, Cisbani E, Giardina I, Orlandi A, Parisi G, Procaccini A, Viale M, Zdravkovic V (2008b) Empirical investigation of starling flocks: a benchmark study in collective animal behavior. Anim Behav 76:201–215

Barberis L, Albano EV (2014) Evidence of a robust universality class in the critical behavior of self-propelled agents: metric versus topological interactions. Phys Rev E 89:012139

Barnsley M (1993) Fractals Everywhere. Morgan Kaufmann Publishers, San Francisco

Baskaran A, Marchetti MC (2009) Statistical mechanics and hydrodynamics of bacterial suspensions. Proc Natl Acad Sci U S A 106(37):15567–15572

Battersby S (2015) The cells that flock together. Proc Natl Acad Sci U S A 112(26):7883–7885

Bazazi S, Buhl J, Hale JJ, Anste ML, Sword GA, Simpson SJ, Couzin ID (2008) Collective motion and cannibalism in locust migratory bands. Curr Biol 18:735–739

Beekman M, Sumpter DJT, Ratnieks FLW (2001) Phase transition between disordered and ordered foraging in Pharaoh's ants. Proc Natl Acad Sci U S A 98:9703–9706

Benisty S, Ben-Jacob E, Ariel G, Be'er A (2015) Antibiotic-induced anomalous statistics of collective bacterial swarming. Phys Rev Lett 114(1):018105

Bialek W, Cavagna A, Giardina I, Mora T, Silvestri E, Viale M, Walczak A (2012) Statistical mechanics for natural flocks of birds. Proc Natl Acad Sci U S A 109(13):4786–4791

Biro D, Sumpter DJ, Meade J, Guilford T (2006) From compromise to leadership in pigeon homing. Curr Biol 16(21):2123–2128

Brugués J, Needleman D (2014) Physical basis of spindle self-organization. Proc Natl Acad Sci U S A 111(52):18496–18500

Brush SG (1967) History of the Lenz-Ising model. Rev Mod Phys 39(4):883–895

Buhl J, Sumpter DJT, Couzin ID, Hale JJ, Despland E, Miller ER, Simpson SJ (2006) From disorder to order in marching locusts. Science 312:1402–1406

Butler MT, Wang Q, Harshey RM (2010) Cell density and mobility protect swarming bacteria against antibiotics. Proc Natl Acad Sci U S A 107(8):3776–3781

Butt T, Mufti T, Humayun A, Rosenthal PB, Khan S, Khan S, Molloy JE (2010) Myosin motors drive long range alignment of actin filaments. J Biol Chem 285(7):4964–4974

Camperi M, Cavagna A, Giardina I, Parisi G, Silvestri E (2012) Spatially balanced topological interaction grants optimal cohesion in flocking models. Interface Focus 2:715–725

Cavagna A, Cimarelli A, Giardina I, Orlandi A, Parisi G, Procaccini A, Santagati R, Stefanini F (2008a) New statistical tools for analyzing the structure of animal groups. Math Biosci 214:32–37

Cavagna A, Giardina I, Orlandi A, Parisi G, Procaccini A, Viale M, Zdravkovic V (2008b) The STARFLAG handbook on collective animal behavior. 1: empirical methods. Anim Behav 76:217–236

Cavagna A, Giardina I, Orlandi A, Parisi G, Procaccini A (2008c) The STARFLAG handbook on collective animal behavior. 2: three-dimensional analysis. Anim Behav 76:237–248

Cavagna A, Cimarelli A, Giardina I, Parisi G, Santagati R, Stefanini F, Viale M (2010) Scale-free correlations in starling flocks. Proc Natl Acad Sci U S A 107:11865–11870

Cavagna A, Duarte Queriós SM, Giardina I, Stefanini F, Viale M (2013) Diffusion of individual birds in starling flocks. Proc R Soc B 280(1756):20122484

Chang WK, Carmona-Fontaine C, Xavier JB (2013) Tumour-stromal interactions generate emergent persistence in collective cancer cell migration. Interface Focus 3(4):20130017

Chaté H, Ginelli F, Grégoire G (2007) Comment on "Phase transitions in systems of self-propelled agents and related network models". Phys Rev Lett 99:229601

Chen X, Dong X, Be'er A, Swinney HL, Zhang HP (2012) Scale-invariant correlations in dynamic bacterial clusters. Phys Rev Lett 108(14):148101

Coburn L, Cerone L, Torney C, Couzin ID, Neufeld Z (2013) Tactile interactions lead to coherent motion and enhanced chemotaxis of migrating cells. Phys Biol 10(4):046002

Couzin ID (2009) Collective cognition in animal groups. Trends Cogn Sci 13(1):36–43

Couzin ID, Krause J, James R, Ruxton GD, Franks NR (2002) Collective memory and spatial sorting in animal groups. J Theor Biol 218:1–11

Couzin ID, Krause J, Franks NR, Levin SA (2005) Effective leadership and decision-making in animal groups on the move. Nature 433:513–516

Czirók A, Stanley HE, Vicsek T (1997) Spontaneously ordered motion of self-propelled particles. J Phys A Math Gen 30:1375–1385

Das T, Safferling K, Rausch S, Grabe N, Boehm H, Spatz JP (2015) A molecular mechanotransduction pathway regulates collective migration of epithelial cells. Nat Cell Biol 17(3):276–287

Deisboeck TS, Couzin ID (2009) Collective behavior in cancer cell populations. BioEssays 31(2):190–197

Deneubourg J-L, Aron S, Goss S, Pasteels J-M (1990) The self-organizing exploratory pattern of the Argentine ant. J Insect Behav 3:159–168

Dombrowski C, Cisneros L, Chatkaew S, Goldstein RE, Kessler JO (2004) Self-concentration and large-scale coherence in bacterial dynamics. Phys Rev Lett 93(9):098103

Dorigo M, Maniezzo V, Colorni A (1996) The ant system: optimization by a colony of cooperating agents. IEEE Trans Syst Man [sic!] Cybern B 26(1):1–13

Dussutour A, Fourcassie V, Helbing D, Deneubourg J-L (2004) Optimal traffic organization in ants under crowded conditions. Nature 428:70–73

Dyer JRG, Ioannou CC, Morrell LJ, Croft DP, Couzin ID, Waters DA, Krause J (2008) Consensus decision making in human crowds. Anim Behav 75:461–470

Giardina I (2008) Collective behavior in animal groups: theoretical models and empirical studies. HFSP J 2(4):205–219

Ginelli F, Peruani F, Bär M, Chaté H (2010) Large-scale collective properties of self-propelled rods. Phys Rev Lett 104:184502

Grégoire G, Chaté H (2004) Onset of collective and cohesive motion. Phys Rev Lett 92(2):025702

Grégoire G, Chaté H, Tu Y (2003) Moving and staying together without a leader. Phys D 181:157–170

Hamilton WD (1971) Geometry for the selfish herd. J Theor Biol 31:295–311

Helbing D, Farkas I, Vicsek T (2000) Simulating dynamical features of escape panic. Nature 407:487–490

Hemelrijk CK, Kunz H (2005) Density distribution and size sorting in fish schools: an individual-based model. Behav Ecol Sociobiol 16:178–187

Hussain S, Molloy JE, Khan SM (2013) Spatiotemporal dynamics of actomyosin networks. Biophys J 105(6):1456–1465

Kraikivski P, Lipowsky R, Kierfeld J (2006) Enhanced ordering of interacting filaments by molecular motors. Phys Rev Lett 96:258103

Lai S, Tremblay J, Déziel E (2009) Swarming motility: a multicellular behaviour conferring antimicrobial resistance. Environ Microbiol 11(1):126–136

Loose M, Mitchison TJ (2014) The bacterial cell division proteins FtsA and FtsZ self-organize into dynamic cytoskeletal patterns. Nat Cell Biol 16(1):38–46

Mayor R, Carmona-Fontaine C (2010) Keeping in touch with contact inhibition of locomotion. Trends Cell Biol 20(6):319–328

Ordemann A, Balázsi G, Caspari E, Moss F (2003) Daphnia swarms: from single agent dynamics to collective vortex formation. In: Bezrukov SM, Frauenfelder H, Moss F (eds) Fluctuations and Noise in Biological, Biophysical, and Biomedical Systems. Proc. SPIE 5510:172–179

Partridge BL (1982) The structure and function of fish schools. Sci Am 246:114–123

Partridge BL, Pitcher T, Cullen JM, Wilson J (1980) The three-dimensional structure of fish schools. Behav Ecol Sociobiol 6:277–288

Peterson GD (1993) Animal aggregation: experimental simulation using vision-based behavioural rules. In: Nadel L, Stein DL (eds) 1992 Lectures in Complex Systems. Addison-Wesley, Reading

Pimentel JA, Aldana M, Huepe C, Larralde H (2008) Intrinsic and extrinsic noise effects on phase transitions of network models with applications to swarming systems. Phys Rev E 77:061138

Reynolds CW (1987) Flocks, herds, and schools: a distributed behavioral model. Comput Graph 21(4):25–34

Romey WL (1996) Individual differences make a difference in the trajectories of simulated fish schools. Ecol Model 92:65–77

Schaller V, Weber C, Semmrich C, Frey E, Bausch AR (2010) Polar patterns of driven filaments. Nature 467(7311):73–77

Silverberg JL, Bierbaum M, Sethna JP, Cohen I (2013) Collective motion of humans in mosh and circle pits at heavy metal concerts. Phys Rev Lett 110:228701

Simha RA, Ramaswamy S (2002) Hydrodynamics fluctuations and instabilities in ordered suspensions of self-propelled particles. Phys Rev Lett 89(5):058101

Simpson SJ, Sword GA, Lorch PD, Couzin IA (2006) Cannibal crickets on a forced march for protein and salt. Proc Natl Acad Sci U S A 103(11):4152–4156

Still GK (2014) Introduction to Crowd Science. Taylor and Francis Group, CRC Press, Boca Raton

Strefler J, Erdmann U, Schimansky-Geier L (2008) Swarming in three dimensions. Phys Rev E 78:031927

Strogatz SH (1994) Nonlinear Dynamics and Chaos with Applications to Physics and Engineering. Addison-Wesley, Reading

Sumino Y, Nagai KH, Shitaka Y, Tanaka D, Yoshikawa K, Chaté H, Oiwa K (2012) Large-scale vortex lattice emerging from collectively moving microtubules. Nature 483(7390):448–452

Szabó B, Szöllösi GJ, Gönci B, Jurányi ZS, Selmeczi D, Vicsek T (2006) Phase transition in the collective migration of tissue cells: experiment and model. Phys Rev E 74:061908

Toner J, Tu Y (1995) Long-range order in a two-dimensional dynamical XY model: how birds fly together. Phys Rev Lett 75(23):4326–4329

Toner J, Tu Y (1998) Flocks, herds and schools: a quantitative theory of flocking. Phys Rev E 58:4828–4858

Toner J, Tu Y, Ramaswamy S (2005) Hydrodynamics and phases of flocks. Ann Phys 318:170–244

Vicsek T (2012) Swarming microtubules. Nature 483(7390):411–412

Vicsek T, Czirók A, Ben-Jacob E, Cohen I, Schochet O (1995) Novel type of phase transition in a system of self-driven particles. Phys Rev Lett 75(6):1226–1229

Volfson D, Cookson S, Hasty J, Tsimring LS (2008) Biomechanical ordering of dense cell populations. Proc Natl Acad Sci U S A 105(40):15346–15351

Vollmer J, Vegh AG, Lange C, Eckhardt B (2006) Vortex formation by active agents as a model for *Daphnia* swarming. Phys Rev E 73:061924

Wioland H, Woodhouse FG, Dunkel J, Kessler JO, Goldstein RE (2013) Confinement stabilizes a bacterial suspension into a spiral vortex. Phys Rev Lett 110:268102

Yeomans JM (1992) Statistical Mechanics of Phase Transitions. Oxford University Press (Clarendon Press), Oxford

Zhang HP, Be'er A, Florin EL, Swinney HL (2010) Collective motion and density fluctuations in bacterial colonies. Proc Natl Acad Sci U S A 107(31):13626–13630

Chapter 9
Biofilms

...and not quite enough time.

Leonard Bernstein

BACTERIAL BIOFILMS[1] have recently risen to prominence in the scientific and medical literature in large part because of the dramatic increase in antibiotic-resistant infections in hospital settings. Yet biofilms have been around for 3.25 billion years, emerging at the same time as multicellular organisms like filamentous prokaryotes and cyanobacterial mats (de la Fuente-Núñez et al. 2013). Experimental observations of biofilms also date back a considerable time as well, though far less than 3.25 billion years. The material Leeuwenhoek famously scraped off his teeth and examined under a microscope in one of his studies of "animalcules" was none other than a biofilm. J. B. Burton-Sanderson noted in 1870, in the *13th Report of the Medical Officer of the Privy Council*, that he had observed rods at the surface of a liquid that "adhere together by their sides after the manner of the elements of a columnar epithelium". He noted, however, that there was "strong reason to believe that this adhesion is not direct, that is, that they are not in actual contact but glued together by a viscous intermediary substance" (Vlamakis et al. 2013). This substance is now known to be an extracellular matrix, composed of a variety of molecular components, which can be assembled – and disassembled – by bacteria depending on the stress conditions they experience.

Modern studies of biofilms were inaugurated by Canadian microbiologist Bill Costerton. He demonstrated that biofilms account for the majority of naturally occurring bacterial populations. Costerton recalls the initial discovery of the biofilms in Canadian mountain streams in his 2007 book *The Biofilm Primer.* Around 1977, two young colleagues

> took advantage of their outstanding physical condition to gallop tens of miles into the alpine zones of the Absorka and Bugaboo mountains, where they plated and cultured water from icy streams crashing down boulder fields... These cultures yielded only ±10 bacterial cells per milliliter, but it soon became obvious that rocks in these streams were covered with

---

[1] While bacterial biofilms are the best known, archaea, as well as eukaryotic organisms such as fungi, can also form biofilms (de la Fuente-Núñez et al. 2013).

© Springer Science+Business Media B.V. 2018
S. Bahar, *The Essential Tension*, The Frontiers Collection,
DOI 10.1007/978-94-024-1054-9_9

slippery biofilms, and direct examination of these clear slime layers showed the presence of millions of bacterial cells...encased in transparent matrices. (Costerton 2007, pp. 5–7)

Imaging biofilms at the cellular level presented significant technical problems until the early 1990s. Initial studies using transmission electron microscopy (TEM) and scanning electron microscopy (SEM) produced dehydration artifacts as a result of sample preparation, and thus the viscous extracellular matrix holding the biofilm together could only be visualized in distorted form using these techniques (Costerton 2007, p. 13). In 1992, John Lawrence produced confocal laser scanning microscope (CLSM) images of biofilms that set the community "literally buzzing with excitement" when they were first revealed at a conference in Kyoto. As Costerton describes,

> [s]essile cells could be seen to be embedded in a transparent viscous matrix, but the most significant revelations were that biofilms are composed of microcolonies of these matrix-enclosed cells...and that the community is intersected by a network of open water channels... The microcolonies were seen to take the form of simple towers, or of mushrooms, and the water channels were devoid of cells and appeared to constitute a primitive circulatory system that one could imagine being responsible for the delivery of nutrients and removal of wastes... As the delegates returned home from Kyoto...it was clear that bacteria had taken a very significant step upwards on the ladder of evolution and that these organisms were capable of forming very complex and highly structured multicellular communities. (Costerton 2007, p. 15).

The channels identified in the 1992 images were experimentally shown several years later to facilitate oxygen delivery and to permit the flow of water and other small molecules. Two decades on, Wilking et al. (2013) noticed a connection between a wrinkled biofilm surface and a possible mechanism of nutrient transport (Fig. 9.1). By injecting a solution of fluorescent colloidal particles into the wrinkles, they showed that the wrinkles formed a network of interconnected channels with low resistance to liquid flow. Scanning electron microscopy showed that the channel floor could be formed by the agar substrate in laboratory-grown biofilms. Further experiments identified a mechanism that promoted flow through the channels. Pressure within the channels was less than the atmospheric pressure outside, and this pressure gradient could drive flow through the channels. Wilking et al. showed that this pressure gradient is maintained by evaporation from the biofilm surface, facilitating upward nutrient flow through the biofilm. Transverse flow along the channels is driven by a more complex process involving spatial variation in the evaporation rate, and hence in the pressure gradient, related to temperature variations throughout the biofilm. The structure of the channels changes during the life cycle of a biofilm, and cells eventually cover over the agar floor (Wilking et al. 2013). Cairns et al. (2014) speculate that the channels are lined by a hydrophobic protein, BslA, "to allow wicking of fluids into the deeper parts of the biofilm", facilitating flow.

The initial formation of the channels may derive from buckling instability of the matrix, with wrinkles being formed when the matrix presses against a hard surface. A study by Asally et al. (2012) showed that "cell death at the base of the biofilm was linked with buckling in the vertical plane and thus [with] wrinkles" (Cairns et al.

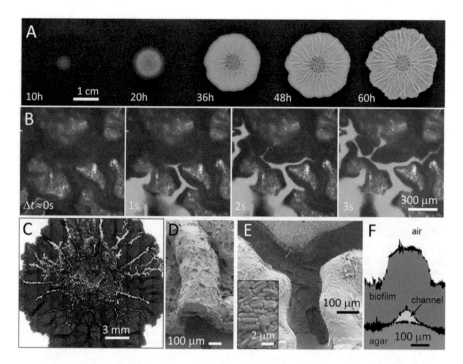

Fig. 9.1 The 2013 study by Wilking et al. demonstrates the formation of fluid-filled channels visible in the biofilm as it grows over time (panel **a**). Microscopy images show the movement of an aqueous dye (*green*) through the channels (panel **b**). Injection of fluorescent beads (panel **c**) shows the connectivity of the channels. Panel **d** shows an SEM image of a cross-section of a "wrinkle"; seen from below, while another SEM image (panel **e**) reveals the highly structured channels. A side-view of the channel structure, reconstructed from a plastic mold, is shown in panel **f**. Figure reproduced with permission from Wilking et al., *Proc. Natl. Acad. Sci. U.S.A.* 110(3), 848–852, 2013

2014). Asally et al. showed that the deletion of genes for certain proteins involved in extracellular matrix synthesis led to a more homogeneous distribution of cell death. In this case, wrinkles failed to form. Conversely, by artificially increasing cell density, Asally and colleagues induced localized cell death and found that "wrinkles were formed in patterns that mirrored cell death zones" (Cairns et al. 2014).

Water-filled channels give biofilms a structure at a comparatively large spatial scale. But bacteria in biofilms can also deposit honeycomb-like grids of hexagonal "cells", each with a diameter on the order of 8 μm, several times the size of a single bacterial cell. These grids, which have only been observed in multi-species biofilms, exhibit such structural regularity that they were initially assumed to be decayed plant structures (Costerton 2007, p. 27). The mechanism by which these structures are formed remains obscure. As Costerton writes, "[t]he ability to form regular tissue-like structures has always been a property reserved for eukaryotic cells, and we have not yet developed an intellectual rubric into which to place the fact that prokaryotic cells can control large-scale activities of this kind" (Costerton 2007, p. 28). Equally surprising is the fact that bacteria "abandon" the honeycomb

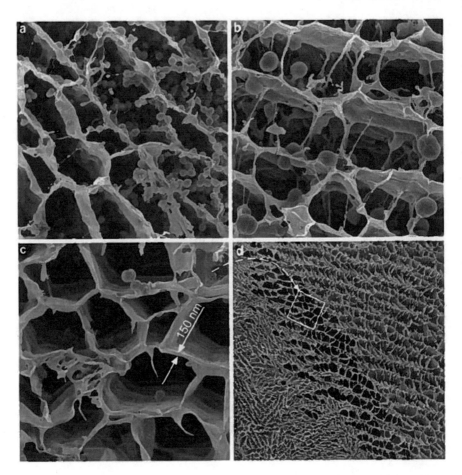

Fig. 9.2 Scanning electron microscope (SEM) images of honeycomb structures in biofilms. Panel **a** shows the honeycomb structure partially abandoned by the bacterial cells, and panel **b** shows the structure essentially devoid of bacteria. Panel **c** shows a close-up image of the area enclosed by the *white box* in panel **d**. Note the high level of structure in the plates forming the honeycomb (*arrows* in panel **c**) and the overall structure (panel **d**). From Costerton, *The Biofilm Primer* (Figure 17)

structure after having completed it (Fig. 9.2). The mechanisms regulating this movement away from the structure, let alone the biological reasons for its formation and subsequent abandonment, remain mysterious.

In addition to honeycomb matrices and water-filled channels, some biofilms produce transitory structures as complex as those of the "social amoeba" *Dictyostelium*, which we have mentioned briefly above and will return to in more depth in the next chapter. Costerton describes these structures with barely contained excitement:

> Kjelleberg's group has shown that the marine organism *Serratia liquefaciens* strain MG 1 forms biofilms in which the organism's cells are arranged into vertical stalks that bear rosettes of cells connected to other rosettes by long chains of cells and that this architectural

marvel is controlled by specific genes... Tim Tolker-Nielsen's group has shown that one clone of *P. aeruginosa* forms stumplike pedestals on colonized surfaces and that mobile cells of a second clone crawl up the pedestals and form the 'caps' of the mushrooms that are such a prominent feature of the biofilms formed by this organism. At the recent 11th meeting of the International Society for Microbial Ecology (ISME) in Vienna (August 2006) Tim presented evidence...that the cells of the second clone may actually form the mushroom caps on templates of DNA produced by programmed apoptosis of specialized cells in the tops of these pedestals. (Costerton 2007, p. 19)

Naturally occurring biofilms typically contain individual cells separated by some distance (up to 10 μm) from their neighbors. Sister cells in biofilms separate shortly after fission. However, cells in a biofilm are often connected by pili, the hairlike bacterial appendages used by bacteria for attachment to surfaces, gliding motility, and horizontal gene transfer. Costerton has suggested that the pili may play a structural role in biofilms akin to that of microtubules in eukaryotic cell structure (Costerton 2007, p. 26).

The presence of pili on bacteria within biofilms suggests a high rate of horizontal gene transfer, leading to rich genetic diversity within the population, and enhancing the biofilm's ability to rapidly adapt to external stresses, whether from environmental changes such as fluctuations in oxygenation or acidity, or from the application of antimicrobial agents by humans. Further, "enhanced horizontal gene transfer in biofilms allows some clones within certain species to jettison certain 'high maintenance' genes, not required for current activities, and to reacquire these genes from their clonal partners at a later time and in different circumstances" (Costerton 2007, p. 75). Thus, a temporary division of labor facilitates metabolic efficiency. Indeed, the up-regulation of genes involved in recombination has been observed in bacterial cells in a biofilm environment (Costerton 2007, p. 58; Boles et al. 2004).

The degree of horizontal gene transfer in biofilms has been described as producing no less than a "distributed genome" (Ehrlich et al. 2010). In this model,

we can conceive of a situation in which the complex and genetically expensive machinery required for the degradation of certain complex pollutants could be jettisoned by cells intent on degrading more amenable nutrients. However, as long as at least some cells have retained all of the genetic elements of the degradative pathway, the community as a whole will be ready to mobilize the genome from its distributed sources and swing into action if this complex substrate should suddenly become available. (Costerton 2007, p. 74)

The genetic diversity enabled by horizontal gene transfer in biofilms is amplified further by the fact that most naturally occurring biofilms are consortia of multiple species. As Costerton emphasizes repeatedly in *The Biofilm Primer*, this means that the majority of laboratory studies of biofilms, typically performed using only a single species, differ radically from natural populations. Even more, most studies of bacteria, and development of antibiotics, are performed on individual "planktonic" (free-living) cells, and on clones of single planktonic cells, reducing the natural genetic diversity even further. The bias toward planktonic cells was called the "bottle effect" by ZoBell in the 1940s; as Costerton explains, "when we transfer liquid bacterial cultures, we always take a loop from the bulk fluid to inoculate the next (sterile) tube. In doing so, we leave behind all of the bacterial cells that have adhered to the surfaces of the test tube...*and we gradually select for mutants that are defective in biofilm formation*" (Costerton 2007, p. 37, my italics). Indeed, ten serial culture transfers of wild-type *E. coli* will result in a loss of 37.5% of the genome (Costerton 2007, p. 37). The retained genes are hardly likely to be critical for biofilm formation, and thus by the time a bacterial cell is studied, it is an exemplar of one particular type of bacterial lifestyle, but not of the predominant form in which bacteria exist in nature. Since most development of antibiotics is performed using such pacified strains, it is no surprise that antibiotics are more effective against planktonic bacteria than against biofilms. This is clearly a problem, since over 99% of bacteria in natural populations – including in human patients – occur in biofilms (Ehrlich et al. 2005).

One might initially assume that the very fact of being a member of a biofilm affords a cell protection from external threats, as in the selfish herd model. Indeed, this is true, up to a point. Grazing amoebae can ingest (and then digest) free-living planktonic cells, but not groups of bacteria enclosed in a biofilm matrix. From this point of view, biofilms afford a purely mechanical protection. However, cells need access to nutrients, and thus exchange of materials between the exterior and interior of the biofilm is essential. Nutrients can flow through water-filled channels to access the biofilm interior, but this poses another problem. If nutrients can gain access to the interior of a biofilm, so can toxins. Moreover, experiments have shown that external toxins can penetrate through the extracellular matrix (Costerton 2007, p. 58).

Antibiotic resistance clearly demands more than a physical barrier. In fact, it derives from changes in gene expression between the planktonic and biofilm states. For example, membrane proteins that allow transport from the extracellular to the intracellular space may cease to be produced in the biofilm state, and pumps that mediate toxin efflux may be expressed instead.[2] Studies suggest that gene expression in cells in the planktonic vs. the biofilm state may differ by up to 70% (Costerton 2007, p. 58).

While the biofilm matrix does not create a physical barrier to penetration by toxins, it still generates a protective environment. For example, it may provide anaerobic regions with high proton concentration, in which some toxins may have difficulty functioning (Costerton 2007, p. 58). The matrix is highly hydrophobic, which helps provide resistance to penetration by toxins, including antimicrobial agents (Vlamakis et al. 2013). There is evidence that the matrix accumulates enzymes like β-lactamase that actively degrade antibiotics (de la Fuente-Núñez et al. 2013).

Biofilms restrict diffusion of metabolic products away from the cells that produce them. Bacteria depend on the re-influx of extruded protons in order to generate ATP. In the turbulent water of a stream, these protons would be easily swept away, as would be any small organic nutrient molecules. In the case of protons specifically, many of the polymers that compose the matrix are negatively charged, and

---

[2] Recent studies suggest that efflux pumps may also be involved in the process of biofilm assembly (de la Fuente-Núñez et al. 2013)

thus bind positive cations; these cations then repel the protons, forcing them to remain near the membranes of the bacteria that extruded them (Costerton 2007, p. 72).

Biofilms provide a means for bacteria to remain in one place in order to feed. Many nutrient sources in turbulent aquatic environments are themselves transitory: "the death of a fish would provide few benefits to planktonic bacteria passing in the water, while the nutrients from the vertebrate disaster can (and do) nourish billions of bacteria that live in the biofilms that soon form right on the nutrient source" (Costerton 2007, pp. 72–73). Other benefits accrue from this group lifestyle. Organic material released from dead bacteria is immediately available for consumption by the survivors. Costerton calls it "a perfect nutrient paradise,…a puree of the bodies of their neighbors" and notes that, as a result of this banquet, "regrowth rates of stressed biofilms are truly phenomenal when the stress is removed" (Costerton 2007, p. 58). A further benefit may arise from secretion of "public good" metabolites. The uptake of secreted metabolites by neighboring cells might prevent feedback inhibition of the cells that produced them (Costerton 2007, p. 72). In other words, the collective milks itself for maximum production by preventing feedback inhibition.

Due to channels within the biofilm, a toxin such as bleach can kill cells even deep within the structure. Yet bleach fails to kill a small subset of cells. Since these cells are not localized within a particular region of the biofilm, their survival has been assumed to be the result of some mutation (Costerton 2007, p. 58). Dubbing these cells "persisters", Lewis (2001) suggested that they might play a major role in biofilm survival and recovery from antibiotic agents, allowing populations to recover (and to feast on their dead cousins in the process) in cases where the biofilm phenotype itself was not sufficient to confer resistance.

The existence of persister cells highlights the genetic diversity within a biofilm. As Gerdes and Semsey (2016) explain, *persistence* is quite different from the development of antibiotic *resistance*. In the latter case, the majority of surviving cells, if not all of them, will be resistant to an antibiotic. In contrast, "regrowth of a persistent population results in the same percentage of drug-sensitive cells as before". Persistence can be seen as a bet-hedging strategy; importantly, the maintenance of subpopulation of persister cells is a population-level trait.

It has been known for some time that persister cells can enter a nearly dormant state in which their slowed metabolism reduces the update of toxins (Balaban et al. 2004). Dormancy can be induced by the expression of type II toxin-antitoxin (TA) genes, which inhibit translation (Balaban et al. 2004; Maisonneuve et al. 2011, 2013; Helaine et al. 2014; Gerdes and Semsey 2016). Such expression can occur not only when bacteria are exposed to antibiotics, but also when they are internalized by macrophages (Helaine et al. 2014). Additionally, type I TA genes promote the dormancy state by decreasing ATP levels (Verstraeten et al. 2015; Gerdes and Maisonneuve 2015; Gerdes and Semsey 2016).

Persisters are not simply resistant to toxins because of a slow metabolism. They also have less passive defense mechanisms. Pu et al. (2016) showed that *E. coli* can actively export an antibiotic through the efflux protein TolC. Persister cells were found to express more TolC than non-persisters. Further, inhibition of TolC activity or deletion of the *tolC* gene reduced the persister phenotype in the population. Day (2016) suggested that epigenetic mechanisms may mediate bacterial persistence; stochasticity of gene expression may also play a role in maintaining a subpopulation of persister cells (Day 2016; Gerdes and Semsey 2016).

The genetic diversity evidenced by the existence of persister cells is amplified by the fact that naturally occurring biofilms are typically composed of multiple species; mechanisms of interspecies signalling have been identified and shown to play a role in the formation of biofilms. Costerton and his collaborators first demonstrated interspecific (indeed, intergeneric) signaling between *Streptococcus cristatus* and *Porphyromonas gingivalis* in dental plaques (Xie et al. 2000). The quorum sensing molecule autoinducer-2 has been referred to as a "bacterial universal language" (Raut et al. 2013). In quorum sensing, bacteria secrete signalling molecules which, when released into the environment and passing a threshold concentration, are taken up by neighboring bacteria, which are then induced to both produce more of the signaling molecules themselves, and also to begin the process of extracellular matrix formation (Rutherford and Bassler 2012). Just as *Dictyostelium* cells aggregate via cAMP signaling to form a slug (see Chap. 10), bacterial quorum sensing also depends on a highly nonlinear, threshold-driven, positive feedback process.

Quorum sensing serves as a marker of population density, and can be mediated by a variety of biochemical pathways (Hooshangi and Bentley 2008; Ng and Bassler 2009; Rutherford and Bassler 2012; Hawver et al. 2016). Even though some quorum sensing signals can be read as a "universal language" across bacterial species, different types of bacteria do use different quorum-sensing mechanisms. For example, Gram-negative bacteria use small molecules such as acyl-homoserine lactones, known as auto-inducers (AIs), while Gram-positive bacteria use larger auto-inducing peptides (AIPs). In both cases, higher cell density leads to higher concentration of AIs or AIPs, which in turn stimulates their secretion. The quorum sensing process also regulates bacterial metabolism. Quorum sensing triggers the production of oxalate, which serves as a nutrient, in order to promote survival of bacteria of the genus *Burkholderia* during starvation conditions (Goo et al. 2015). Not surprisingly, the production of such "public goods" during quorum sensing can lead to cheating, with some bacteria attempting to conserve their energetic resources and profit, for example, from oxalate production, while not expending the energy to produce it themselves (Diggle et al. 2007; Dunny et al. 2008; Popat et al. 2012, 2015; García-Contreras et al. 2015; Katzianer et al. 2015; Zhang et al. 2015).

*** *** ***

The genetic differences between planktonic bacterial cells and their biofilm-inhabiting cohorts arise as a result of complex gene expression cascades triggered not just by population density, by also by external stress in the environment. Under adverse conditions, free-living bacterial cells adhere to a surface and begin to express a set of genes that lead to biofilm formation.

Biofilm formation is a reversible process: when conditions improve, the biofilm structure is dissolved and the bacteria return to their planktonic lifestyle. Since cells within the biofilm structure take on different roles, biofilm formation involves a reversible cellular differentiation process. Some of this differentiation is driven by the varying spatial environment at different locations within the biofilm (van Gestel et al. 2015). At deeper layers, cells have less access to oxygen and other nutrients and thus grow more slowly and are less metabolically active[3] than cells closer to the surface (de la Fuente-Núñez et al. 2013).

Cell activity is not only regulated by location within the biofilm structure, however. It is also determined by the temporal expression of particular genes, and by the local concentration of key proteins. Richly complex regulatory pathways control the cascades of gene expression that trigger biofilm formation and disassembly. This may explain why individual cells can only be recruited into biofilms during a limited time window: it is much harder for free-living cells to join an established biofilm once it is metabolically integrated (Costerton 2007, p. 51).

While it may be difficult for individual cells to join an established biofilm, individual cells can detach and act as "planktonic scouts", flowing down a mountain stream, or through a human's lung, until they attach to a favorable new site for colonization. In order for this to occur, however, these cells must

> return to the planktonic phenotype and disentangle themselves from matrix components before they can leave the community. This process must include a signal that triggers the synthesis or release of enzymes that can degrade the basic polymers that constitute the biofilm matrix as well as the conversion of the cells to the planktonic phenotype. In *P. aeruginosa*, the lyase enzyme that digests alginate [an exopolysaccharide matrix constituent] is continuously synthesized and stored in the periplasmic space…and is released to digest the matrix when detachment is initiated. (Costerton 2007, pp. 53–54)

This process is closely related to full-scale biofilm disassembly. When cells begin to emerge from the trance of the biofilm Borg, those closer to the center awaken first. In the mushroom-like structures formed by *P. aeruginosa*,

> the return to the planktonic phenotype gradually spreads from the center …, so that the swarm of swimming planktonic cells finds a breach in the dissolving microcolony, and the individual cells swim to freedom as the walls of [their] erstwhile home collapse." (Costerton 2007, p. 54)

The pathways involved in biofilm formation have been studied in great detail for species such as *Bacillus subtilis*. This bacterium is Gram-positive[4] and

---

[3]This metabolic gradient also has an effect on the bacterial response to various antimicrobial drugs; tetracycline and fluoroquinolones are able to kill metabolically active cells near the surface of a biofilm of *P. aeruginosa*, while colistin, a lipopeptide, is more effective against the slower-growing cells in the deeper layers of the biofilm. Actively growing cells upregulate an operon encoding genes that confer resistance to colistin (de la Fuente-Núñez et al. 2013).

[4]Gram-positive bacteria can be stained with crystal violet, a method developed in the 1880s by H. C. Gram. Gram-positive bacteria appear violet after the stain has been washed away, since their thicker peptidoglycan layer retains the stain. Gram-negative bacteria have a thinner peptidoglycan layer, and do not retain the crystal violet stain (they can be stained pink with a "counterstain", however). Gram-positive bacteria are more susceptible to antibiotics since, despite their thicker

non-pathogenic, and therefore not as medically important as species like *Pseudomonas aeruginosa*. However, it is well studied in the laboratory and provides a useful model for the underlying genetic processes of single-species biofilm formation.

In its planktonic, free-living state, *B. subtilis* is self-propelled, using flagella to navigate through the environment. In response to external stress or quorum sensing signals, bacteria lose their flagellar motility. In *B. subtilis*, this process is regulated by a protein called EpsE, which is also involved in producing the principal extracellular matrix protein, an extracellular polysaccharide (EPS). In a series of delicate experiments combining biophysics and genetics, Blair et al. (2008) showed that EpsE acts as a "molecular clutch" rather than a "brake". During normal flagellar operation, the subunits of the protein FliG "polymerize into a wheel-like rotor attached to the flagellar basal body and transduce the energy of proton flux through the MotA-MotB proton channel into the rotational energy of the flagellum" (Blair et al. 2008).

When the clutch is thrown, EpsE interacts with FliG to decouple it from MotA-MotB (Fig. 9.3). The flagella are thus not immobilized, but rather *uncoupled from their power source*, rendering rotation impossible (Blair et al. 2008; Guttenplan and Kearns 2013). As Vlamakis et al. (2013) phrase it, this is a "beautiful example of the multiple levels of regulation which ensure that matrix-producing cells are non-motile". As Blair et al. (2008) emphasize, the disabling of flagellar motility is entirely reversible,[5] so that "if biofilm formation is prematurely aborted, flagella once disabled by the clutch might be reactivated, allowing cells to bypass fresh investment in flagellar synthesis". Reversible control of flagellar motility is a much more rapid and flexible process than down-regulation of flagellar protein expression, which would take generations to be fully effective.

After losing their flagella, cells

> form long chains of non-motile cells that adhere to each other and to the surface by secreting an extracellular matrix... As the biofilm matures, the cell clusters enlarge and the community is protected and organized by the extracellular matrix. In addition to matrix producers, motile cells and spores are present and are spatially organized within the maturing biofilm... The presence and localization of the different cell types is dynamic and there seems to be an ordered sequence of differentiation such that motile cells become matrix-producing cells, which go on to become spores... Importantly, this process of differentiation is not terminal; as environmental conditions change, it is possible for cells to alter their gene expression (in the case of motile or matrix-producing cells) or to germinate (in the case of spores). Phenotypic heterogeneity in *B. subtilis* is not limited to these three cell types... In laboratory conditions, biofilms have a limited lifespan, and they eventually disassemble in response to self-generated signals... As a biofilm disassembles, spores are released from the matrix, giving them the potential to disperse and encounter environmental conditions that are propitious for germination. (Vlamakis et al. 2013)

---

peptidoglycan layer, they lack the tough outer protein/lipopolysaccharide layer that protects Gram-negative species.

[5] While biofilm formation is a reversible process, de la Fuente-Núñez et al. (2013) point out that "growth in a biofilm can favor the occurrence of processes that lead to the acquisition of inheritable resistance, such as horizontal gene transfer...or adaptive mutations."

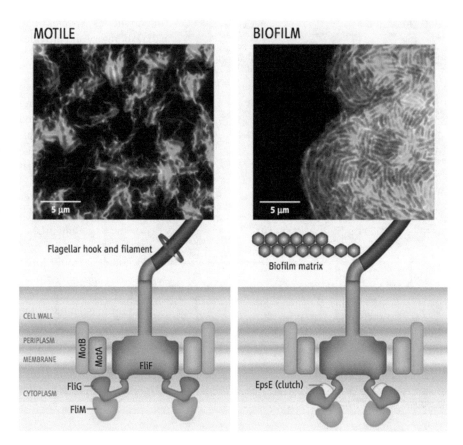

Fig. 9.3 Schematic diagram of the *B. subtilis* rotary flagellar motor, in which EpsE acts as a molecular clutch (*bottom panels*). The *top panels* show fluorescence microscopy images of *B. subtilis* in motile (*top left*) and biofilm (*top right*) form, with bacterial membranes stained *red* and flagella stained in *green*. Figure reproduced from R.M. Berry and J.P. Armitage, How bacteria change gear, *Science* 320: 1599–1600, 2008. *Top panels* reproduced with the permission of D. Kearns. *Bottom panels* reprinted with permission from AAAS

The temporal programme of cell differentiation is controlled by a number of interlinked regulatory processes. Figure 3a in Vlamakis et al. (2013), reproduced here as Fig. 9.4, illustrates the complexity of these regulatory networks in *B. subtilis*. One key regulatory protein, whose activity affects the transcription of over 100 other genes, is Spo0A. This protein's activation depends on the phosphorylation of a particular aspartate amino acid residue, and the relative concentration of phosphorylated to unphosphorylated Spo0A in a cell determines which genes are expressed, and hence the particular role played by that cell within the biofilm. An intermediate level of phosphorylated Spo0A (Spo0A~P) leads cells to produce extracellular matrix, while a higher level triggers the spore-forming developmental pathway. Gradual accumulation of Spo0A may lead to the formation of spores later

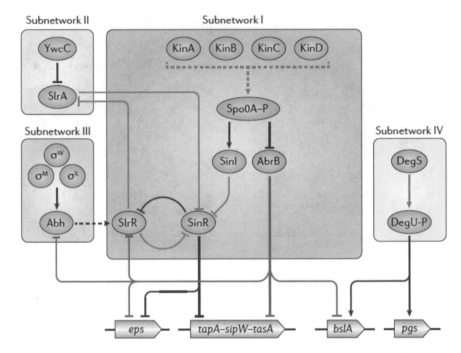

Fig. 9.4 A schematic diagram of just some of the gene regulatory network controlling biofilm formation in *B. subtilis*. *T-bars* indicate repression; arrows indicate activation. Reprinted by permission from Macmillan Publishers Ltd: *Nature Reviews Microbiology* (Vlamakis et al. *Nat. Rev. Microbiol.* 11(13), 157-168, 2013), copyright 2013

in the biofilm's life cycle (Vlamakis et al. 2013). The SpoOA protein has been implicated in regulatory processes in other bacteria, such as *C. difficile* (Dawson et al. 2012).

Since matrix synthesis is energetically expensive, tight regulation is essential in order to avoid wasting precious resources. As summarized by Cairns et al. (2014), phosphorylated SpoOA~P functions as a transcription factor in several different pathways controlling this process. It can block transcription of a gene encoding a repressor (AbrB) of extracellular matrix synthesis genes. It also promotes the expression of a protein, AbbA, which binds to AbrB, keeping this repressor sequestered away from its DNA target. SpoOA~P promotes the expression of SinI, which irreversibly binds to SinR, a protein that represses the *eps* operon, which encodes 15 genes essential for extracellular matrix production, and the *tapA-sipW-tasA* operon, which is also essential to matrix formation. Other proteins, SlrA and SlrR, have been identified, which act as homologues of SinI and SinR, and promote the expression of matrix synthesis genes (Kobayashi 2008). SlrR, for example, promotes expression of genes in the *eps* and *tapA* operons by binding directly to SinR with high affinity, preventing it from carrying out its repressor role (Vlamakis et al. 2013; Cairns et al. 2014).

The pathways just described may seem to be a wild tangle of double negatives, yet these subtle regulatory mechanisms are delicate, tightly controlled, and quite typical of gene regulation in many species. I mention them here in order to give a flavor of the complex regulatory processes involved in biofilm formation. Other proteins, such as RemA and RemB, are involved in the activation of the *eps* and *tapA* operons. These proteins seem to be activated by environmental stress since "*remA*[6] is situated alongside genes connected to the stringent[7] response, suggesting a link between *remA* and the nutrient status of the cell" (Cairns et al. 2014). Another transcription factor, DegU, is involved in multiple aspects of biofilm formation as well as bacterial locomotion in the planktonic state. The concentration of DegU has subtle – and self-limiting – effects.

> The level of phosphorylated DegU (DegU~P) in the cell dictates which behaviour manifests. For example, activation of biofilm formation requires intermediate levels of DegU~P and inhibition of biofilm formation requires high levels of DegU~P. Biofilm activation occurs when DegU~P indirectly promotes transcription of *bslA*, which encodes a hydrophobic biofilm coat protein… However, under conditions where DegU~P levels in the cell are high, biofilm formation is inhibited due to a lack of transcription from the *eps* and *tapA* operons… Recent work has indicated that the DegS-DegU pathway [DegS is a histidine kinase that phosphorylates DegU] is activated by inhibition of flagellar rotation, as may conceivably occur when a cell senses a surface prior to adherence. (Cairns et al. 2014)

The *eps* operon codes for proteins involved in the synthesis of polysaccharide constituents of the extracellular matrix. The composition of these polysaccharides varies with the availability of the raw materials. When grown in glutamic acid and glycerol, bacteria will synthesize galactose, glucose and N-acetyl-galactose. Grown in TY broth, bacteria produce polysaccharides primarily composed of mannose. *B. subtilis* can synthesize other polysaccharides using a pathway independent of *eps*, when grown in the presence of sucrose; this form of the matrix may be more prevalent when biofilms grow in natural conditions such as on the roots of plants (Cairns et al. 2014).

In addition to polysaccharides, the extracellular matrix contains proteins. One such protein, TasA, polymerizes in a fashion similar to aggregating prions or to the β-amyloid protein found in the neural tissue of Alzheimer's patients. TasA can shift from an oligomeric form rich in α-helices to a fiber-like state rich in β-sheets. This transition appears to be triggered by contact with a surface, and is regulated by the acidity of the medium; it is also stabilized by another protein, TapA (Cairns et al. 2014). The protein BslA is also essential for biofilm assembly. It forms "rafts" below

---

[6]A reminder for non-biologists: the usual convention is to write the names of proteins in regular type, beginning with a capital letter, as RemA, and to write the gene that codes for the protein in italics, beginning with lower case, as *remA*. However, this convention does not apply for all species; genes in yeast are typically written with capital non-italic letters, so the gene coding for the sucrose-hydrolyzing enzyme invertase in *S. cerevisiae* is SUC2 (see Chap. 13).

[7]The *stringent response* occurs following external stresses such as heat shock, iron or fatty acid depletion, and amino acid starvation. See Boutte and Crosson (2013) for a recent review of the diversity of stringent responses across bacterial species.

biofilms called pellicles, existing at a liquid-air interface. BslA contains a hydrophobic cap comparable to the hydrophobin proteins in fungi (Cairns et al. 2014).

All the molecules involved in the extracellular matrix, from polysaccharides to TasA, TapA and BslA, provide a benefit to any bacterium in the biofilm, regardless of whether or not that bacterium is actively involved their production. These products thus act as "public goods", and create the possibility for cheaters to arise within the population, free-riding on the rafts, so to speak, produced by others. We will revisit this below.

Perhaps the most surprising constituent of the extracellular matrix is DNA. The presence of DNA in the matrix was first observed by Whitchurch et al. (2002), who found that the addition of DNase I to the culture medium inhibited the formation of *P. aeruginosa* biofilms. Moreover, DNase I dissolved established biofilms as long as they had not reached the mature age of ~84 h. ("[T]he matrix in mature biofilms may be strengthened by other substances," suggested Whitchurch and colleagues, or "mature biofilms may produce sufficient proteolytic exoenzymes to locally inactivate the DNase I".) Could extracellular DNA play a role in the matrix structure? It is, after all, a polymer!

Where does this extracellular DNA come from? It could conceivably be deposited in the extracellular space as a result of the lysis of dying bacteria. However, Whitchurch et al. noted that the DNA they observed "is presumably derived from membrane vesicles rather than cell lysis as we saw no evidence of the latter during biofilm formation". More recent studies have confirmed that extracellular DNA can indeed be released via vesicles, and that its production is upregulated during biofilm production in *Streptococcus mutans*, a bacterium that forms biofilms on the surface of teeth (Liao et al. 2014). Intriguingly, Liao et al. observed that the extracellular DNA released from vesicles was distributed in a structured network throughout the biofilm, in contrast to the randomly dispersed extracellular DNA resulting from cell lysis. Some extracellular DNA does appear to be produced from programmed cell death, however. Tolker-Nielsen showed that cells at the tips of the "mushroom stalks" formed in *P. aeruginosa* biofilms appear undergo apoptosis *in order* to release their DNA, which then serves as the structural basis for the "mushroom cap" formed by other cells (Costerton 2007, p. 23).[8]

Factors that promote biofilm disassembly are of clear medical importance, since they could be exploited to disrupt potentially fatal biofilms that form, for example, on implanted devices or in the lungs of cystic fibrosis patients. The medical strategy is to divide and conquer. Once a biofilm is disrupted, individual cells (except persisters) regain their vulnerability to conventional antibiotics.

A slower means of disassembling a biofilm is to silence the genetic pathways involved in matrix formation. For example, degradation of a protein such as SlrR could push bacteria back into a state where SinR can once again act as an effective

---

[8]This is reminiscent of the death of cells in the *Dictyostelium* stalk, though the role of the corresponding cells in *Dictyostelium* is the production of cellulose rather than the release of DNA. See Chap. 10.

repressor of the *eps* operon (Cairns et al. 2014). Consistent with this, Chai et al. (2010) observed that the level of SlrR expression decreases as a biofilm matures.

Biofilm disassembly is likely to involve active degradation of the matrix. Kolodkin-Gal et al. (2010) claimed that bacteria produce D-amino acids[9] during the later stages of the biofilm life cycle. These amino acids have also been shown to inhibit biofilm formation (de la Fuente-Núñez et al. 2013). D-amino acids were suggested to play a variety of roles in matrix disassembly, such as aiding in the release of amyloid fibers from the matrix structure (Kolodkin-Gal et al. 2010). However, a study by Leiman et al. (2013) suggested that the disassembly-promoting effect of these D-amino acids was due to their incorporation into proteins, rendering them incapable of properly performing their normal tasks, and thus slowing cell growth. The initial strain of *B. subtilis* in which the D-amino acid effect was observed had a mutation in a gene coding for an enzyme that removed D-amino acids from tRNAs; a strain without this mutation showed no D-amino acid effect on matrix disassembly (Leiman et al. 2013; Cairns et al. 2014). Other studies, however, continue to support a role for D-amino acids in the disassembly process (Yu et al. 2016). Complicating the picture further, a study in *P. aeruginosa* found reduced cell viability in a biofilm following treatment with a D-amino acid mixture, accompanied by a 30% *increase* in extracellular matrix production, rather than disassembly (Sanchez et al. 2013).

A 2012 study by Kolodkin-Gal et al. suggested that the bacterial product norspermidine might be involved in matrix disassembly. However, *B. subtilis* was subsequently shown to be incapable of producing norspermidine; in fact, Hobley et al. (2014) showed that exogenous norspermidine promoted biofilm *formation*. In 2015, after unsuccessful attempts to reproduce their original norspermidine results, the Kolodkin-Gal et al. paper was retracted. To date, the scientific understanding of the mechanisms of biofilm disassembly remains in a state of disarray. Cairns et al. summarize the current state of knowledge as follows:

> the issue of biofilm disassembly by *B. subtilis* remains a much-debated topic within the field. Further investigation will be required to determine whether the reduction in biofilm biomass observed in late-stage biofilms is the result of an organized disassembly process or simply the result of the onset of sporulation by the majority of the population after exhaustion of the nutrient supply. (Cairns et al. 2014)

*** *** ***

Between assembly and disassembly (Fig. 9.5), a biofilm exists as a complex "city of microbes" (Watnick and Kolter 2000). With cells exhibiting some level of differentiation, and the presence of "public goods", it is not surprising that the possibility exists for cheaters to arise within the population. Experiments such as those of Popat et al. (2012) clearly show the tension between cooperation and competition in biofilms: *cheaters do better individually, but the population as a whole does better*

---

[9] Amino acids can exist as L or D enantiomers (with the same chemical formula, but mirror-image structures). It is the L-form that is typically produced in living cells, though some D-amino acids are observed in the bacterial cell wall.

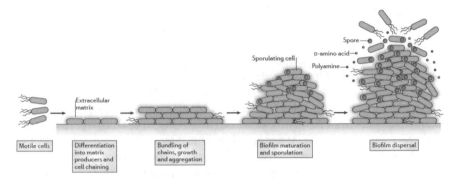

Fig. 9.5 Schematic representation of the life-cycle of a *B. subtilis* biofilm, showing (from *left*) motile cells with flagella, biofilm assembly, generation of sporulating cells, and sporulation from the mature biofilm. Reprinted by permission from Macmillan Publishers Ltd: *Nature Reviews Microbiology* (Vlamakis et al. *Nat. Rev. Microbiol.* 11(13), 157–168, 2013), copyright 2013

without cheaters. This is a clear conflict between fitness at the individual level and fitness at the group level. In biofilms, the conflict seems typically to be resolved in favor of the group. What tips the scales in favor of cooperators? They benefit ultimately, as members of a surviving group, but must put their metabolic stakes on the table before raking in the rewards.

Popat and colleagues investigated wild-type and *lasR* mutant strains of *P. aeruginosa*. The *lasR* mutants fail to respond to quorum-sensing signals, and thus could putatively play the role of "cheaters" in a mixed population with wild-type bacteria. In this case, cheaters are specifically defined as cells that cannot directly use a nutrient source but can exploit the extracellular digestion of that source by other cells. It had previously been shown that *lasR* mutants could act as social cheats in well-mixed planktonic cultures (Diggle et al. 2007; Sandoz et al. 2007).

In order to investigate cheating in biofilms, rather than planktonic cultures, Popat and colleagues used several different experimental designs. First, they compared the relative growth of wild-type and mutant strains in monoculture, and then in co-cultures. Biofilms were grown in flow cells, through which both regular medium, and minimal medium to induce quorum sensing, could alternately be pumped. Wild-type and mutant cells were infected with plasmids expressing different fluorescent proteins, so that imaging of co-cultures would reveal the proportion of each cell type. When grown in monoculture, the wild-type cells were significantly better at producing biofilms than the mutant cells. In mixed cultures, however, the wild-type cells formed a comparatively reduced percentage of the biofilm's biomass compared to the mutant, cheating cells. These studies can be compared with the investigations of cheating in yeast performed by Jeff Gore and colleagues (2009), Celiker and Gore (2012), Sanchez and Gore (2013), Datta et al. (2013). While yeast do not form biofilms, they do secrete enzymes that act as public goods, and are therefore susceptible to the development of cheater strains; see Chap. 13 for a detailed discussion of these studies.

Not only did Popat et al. demonstrate that cheaters did well in comparison to wild-type cells, but they also found that co-cultures including mutants produced biofilms with reduced density and thickness compared to wild-type monocultures. The detrimental effect of cheaters was greater in co-cultured biofilms than in planktonic co-cultures. In other word, cheating was more detrimental in a more interdependent community. Inclusion of cheaters also increased biofilm susceptibility to antibiotics. In other words, the cheaters did better *within* a co-culture, but damaged the health of the community as a whole.

Hypothesizing that "the negative consequences of cheats at the population level could be partially compensated for if the cooperators increased their rate of cooperation", Popat et al. added a synthetic quorum-sensing signal. This increased the population productivity when the initial cheat population was low. The effect was mitigated at higher cheat populations, since, given the fact the mutant cells were incapable of responding to the quorum sensing signal, a larger percentage of mutants naturally reduced the overall population response.

What tips the balance in favor of cooperators? Once cooperation has begun, a group advantage clearly emerges. But what gives cooperating cells a fitness edge over cheats to begin with? Might the answer be as prosaic as a random mutation conferring an immediate group benefit that outweighed the sum of individual disadvantages? In the case of biofilms, the question is far from resolved. However, several directions of research point toward possible mechanisms of cheating *suppression*. For example, there is evidence that biofilms can spatially segregate cooperating and cheating cells, making it harder for cheaters to exploit cooperators (Kreft 2004; Xavier and Foster 2007). When found in the lungs of cystic fibrosis patients, *lasR* mutants are isolated, suggesting that such segregation can be enforced in naturally occurring biofilms (Popat et al. 2012). Other studies have identified "metabolic incentives" that reward cooperation (Dandekar et al. 2012), as well as metabolic disincentives to cheating, such as production of cyanide by wild-type *P. aeruginosa*, to which *lasR* mutants are more susceptible than wild-type cells (Wang et al. 2015). Other studies in *P. aeruginosa* have shown that oxidative stress selects for bacterial strains able to mount a quorum-sensing response, and thus promotes selection against cheater mutants (García-Contreras et al. 2015). However, despite all these examples of cheater-suppression, there are bacterial species in which cheater strains arise frequently in natural populations, such as a "quorum non-sensing" strain of *Vibrio cholerae* (Katzianer et al. 2015).

Another striking set of experiments, from Gürol Süel's laboratory at UCSD, has shown that cells can experience a private benefit from the production of public goods (Zhang et al. 2015). This study not only addresses the problem of how cheating is prevented, but an even more important one: how division of labor is sustained in biofilms. Even more recently, Süel and colleagues showed that different *B. subtilis* biofilm communities can synchronize their growth dynamics when competing for resources, essentially "time-sharing" resources like two families sharing a beach house on the Jersey Shore (Liu et al. 2017; see Gordon 2017 for commentary).

Chai et al. (2008) found that, as a result of bistable gene regulation controlled by Spo0A, while most cells express *sinR*, only a small subpopulation expresses the antirepressor *sinI*. (Recall that *sinI* counteracts *sinR* repression of operons coding for extracellular matrix proteins and enzymes that synthesize extracellular polysac-charides.) The population was essentially separated into subgroups expressing different genes and performing different tasks within the community. Only about 10% of the cells took on the metabolically costly role of synthesizing the extracellular matrix. Kearns notes several caveats that should be attached to these conclusions, however.

> The work by Chai et al. raises the intriguing possibility that only a subset of cells within a given population is responsible for producing the biofilm matrix components. Matrix production is energetically costly and this strategy might relegate the cost to a subpopulation that provides protection for the entire community. With respect to this idea, the results obtained by Chai et al. have two caveats. First, while the experiments were conducted in a medium that promotes biofilm formation, the well-agitated conditions used did not actually allow stable biofilms to form. Second, the cytological images were essentially snapshots of gene expression. If activation of *sinI* is transient and/or dynamic, a larger subpopulation than was reported might activate SinI during biofilm formation. Regardless of these caveats, it is clear that at least under conditions that are likely similar to those experienced early in biofilm formation, production of the matrix components was confined to a subpopulation. (Kearns 2008)

While studies of biofilm formation are still in their infancy, it is clear that communication can be performed within these massive ensembles of bacterial cells through quorum sensing, complex structures can be formed, and that, through simple genetic switches, cells can effectively differentiate. Even for prokaryotes, then, the transition from individual to collective, with its delicately negotiated balance between competition and cooperation, occurs with comparative ease.

References

Asally M, Kittisopikul M, Rué P, Du Y, Hu Z, Çağatay T, Robinson AB, Lu H, Garcia-Ojalvo J, Süel GM (2012) Localized cell death focuses mechanical forces during 3D patterning in a biofilm. Proc Natl Acad Sci U S A 109(46):18891–18896

Balaban NQ, Merrin J, Chait R, Kowalik L, Leibler S (2004) Bacterial persistence as a phenotypic switch. Science 305(5690):1622–1625

Berry RM, Armitage JP (2008) How bacteria change gear. Science 320:1599–1600

Blair KM, Turner L, Winkelman JT, Berg HC, Kearns DB (2008) A molecular clutch disables flagella in the *Bacillus subtilis* biofilm. Science 320:1636–1638

Boles BR, Thoendel M, Singh PK (2004) Self-generated diversity produces "insurance effects" in biofilm communities. Proc Natl Acad Sci U S A 101(47):16630–16635

Boutte CC, Crosson S (2013) Bacterial lifestyle shapes stringent response activation. Trends Microbiol 21(4):174–180

Cairns LS, Hobley L, Stanley-Wall N (2014) Biofilm formation by *Bacillus subtilis*: new insights into regulatory strategies and assembly mechanisms. Mol Microbiol 3(4):587–598

Celiker H, Gore J (2012) Competition between species can stabilize public-goods cooperation within a species. Mol Syst Biol 8:621

Chai Y, Chu F, Kolter R, Losick R (2008) Bistability and biofilm formation in *Bacillus subtilis*. Mol Microbiol 67:254–263

Chai Y, Kolter R, Losick R (2010) Reversal of an epigenetic switch governing cell chaining in *Bacillus subtilis* by protein instability. Mol Microbiol 78:218–229

Costerton JW (2007) The Biofilm Primer. Springer, Berlin/Heidelberg

Dandekar AA, Chugani S, Greenberg EP (2012) Bacterial quorum sensing and metabolic incentives to cooperate. Science 338:264–266

Datta MS, Korolev KS, Cvijovic I, Dudley C, Gore J (2013) Range expansion promotes cooperation in an experimental microbial metapopulation. Proc Natl Acad Sci U S A 110(18):7354–7359

Dawson LF, Valiente E, Faulds-Pain A, Donahue EH, Wren BW (2012) Characterisation of *Clostridium difficile* biofilm formation, a role for Spo0A. PLoS ONE 7(12):e50527

Day T (2016) Interpreting phenotypic antibiotic tolerance and persister cells as evolution via epigenetic inheritance. Mol Ecol 25(8):1869–1882

de la Fuente-Núñez C, Reffuveille F, Fernández L, Hancock REW (2013) Bacterial biofilm development as a multicellular adaptation: antibiotic resistance and new therapeutic strategies. Curr Opin Microbiol 16:580–589

Diggle SP, Griffin AS, Campbell GS, West SA (2007) Cooperation and conflict in quorum-sensing bacterial populations. Nature 450:411–414

Dunny GM, Brickman TJ, Dworkin M (2008) Multicellular behavior in bacteria: communication, cooperation, competition and cheating. BioEssays 30:296–298

Ehrlich GD, Hu FZ, Shen K, Stoodley P, Post JC (2005) Bacterial plurality as a general mechanism driving persistence in chronic infections. Clin Orthop Relat Res 437:20–24

Ehrlich GD, Ahmed A, Earl J, Hiller NL, Costerton JW, Stoodley P, Post JC, DeMeo P, Hu FZ (2010) The distributed genome hypothesis as a rubric for understanding evolution in situ during chronic bacterial biofilm infectious processes. FEMS Immunol Med Microbiol 59(3):269–279

García-Contreras R, Nuñez-López L, Jasso-Chávez R, Kwan BW, Belmont JA, Rangel-Vega A, Maeda T, Wood TK (2015) Quorum sensing enhancement of the stress response promotes resistance to quorum quenching and prevents social cheating. ISME J 9:115–125

Gerdes K, Maisonneuve E (2015) Remarkable functional convergence: alarmone ppGpp mediates persistence by activating type I and II toxin-antitoxins. Mol Cell 59(1):1–3

Gerdes K, Semsey S (2016) Pumping persisters. Nature 534:41–42

Goo E, An JH, Kang Y, Hwang I (2015) Control of bacterial metabolism by quorum sensing. Trends Microbiol 23(9):567–576

Gordon V (2017) Coupling and sharing when life is hard. Science 356(6338):583–584

Gore J, Youk H, van Oudenaarden A (2009) Snowdrift game dynamics and facultative cheating in yeast. Nature 459(7244):253–256

Guttenplan SB, Kearns DB (2013) Regulation of flagellar motility during biofilm formation. FEMS Microbiol Rev 37:849–871

Hawver LA, Jung SA, Ng W-L (2016) Specificity and complexity in bacterial quorum-sensing systems. FEMS Mircobiol Rev 40:738–752

Helaine S, Cheverton AM, Watson KG, Faure LM, Matthews SA, Holden DW (2014) Internalization of *Salmonella* by macrophages induces formation of nonreplicating persisters. Science 343(6167):204–208

Hobley L, Kim SH, Maezato Y, Wyllie S, Fairlamb AH, Stanley-Wall NR, Michael AJ (2014) Norspermidine is not a self-produced trigger for biofilm disassembly. Cell 156:844–854

Hooshangi S, Bentley WE (2008) From unicellular properties to multicellular behavior: bacteria quorum sensing circuitry and applications. Curr Opin Biotechnol 19:550–555

Katzianer DS, Wang H, Carey RM, Zhu J (2015) "Quorum non-sensing": social cheating and deception in *Vibrio cholerae*. Appl Environ Microbiol 81(11):3856–3862

Kearns DB (2008) Division of labour during *Bacillus subtilis* biofilm formation. Mol Microbiol 67(2):229–231

Kobayashi K (2008) SlrR/SlrA controls the initiation of biofilm formation in *Bacillus subtilis*. Mol Microbiol 69(6):1399–1410

Kolodkin-Gal I, Romero D, Cao S, Clardy J, Kolter R, Losick R (2010) D-amino acids trigger biofilm disassembly. Science 238:627–629

Kolodkin-Gal I, Cao S, Chai L, Böttcher T, Kolter R, Clardy J, Losick R (2012) A self-produced trigger for biofilm disassembly that targets exopolysaccharide. Cell 149:684–692. RETRACTED

Kreft J-U (2004) Biofilms promote altruism. Microbiology 150:2751–2760

Leiman SA, May JM, Lebar MD, Kahne D, Kolter R, Losick R (2013) D-amino acids indirectly inhibit biofilm formation in *Bacillus subtilis* by interfering with protein synthesis. J Bacteriol 195:5391–5395

Lewis K (2001) Riddle of biofilm resistance. Antimicrob Agents Chemother 45(4):999–1007

Liao S, Klein MI, Heim KP, Fan Y, Bitoun JP, Ahn SJ, Burne RA, Koo H, Brady LJ, Wen ZT (2014) *Streptococcus mutans* extracellular DNA is upregulated during growth in biofilms, actively released via membrane vesicles, and influenced by components of the protein secretion machinery. J Bacteriol 196(13):2355–2366

Liu J, Martinez-Corral R, Prindle A, Lee DD, Larkin J, Gabalda-Sagarra M, Garcia-Ojalvo J, Süel GM (2017) Coupling between distant biofilms and emergence of nutrient time-sharing. Science 356(6338):638–642

Maisonneuve E, Shakespeare LJ, Jørgensen MG, Gerdes K (2011) Bacterial persistence by RNA endonucleases. Proc Natl Acad Sci U S A 108(32):13206–13211

Maisonneuve E, Castro-Camargo M, Gerdes K (2013) (p)ppGpp controls bacterial persistence by stochastic induction of toxin-antitoxin activity. Cell 154(5):1140–1150

Ng W-L, Bassler BL (2009) Bacterial quorum-sensing network architectures. Annu Rev Genet 43:197–222

Popat R, Crusz SA, Messina M, Williams P, West SA, Diggle SP (2012) Quorum-sensing and cheating in bacterial biofilms. Proc R Soc B 279(1748):4765–4761

Popat R, Cornforth DM, McNally L, Brown SP (2015) Collective sensing and collective responses in quorum-sensing bacteria. J R Soc Interface 12:20140882

Pu Y, Zhao Z, Li Y, Zou J, Ma Q, Zhao Y, Ke Y, Zhu Y, Chen H, Baker MA, Ge H, Sun Y, Xie XS, Bai F (2016) Enhanced efflux activity facilitates drug tolerance in dormant bacterial cells. Mol Cell 62(2):284–294

Raut N, Pasini P, Daunert S (2013) Deciphering bacterial universal language by detecting the quorum sensing signal, autoinducer-2, with a whole-cell sensing system. Anal Chem 85(20):9604–9609

Rutherford ST, Bassler BL (2012) Bacterial quorum sensing: its role in virulence and possibilities for its control. Cold Spring Harb Perspect Med 2:a012427

Sanchez A, Gore J (2013) Feedback between population and evolutionary dynamics determines the fate of social microbial populations. PLoS Biol 11(4):e1001547

Sanchez Z, Tani A, Kimbara K (2013) Extensive reduction of cell viability and enhanced matrix production in *Pseudomonas aeruginosa* PAO1 flow biofilms treated with a D-amino acid mixture. Appl Environ Microbiol 79(4):1396–1399

Sandoz KM, Mitzimberg SM, Schuster M (2007) Social cheating in *Pseudomonas aeruginosa* quorum sensing. Proc Natl Acad Sci U S A 104:15876–15881

van Gestel J, Vlamakis H, Kolter R. Division of labor in biofilms: the ecology of cell differentiation. Microbiol Spectrum 3(2): MB-0002-2014, 2015

Verstraeten N, Knapen WJ, Kint CI, Liebens V, Van den Bergh B, Dewachter L, Michiels JE, Fu Q, David CC, Fierro AC, Marchal K, Beirlant J, Versées W, Hofkens J, Jansen M, Fauvart M, Michiels J (2015) Obg and membrane depolarization are part of a microbial bet-hedging strategy that leads to antibiotic tolerance. Mol Cell 59(1):9–21

Vlamakis H, Chai Y, Beuaregard P, Losick R, Kolter R (2013) Sticking together: building a biofilm the *Bacillus subtilis* way. Nat Rev Microbiol 11:157–168

Wang M, Schaefer AL, Dandekar AA, Greenberg EP (2015) Quorum sensing and policing of *Pseudomonas aeruginosa* social cheaters. Proc Natl Acad Sci U S A 112(7):2187–2191

Watnick P, Kolter R (2000) Biofilm, city of microbes. J Bacteriol 182:2675–2679

Whitchurch CB, Tolker-Nielsen T, Ragas PC, Mattick JS (2002) Extracellular DNA required for bacterial biofilm formation. Science 295:1487

Wilking JN, Zaburdaev V, De Volder M, Loscik R, Brenner MP, Weitz DA (2013) Liquid transport facilitated by channels in *Bacillus subtilis* biofilms. Proc Natl Acad Sci U S A 110(3):848–852

Xavier JB, Foster KR (2007) Cooperation and conflict in microbial biofilms. Proc Natl Acad Sci U S A 104:876–881

Xie H, Cook GS, Costerton JW, Bruce G, Rose TM, Lamont RJ (2000) Intergeneric communication in dental plaque biofilms. J Bacteriol 182(24):7067–7069

Yu C, Li X, Zhang N, Wen D, Liu C, Li Q (2016) Inhibition of biofilm formation by D-tyrosine: effect of bacterial type and D-tyrosine concentration. Water Res 92:173–179

Zhang F, Kwan A, Xu A, Süel GM (2015) A synthetic quorum sensing system reveals a potential private benefit for public good production in a biofilm. PLoS ONE 10(7):e0132948

Chapter 10
Multicellularity: *Dictyostelium*

> *A good artist understands painting and agrees with Caravaggio about everything.*
>
> Caravaggio

THE SLIME mold *Dictyostelium discoideum* and its relatives (*D. mucuroides*, *D. purpureum*, and members of the genus *Polysphondylium*) provide a mesmerizing example of differentiation, aggregation, and the coalescence of an ensemble of individuals into a new organism.[1] As recounted in Richard Kessin's comprehensive study of *Dictyostelium* biology, *D. mucuroides* was first observed by Oskar Brefeld in the 1860s in horse dung. He named the organism after its net-like (dicty) aggregation patterns and its tower-like (stelium) stalk. Brefeld initially thought that the amoebae gave up their identity as they aggregated, forming a syncytial mass. Studies by Philippe van Tieghem a decade later showed that the amoebae did not fuse, and Brefeld followed up quickly with a detailed paper in 1884 in which he characterized the aggregation process in detail.

Early studies of the *Dictyostelium* life cycle were hampered by the fact that phagocytosis was unknown at the time of Brefeld and van Tieghem's original studies (Kessin 2001, p. 10). The fact that *Dictyostelium* could engulf and then internally digest bacteria was not even considered. It was not even known that the amoebae needed bacteria as a food source, and no one suspected that the trigger for *Dictyostelium* aggregation was starvation in the absence of bacteria.

Early studies used dung as a culture medium. This proved problematic, since it was often contaminated with bacteria. This contamination, however, led G. A. Nadson to realize that *Dictyostelium* growth was actually enhanced in the presence of bacteria, and he suggested in 1899 that they had a symbiotic relationship. With more careful use of culture media in the early 1900s, G. Potts showed that bacteria appeared to play the role of a food source, and that provision with ample bacteria kept the amoebae in a vegetative state. However, he did not suggest that *Dicty*

---

[1] Other species form multicellular collectives that sporulate, such as the bacterium *Myxococcus xanthus*, and fuse, such as the much more complex colonial protochordate *Botryllus schlosseri*, also known as a star ascidian or a golden star tunicate. These organisms also exhibit individuals that "cheat" within the collective, as we will discuss in the case of *Dictyostelium* below (Kessin 2001).

© Springer Science+Business Media B.V. 2018
S. Bahar, *The Essential Tension*, The Frontiers Collection,
DOI 10.1007/978-94-024-1054-9_10

digested the bacteria intracellularly. E. W. Olive observed ingestion of bacteria by *Dicty* in 1902, but held that *Dicty* did not derive any nutritional benefit from the engulfment (Fig. 10.1). The following year, Paul Vuillemin suggested that amoebae digested the bacterial after engulfment, and described the amoebae as "bacterio-phages" (in the sense of "bacteria-eaters", rather than the more technical use of the term bacteriophage, which today refers to viruses that infect bacteria). It was not until several decades later that R. A. Harper made some of the next major advances in the study of *Dicty* and related organisms, publishing studies which focused on the maintenance of proportionality of cell populations in the growing aggregate, and on the light-sensitivity of the slug. In his 1932 study of "Organization and Light Relations in *Polysphondylium*", a relative of *Dicty* that produces multiple sori budding from a single stalk, Harper also highlighted an important aspect of slime mold biology, emphasizing that in these species "the processes of growth and cell multiplication are separated from those of morphogenesis and differentiation" (Harper 1932).[2]

The species *Dictyostelium discoideum*, now the best known and most well studied of the *Dictyostelids* (and often affectionately known as just "*Dicty*") was discovered by Kenneth Raper in the 1930s among decaying forest leaves in North Carolina (Raper 1935). Raper developed a method for growing *D. discoideum* on a lawn of bacteria rather than on dung, and used grafting experiments to identify the fate map of cells in a developing *Dictyostelium* aggregate. He then "stained" certain populations by growing them on *Serratia marcesens*, a type of bacteria with red pigment that is not degraded during digestion. Having previously shown that portions of two pseudoplasmodia could be grafted together, Raper wrote that it then occurred to him

> ...that by grafting the anterior fraction of a red pseudoplasmodium,[3] consisting so to speak of ear-marked myxamoebae, upon the body of a colorless pseudoplasmodium previously decapitated, additional information might be gained regarding the movement and organization of the migrating pseudoplasmodium and the formation of the fruiting structure. Apical fractions were removed from colorless pseudoplasmodia...produced in culture with *Escherichia coli* upon lactose-peptone, dextrose-peptone, or carrot-peptone agar; and in their stead were immediately placed comparable fractions of red pseudoplasmodia...produced in cultures with *Serratia marcescens*, upon buffered carrot-peptone agar... As in earlier experiments with uncolored pseudoplasmodia, wherever a foreign anterior fraction was placed in close contact with a decapitated body without being crushed or otherwise mutilated the two fractions promptly merged. (Raper 1940)

Using this grafted configuration (which he described as "somewhat simulating in character the joining of a red and a white brick wall"), as well as a configuration in which the posterior portion, rather than the anterior, was stained red, Raper carefully followed the subsequent development of the pseudoplasmodium as it proceeded to culmination (Fig. 10.2). He was surprised to observe that the

---

[2] Harper refers to *Polysphondylium* as a plant. A more recent view of the evolutionary history of *Polysphondylium* and *Dictyostelium* suggests that these species may share a more recent common ancestor with animals and fungi; see Kessin (2001), Chap. 3.

[3] So-called because of the slug's similarity in appearance to Plasmodia such as the malaria parasite.

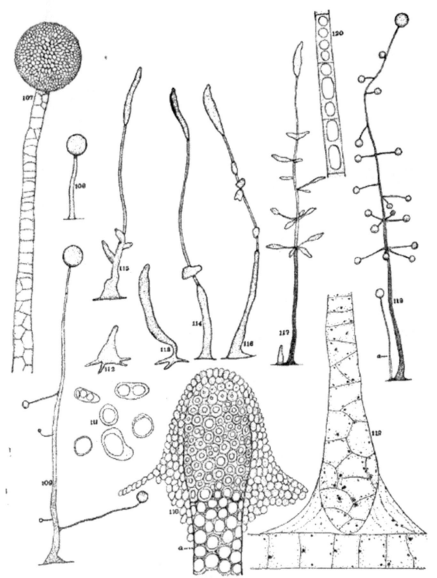

Fig. 10.1 A plate from Edgar W. Olive's 1902 *Monograph of the Acrasiae*, showing various species of *Dictyostelium* in various stages of development. (Harvard Library Open Metadata licensed under CC0 1.0)

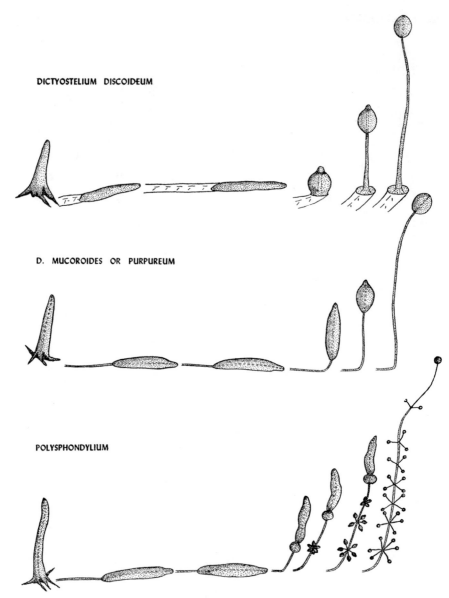

DICTYOSTELIUM DISCOIDEUM

D. MUCOROIDES OR PURPUREUM

POLYSPHONDYLIUM

Fig. 1. The Migration and Culmination Stages in Three Different Types of Cellular Slime Molds

Fig. 10.2 Illustration from a study by John Tyler Bonner (1957) shows the stages of development in various species of slime mold. Similar images can be found in the classic works by Kenneth Raper. Reproduced from Bonner (1957) with the permission of the University of Chicago Press

pseudoplasmodium essentially turned itself inside out, a process reminiscent of gastrulation in embryological development. The anterior (apical) portion of the slug (Fig. 10.3) slowed down its forward movement and became the stalk, while the posterior portion continued moving, crawled up the stalk, and became the sorus (the

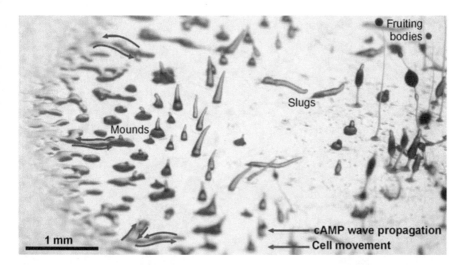

Fig. 10.3 Images of *Dictyostelium* growing on an agar plate beautifully illustrate the concept of "inferring history from a series". At the *left*, cells are in the vegetative state and feed on bacteria. Further to the *right*, at the edge of the "feeding front", cells begin to aggregate and develop into mounds, slugs, and fruiting bodies. *Blue arrows* show the direction of propagation of the cAMP wave; *red arrows* show the direction of cell movement. From Dormann and Weijer, 2006. Figure © 2006 European Molecular Biology Organization

bulblike portion of the fruiting body containing spores, Fig. 10.4). Here is his description of the process (in the case with the anterior portion stained red):

> Marking the onset of sorocarp formation, the red anterior portion gradually ceased forward movement, became raised above the agar surface, and pointed upward. Meanwhile, the myxamoebae comprising the main body of the pseudoplasmodium continued to move forward, crowding around and beneath the colored anterior part. This came to occupy an axial position in the bulbous mass of colorless myxamoebae, projecting conspicuously above it a rounded, nipple-like, apical tip and continuing downward into the center of the mass. The first evidence of sorocarp formation now appeared as the myxamoebae comprising the lower portion of this column became vacuolated and compacted together forming the stalk initial. Subsequently the stalk was built upward by the progressive vacuolation of red colored myxamoebae in this axial region. The basal disk meanwhile was formed by the similar vacuolation of colorless myxamoebae surrounding the base of the stalk and in contact with the substratum. As the stalk lengthened the main body of colorless myxamoebae ascended it *en masse*, ... while becoming differentiated into spores progressively from the periphery of the mass toward its center. When all of the myxamoebae had become transformed either into stalk cells or spores the fruiting structure was complete. The red coloration of the stalk confirmed its origin from the anterior portion of the migrating pseudoplasmodium, while the absence of color in the sorus and in the basal disk indicated their origin from the posterior portion... (Raper 1940)

Another crucial finding in Raper's 1940 paper was that the anterior portion of the slug exerted control over the developmental process. The grafting of multiple tips onto a single slug of *D. discoideum*, for example, resulted in breakup into a number

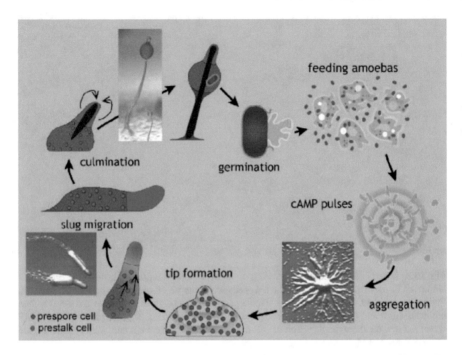

Fig. 10.4 *Dictyostelium* life cycle, showing the fate of prespore (*red*) and prestalk (*blue*) cells. From Schaap, BioEssays 29.7, 635-644, 2007. © 2007 Wiley Periodicals

of individual slugs.[4] Severing a slug produced a front end that generated defective fruiting bodies, while the posterior portion still managed to develop normally.[5] The role of the tip has been compared to that of the Mangold/Spemann organizer in embryonic development.

In the years since the *Dictyostelium discoideum* genome was sequenced (Eichinger et al. 2005; see also Loomis 2006), significant progress has been made in unraveling the various gene regulatory pathways that are activated in the tip and the role of these gene products in modulating the activity of other cells in the aggregate. The genomes of other related species have also been sequenced as well, shedding more light on the evolutionary history of the entire lineage.

Long before its genome was sequenced, however, a fundamental step forward in understanding *D. discoideum* was taken by John Tyler Bonner in experiments he described in his paper "A Theory of the Control of Differentiation in the Cellular Slime Molds", which appeared in the *Quarterly Review of Biology* in 1957. Here is Bonner's introduction to the life cycle of *D. discoideum*.

> The spores of *D. discoideum* are covered with a hardwalled capsule. Upon germination each capsule liberates one amoeba. The amoebae divide mitotically and remain entirely

---

[4] Some other species do form multiple tips naturally.

[5] This is now known to be due to a reserve population of "prestalk" cells that migrate forward and form a new tip (Kessin 2001, p. 15).

separate from one another, feeding independently upon bacteria. When they multiply to a population of sufficient numbers, they stream together to form large collections of many cells or pseudoplasmodia, and this aggregation process appears to be largely due to a chemical substance, acrasin, to which the amoebae are chemotactically sensitive. Certain of the amoebae, which form the center of the aggregate, apparently produce it sooner than others, and in this way an acrasin gradient is set up which is effective in orienting the amoebae. The aggregated cell mass assumes a sausage shape and crawls about the substratum for variable periods of time. During this migration phase it is sensitive to light and heat gradients, orienting itself toward light and toward warmer regions. Differentiation begins at this stage. The anterior cells of the sausage are destined to become part of the supporting stalk and the posterior cells will turn into spores. The final fruiting involves a series of morphogenetic movements in which the anterior presumptive stalk cells are pushed down through the spore mass, and in so doing, these stalk cells become large and vacuolated and are permanently trapped in a delicately tapering cellulose cylinder. During this culmination stage, the spore mass is lifted up into the air and each amoeba in the mass becomes encapsulated into a spore. (Bonner 1957)

Other members of the genus differentiate a similar way, though with some differences in tip formation; the genus *Polysphondylium* forms multiple fruiting bodies radiating from a single stalk (see Figs. 10.1 and 10.2).

In developing a model for *D. discoideum*, Bonner identified three fundamental questions. First, since the anterior end of the "sausage" produced the stalk cells and the posterior end the spore cells, *what are the differences between the front and hind ends of the mass?* Secondly, what is the basis for the well-defined ratio between stalk cells and spore cells that remained constant for all *Dicty* cell aggregates, large or small? This constant ratio could be measured as the slope of a linear proportionality between the log of the volume of cells destined to form the sorus and the log of the volume of cells destined to form the stalk (Fig. 10.5). In other words, Bonner asked, *how, in any one species, is it possible that the differentiation remains proportionate irrespective of size?* Lastly, again in Bonner's words, *how can one account for differences in proportionality ratios in different species and strains?*

To attack the first question, Bonner and colleagues performed a range of microscopic and histological studies to identify differences between anterior and posterior cells. They observed that the anterior cells were typically larger than the posterior ones, and that, after the cell mass had been migrating for some time, the nuclei of the anterior cells also became enlarged. Anterior cells secreted non-starch polysaccharides, which ultimately surrounded them, forming a cellulose cylinder that enclosed the stalk. In contrast, posterior cells appeared to collect polysaccharide into small granules. Toward the end of the migration phase, anterior cells stained for the presence of high concentrations of alkaline phosphatase. Gregg et al. (1954) had shown that the anterior cells lost a significant amount of nitrogen during the culmination phase; the nitrogen was presumably used as a source of energy to fuel the synthesis of the polysaccharide stalk. The energetic activity of the anterior cells, however, left them "mostly trapped and dead inside the stalk cylinder". Bonner wrote that, to all appearances,

> ... the stalk cells have thrown themselves into their morphogenetic activities with such energy that they have lost the ability to perpetuate themselves, and... instead they become depleted, exhausted, inert bricks that fill up the inside of the stalk. The presumptive spores,

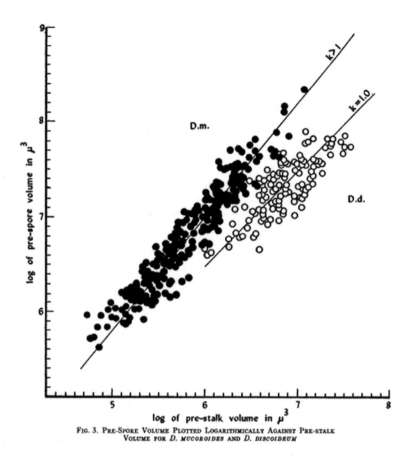

Fig. 3. Pre-Spore Volume Plotted Logarithmically Against Pre-stalk
Volume for *D. mucoroides* and *D. discoideum*

Fig. 10.5 Logarithmic plots of pre-spore vs. pre-stalk volumes in two different species (*D. mucoroides* and *D. discoideum*). Reproduced from Bonner (1957) with the permission of the University of Chicago Press

on the other hand, appear to give off nothing; they neither show a high enzymatic activity (viz., of alkaline phosphatase), nor are they wasteful of their polysaccharides. They would seem to be conserving all for the next generation and they make little or no contribution to our immediate understanding of morphogenesis. *When one considers the activity of these two cell types, it is indeed a striking division of labor.* (Bonner 1957, my italics)

The life cycle of *Dicty*, therefore, is *a living laboratory for the evolution of the division of labor.* Related species undergo generally similar differentiation patterns, though with different sizes of stalk regions (*D. mucoroides* and *D. purpureum*) or with a prolonged period of even staining for metabolic markers followed by a sudden period of differentiation (*Polysphondylium*). The "sacrifice" of the stalk cells should be compared to the reversible differentiation process described for biofilms in the previous chapter, and raises important (and far from answered) questions about the factors which push some species into life-cycles based on terminal differentiation, while others retain the ability to reverse the differentiation process in all cells.

Turning to his second question, Bonner noted that the proportionality found between sorus and stalk volumes was maintained between the volume of cells which would *ultimately* find themselves in the sorus ("presumptive sorus") and the volume of cells which would *ultimately* end up in the stalk ("presumptive stalk"). In other words, the ratio was maintained over the course of development. The pseudoplasmodium's cell volume and stalk volume could be plotted as a coordinate on an allometric plot with constant slope (Fig. 10.5). As the life cycle progressed, the coordinate would move along the line during just as a star moves along the main sequence. But a pseudoplasmodium's measurements would remain on the curve belonging to its own species.

Bonner and others confirmed Raper's observation that cutting migrating pseudoplasmodia produced truncated portions that developed as normal fruiting bodies, complete with spores. This occurred for pieces of the posterior region, that would eventually have developed into the sorus, and also for pieces of the anterior region that would, undisturbed, have "sacrificed" themselves in a somatic role (though the differentiation into a full fruiting body took longer for pieces cut from the anterior region). These experiments led Bonner to conclude that a cell's fate – as suggested by Raper's *Serratia marcescens* experiments – was determined by its position within the pseudoplasmodium. Importantly, this meant that cells must be able to *sense their location within the pseudoplasmodium*.

> To account for proportionality and regulation [i.e., the maintenance of proportionality throughout the life cycle] it must be assumed that one part of the cell mass 'knows' the extent of another. When a pre-stalk tip is isolated, the information that all the posterior prespore cells are missing is somehow registered, so that in a matter of a few hours the deficit is made up and regulation has taken place. A logical conclusion is that there must be some communication between parts. *Let us therefore postulate a polar movement of some key factor which provides the necessary information.* (Bonner 1957, my italics)

To investigate the nature of this "polar messenger", Bonner stained the posterior half of a migrating cell mass with Nile blue sulfate or neutral red, and grafted it onto an unstained anterior portion. Observing these *Dictyostelium* chimeras under the microscope, Bonner observed that cells from the stained posterior portion would migrate into the unstained anterior. Since the entire cell mass was migrating during this time, he concluded so that the stained cells that made it into the anterior portion had a significantly greater velocity than their fellows. These cells were observed to move "as a band", ruling out simple diffusion as an explanation for their behavior. In experiments with a stained anterior half, dyed cells moved "backward" into the posterior region at the same speed that stained cells had moved "forward" in the previous experiment. In other words, the anterior cells were moving at a significantly slower velocity. Bonner concluded that

> there is a polar movement of substances which provides a communication between parts by delivering … essential substances to the anterior end, where there are reactions responsible for the morphogenesis of the slime mold. Furthermore, it is suggested that this polar movement is achieved by especially fast moving cells. (Bonner 1957)

Bonner speculated on the evolutionary origins of the division of labor in *Dictyostelium*. What advantage was there to living as a slime mold rather than an individual amoeba? "This step," he wrote,

> is in many ways the most difficult because we are forced to make the gratuitous assumption that they [slime molds] are adaptively superior; yet there is little or no evidence to show that this is the case. Free living amoebae which are incapable of forming cysts, others which are, and the communal slime molds all live side by side and must have done so for millions of years. (Bonner 1957)

A possible advantage might lie, he considered, in separation between the feeding and reproductive stages. Since

> by lifting the mass away from the region of feeding, a process which is abetted by heat and light tropisms, the spores are in a more favorable position for dispersal. The cellular slime molds seem to have a more effective method of spreading than free-living amoebae. (Bonner 1957)

Bonner did not raise this point, but it is worth noting that spore formation also forces the slime mold's reproductive cycle to pass through a single-cell "bottleneck", a type of reproductive cycle which has a number of advantages, at least for the incipient "organism", as we will explore further below.

Bonner sketched out a hypothetical evolutionary history for the slime molds. Suppose amoeba first had some degree of sensitivity to "substances given off by their own kind", which ultimately could be used as the "polar messenger". Suppose then that the groups of amoebae formed a "pedestal to isolate the spores from the substratum". He noted that some species of solitary amoebae did this, forming single stalked cells. In 1956 Raper had discovered a new genus of slime molds, *Acytostelium,* which formed undifferentiated cell masses and secreted empty cellulose cylinders as anchors. "The point", Bonner wrote, "is that all the cells produce stalk material and then subsequently they all produce spores; there is no division of labor" at this stage. The next step could be driven by simple variability of cell traits, leading to "sorting out of fast, perhaps high-energy, cells, and slow, low-energy cells. With this differential ability to move would come a differential ability to produce or contribute in the stalk or the spore direction." A gradient could lead to entirely different conditions at either end of the cell mass, resulting in full (but still reversible) differentiation. A final step would be a transition from a gradient of cell behaviors to a full division of labor. In Bonner's words,

> in these proposed evolutionary steps the most interesting advance is the idea that the cell variability within an organism may increase, but always necessarily in a continuous fashion. Then, since the variation involves a polar activity (i.e., polar movement), there is an opportunity for reactions to exist at one end of the cell mass that are absent in the other. Therefore, the continuous variation turns into a discontinuous one, and the result is a division of labor, a differentiation. (Bonner 1957)

Dictyostelium clearly straddles the line between single cellular and multicellular life forms. From this standpoint, and because it seems to provide an example of altruistic behavior in the "self-sacrifice" of the stalk cells, it is an endlessly fascinating object of study. *Dictyostelium* also provides a model organism for the study of the evolution of the distinction between self and non-self. For example, Kessin provides a fascinating overview of the parasitic behavior of the aptly-named species *D. caveatum*, which parasitizes other Dicty species, inhibiting their growth and ingesting their cells in a process euphemistically called "nibbling". There exists a cannibalistic mutant of *D. caveatum* that nibbles itself to death; the role of the mutant gene has not yet been identified (Kessin 2001, pp. 33–35).

An adherent of the selfish gene model might assume that the "altruism" of the stalk cells is a result of kin selection. However, as Kessin points out, this interpretation is belied by the fact wild populations have been found to exhibit significant genetic diversity. Amoebae of different *species* do not aggregate; however, the amoebae of a single species that come together to form a slug do not *have* to be clonal. Many slugs are found to be chimeras of several different clones, making the aggregate closer to a colony of social insects of different castes than to a multicellular organism formed from a single zygote (Fortunato et al. 2003a, b). Indeed, if there were not rich genetic diversity within populations, we would "be at a loss to explain the source of the variation that allowed the evolution of a fruiting body from an earlier form" (Kessin 2001, p. 30).

Not only is the phylogenetic tree of *Dictyostelium* quite diverse, but many species exhibit a range of possible life cycles, which may themselves map to different stages in evolutionary development. In addition to the formation of fruiting bodies, there are two other possible responses to starvation: the formation of *microcysts*, or the formation of *macrocysts*. (*D. discoideum* can only form the latter; we will return to this below.)

Microcysts are formed by species such as *P. pallidum* and other species less closely related to *D. discoideum*. Like fruiting bodies, microcyst formation is triggered by starvation, as well as by changes in the ammonia concentration in the soil and environmental changes in osmolarity (Kessin 2001, p. 25). Amoebae individually wrap themselves in a double cellulose layer and become dormant as spores until external conditions improve. "Encystment," as Kessin explains,

> is a solitary form of development and involves no chemotaxis, no multicellularity, and no cell-type proportioning. At least superficially, the encystment process resembles that of many free-living soil amoebae…yet it is part of the developmental repertoire of these simple organisms and the evolutionary innovations of the microcyst, such as cellulose synthesis, are maintained in the other cycles. (Kessin 2001, p. 25)

Macrocyst formation is a multi-cell process. Like fruiting bodies, macrocysts produce spores surrounded by a triple layer of cellulose. As Kessin notes, the commonality of cellulose synthesis has important implications for mapping the evolutionary lineage of *Dicytostelium* and its relatives, since "cellulose biosynthesis and extrusion is a complicated process and it is not likely that it evolved twice" (Kessin 2001, p. 25).

Macrocyst formation likely evolved after microcyst formation, but before the ability to form slugs and fruiting bodies, since macrocysts and slugs use similar chemotactic mechanisms to communicate during aggregation. The formation of a macrocyst is initiated by two cells that belong to different mating types; macrocyst formation is the only sexual reproductive cycle available to *Dictyostelium* and its relatives. When the mating types, called NC4 and V12 cells, encounter each other under conditions of starvation and low ionic strength, they fuse together to form a giant cell, and begin to attract other amoebae.[6] As other amoebae join the aggregate, they are engulfed and digested by the initial pair. The fact that macrocysts digest amoebae as they join a growing aggregate, while developing slugs do not, implies that some form of self-recognition has been activated in the latter case.

In the macrocyst, hundreds of cells are devoured to feed the giant cell. In contrast, when the slug becomes a fruiting body, only 20% of the cells "sacrifice" themselves to form the stalk. For individual cells, then, it is selectively advantageous to participate in the formation of a fruiting body rather than a macrocyst. Both life cycles, however, confer a selective advantage in providing protection from predators that feed on isolated amoebae.

While a cell is more likely to pass on its genes in a fruiting body than in a macrocyst, *within* a fruiting body an amoeba clearly has an individual advantage if it becomes a spore rather than a stalk cell. Yet a population of spore-forming cells alone would be "impaired in their ability to make a fruiting body, and in the wild there would be selection against them because...mechanisms of dispersal depend on a normal fruiting body" (Kessin 2001, p. 31). *Dictyostelium* thus provides a fascinating experimental laboratory within which to study the delicate balance between individual and group selection. Instead of playing prisoner's dilemma games, one can actually investigate the genes whose activation, suppression and mutation control a cell's fate (Fig. 10.6).

Several *Dictyostelium* "cheater" mutants have been identified. These variants preferentially form spores when grown among a population of normal cells. Their impairment when grown clonally, however, is starkly obvious. A cell type identified by Filosa in 1962 formed slugs but no fruiting bodies when grown clonally. Another cell type, studied by Leo Buss in the early 1980s, only formed spores, essentially reverting to the microcyst behavior. Kessin and his colleagues identified a specific mutant strain, *chtA/fbxA*, which forms a very long slug but does not develop into a fruiting body (Ennis et al. 2000). They found that the mutant gene responsible for this behavior codes for a protein, FbxA, involved in proteolysis; the mutation is hypothesized to allow its carrier to "ignore" signals from other cells triggering it to follow the developmental pathway to becoming a stalk cell (Kessin 2001, p. 32). The specific protein targeted by the mutant FbxA protein was later identified as RegA, a cAMP phosphodiesterase (Shaulsky and Kessin 2007).

There are many ways to become a cheater (Fig. 10.7). Rather than ignoring signals telling it to become a stalk cell, for example, an amoeba could, as a result of a "more developed chemotactic mechanism or better motility reach a zone from which spores are more likely to arise" (Kessin 2001, p. 32). Investigation of the particular mutations that drive such cheating may illuminate how inter-amoeba conflict is suppressed.

---

[6] If NC4 and V12 cells encounter one another during the aggregation stage of fruiting body formation, the developmental cycle can be diverted into macrocyst formation.

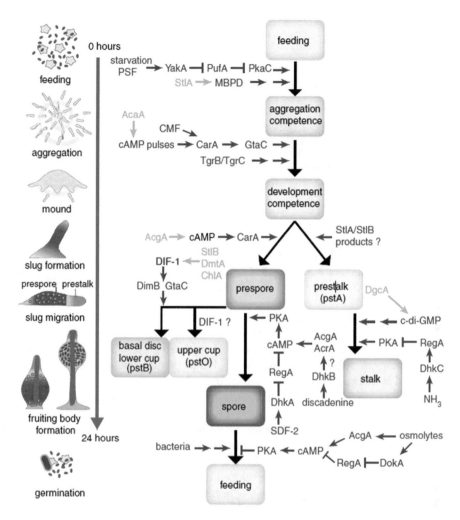

Fig. 10.6 Schematic diagram showing genes activated during *D. discoideum* development. Reproduced from Du et al. (2015) (doi: 10.1016/j.jmb.2015.08.008) under a CC BY 4.0 license (https://creativecommons.org/licenses/by/4.0/)

The *chtA/fbxA* mutant pays a clear fitness cost in that it cannot adequately produce stalk cells when grown clonally. Other cheater mutants, such as the *dimA* strain identified by Strassmann and Queller's group (Foster et al. 2004; Santorelli et al. 2008), and the *csaA* strain identified by Ponte et al. (1998), pay similar fitness costs. Other cheater mutants are successful on their own, not just when mixed with a mutant strain (Strassmann et al. 2000; Kessin 2000). For example, when grown clonally, the *chtB* mutant differentiates successfully, and exhibits a phenotype qualitatively

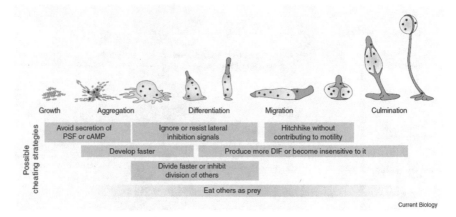

Fig. 10.7 Cheating strategies and the time window during *Dictyostelium* development when they are advantageous. Spore and pre-spore cells are shown in *yellow*, stalk and pre-stalk cells in *blue*, and cheater mutants in *red*. Reprinted from *Curr. Biol.* 17, G. Shaulsky and R.H. Kessin, The cold war of the social amoebae, pp. R684-R692, Copyright (2007), with permission from Elsevier

indistinguishable from its parental strain[7] (Santorelli et al. 2013). When grown as a chimera with the parental strain, however, *chtB* preferentially forms spores, and suppresses the formation of spores by the wild type cells. The *chtB* gene occurs on *Dictyostelium discoideum's* chromosome 5, but its precise function when normally expressed is not yet known. Since the *chtB* mutant can differentiate when grown clonally, it is a *facultative cheater*: its cheating behavior depends on the situation in which it is placed.

Kessin suggested that the evolution of the fruiting-body life cycle may have been facilitated by the genetic diversity provided by the sexual reproduction of the macrocyst cycle (Kessin 2001, p. 28). He and others have proposed an evolutionary sequence in which amoebae first developed the capacity to become microcysts (a stage involving the evolution of the biochemical pathways of cellulose synthesis), and then, after developing chemotaxis, became able to form macrocysts. The addition of cell adhesion mechanisms, differentiation of cell types, and maintenance of proportionality provided a genetic background against which the ability to form fruiting bodies could evolve. Some species, like *D. discoideum*, subsequently abandoned the microcyst lifestyle, but retained key biochemical pathways innovated during the evolution of this stage, such as cellulose synthesis.

In 2006, Schaap and colleagues published a molecular phylogeny of over one hundred *Dictyostelid* species. They concluded that the species could be divided into four groups, the last (and most recently diverged) of which has lost its ability to form microcysts (Fig. 10.8). All members of this fourth group, including

---

[7] Santorelli et al. (2013) note that stalk length in the two populations was not quantitatively compared, so that it remains possible that *chtB* cells might be at some disadvantage with respect to their clonally grown parental population.

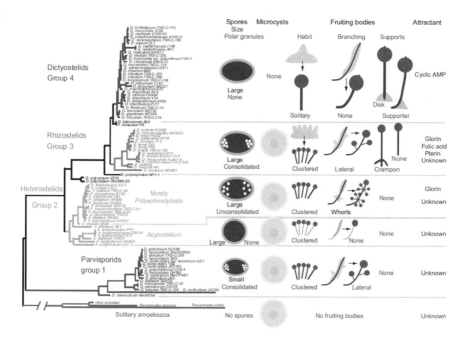

Fig. 10.8 Molecular phylogeny for the *Dictyostelia* and related species. Reproduced with permission from Schaap (2011b)

D. discoideum, use cAMP as a chemoattractant (Konijn et al. 1969), while those in the other groups use glorin, folic acid, pterin or other chemical signals. All of these signals fall under the descriptor "acrasin" in the sense of Bonner's early experiments. For example, folic acid is the acrasin of *D. minutum* (De Wit and Konijn 1983), glorin[8] is the acrasin for *Polysphondylium violaceum* (Shimomura et al. 1982), and pterin is the acrasin for *Dictyostelium lacteum* (Van Haastert et al. 1982).

The use of cAMP as an externally-secreted chemoattractant in *D. discoideum* and its close relatives is quite unusual, since most other organisms use cAMP only *within intracellular* signal transduction pathways (Schaap 2011a). The cAMP signaling strategy in *D. discoideum* is hypothesized to have evolved from a stress response. All *Dictyostelids* use some form of cAMP signaling. The cAMP receptor protein, cAR1, is conserved throughout the phylogeny (Schaap 2011a). However, there are crucial differences between the role of cAMP in groups 1–3 and in the most recently evolved group 4. In the first three groups, cAR1 is only expressed after the individual amoebae aggregate. However, in group 4, cAR1 is expressed before aggregation. Experiments performed in Schaap's laboratory showed that disruption of oscillatory cAMP signaling prevented the formation of both slugs and fruiting bodies in group 4 species, but only blocked fruiting-body formation in groups 1–3 (Alvarez-Curto et al. 2005). Louis et al. (1993) showed that there are

---

[8] N-propionyl-γ-L-glutamyl-L-ornithine-δ-lactam ethyl ester.

two entirely different promoter mechanisms that control expression of the gene coding for cAR1; the mechanisms are activated at entirely different developmental
stages. These results are consistent with the hypothesis that the use of cAMP as a
signal for aggregation evolved *after* cAMP was already in use as a signal for differentiation and morphogenesis.

The role of cAMP in controlling differentiation and morphogenesis can be
revealed by disrupting the expression of the genes that code for cAMP receptors.
Such an experiment in *P. pallidum*, for example, leads to fruiting bodies that make
microcyst-like structures rather than spores. As Schaap points out, encystation and
sporulation involve the interaction of cAMP and phosphokinase A (PKA), but, for
sporulation to occur, cAMP must also interact with cAR1 or other receptor proteins.
When the genes coding for these receptors are mutated or their correct expression is
otherwise prevented, the sporulation process cannot occur. As Schaap explains,

> these results suggest a possible scenario for the evolution of cAMP signaling in *Dictyostelia*
> that starts with cAMP functioning as an intracellular signal that transduces the perception
> of environmental stress into an encystation response in the solitary ancestor... [A]t least
> one species (*D. minutum*) uses the same attractant (folic acid) for food-seeking as it does for
> aggregation.[9] The first colonial amoebas might, therefore, have adapted their food-seeking
> strategy for aggregation, while still using cAMP intracellularly to trigger encystation.
> (Schaap 2011a)

Alvarez-Curto et al. (2005) showed that genes for cAMP receptors already exist
in group 1–3 species such as *D. minutum*, and are used in the formation of fruiting
bodies, though not in the process of aggregation (which in *D. minutum* is driven by
the acrasin folic acid). They suggested that the addition of new regulatory regions to
existing cAMP signaling genes enabled cAMP to be exploited as a chemoattractant
in the group 4 species. In a striking experiment, they expressed a cAR from *D.
minutum* in a mutant strain of *D. discoideum* that lacks cARs with high affinity for
cAMP. When the *D. minutum* protein was expressed, the mutant *D. discoideum* was
able to aggregate virtually as effectively as the wild type.

∗∗∗ ∗∗∗ ∗∗∗

Aggregation is triggered by starvation. The process is mediated through a complex set of biochemical pathways. For example, individual *D. discoideum* amoebae
secrete an autocrine glycoprotein known as PSF ("prestarvation factor") as they
feed on bacteria (Clarke et al. 1992).

> When the ratio of PSF relative to that of bacteria exceeds a certain threshold, cells stop
> proliferating and initiate the expression of genes that are required for their aggregation... A
> second protein, CMF (conditioned medium factor), is secreted during starvation. CMF
> stimulates gene expression in parallel with PSF, and both signals potentiate cAMP signal-
> ling by inducing genes involved in cAMP synthesis and detection. (Schaap 2011a)

---

[9] *D. discoideum* has been observed to do the same: a 1969 paper by Konijn, Bonner and others
showed that cAMP produced by *E. coli* could attract *D. discoideum*.

Once aggregation has begun, the population faces the complex task of negotiating which cells will become stalk and which will become spore. What determines the fate of a cell just joining the aggregate? As even the earliest studies by Raper and others had shown, location of a cell within the aggregate is key to determining its ultimate fate. But what biochemical pathways mediate this process?

As summarized by Jang and Gomer (2011), two prevailing views have dominated the field. According to one hypothesis, the gradient of a differentiation inducing factor (DIF, a lipid-like chlorinated alkyl phenone, often referred to as DIF-1, after its principal species) triggers anterior cells to become prestalk cells and posterior cells to become prespore cells. Town and Stanford (1979) prepared DIF from high-density cell populations, and showed that it could stimulate isolated amoebae to differentiate into stalk cells. Kopachik et al. (1983) showed that mutant *Dictyostelium* strains, impaired in their ability to produce DIF, were also unable to produce stalk cells; when DIF was added to the medium, their ability to produce stalk cells was restored. Kay and Jermyn found that DIF not only promoted formation of stalk cells, but inhibited the formation of spores, and proposed that

> [p]erhaps the simplest known morphogenetic field arises within the multicellular aggregate formed by developing cells of the slime mould *Dictyostelium discoideum*…in *Dictyostelium* it is *almost essential that morphogens should dictate to cells their choice of a differentiation pathway*. We have previously described a crude factor termed DIF which stimulates the differentiation of isolated amoebae into stalk cells. We now show that purified DIF also inhibits spore formation and so switches cells to stalk cell formation. Thus, we believe that DIF is a morphogen which regulates the choice of differentiation pathway of cells in the *Dictyostelium* slug. (Kay and Jermyn 1983, my italics)

Brookman, Jermyn and Kay set out to confirm this hypothesis by measuring the actual DIF-1 gradient throughout the slug.

> Migrating slugs from strain V12M2 were manually dissected into anterior one-third and posterior two-third fragments and the DIF activity extracted. Surprisingly, we found that DIF was not restricted to the prestalk fragment. Instead there appears to be a reverse gradient of DIF in the slug with at least twice the specific activity of total DIF in the prespore region than in the prestalk region. (Brookman et al. 1987)

A mechanistic explanation for this unexpected result was provided when Kay and Thompson (2001) showed that prespore cells actually *secrete* DIF, but seem to be immune to its effects. Prespore cells can thus consign their anterior hapless neighbors to a stalky fate.

A form of DIF appears to be necessary for the formation of the basal disk, which anchors the fruiting body to the soil and is composed of a subpopulation of prestalk cells. Other genes that are markers of the prestalk pathway, such as *ecmA* and *ecmB*, also require the presence of DIF in order to be expressed. However, other prestalk markers are independent of DIF-1, suggesting that "DIF-1 is required only for the differentiation of a subset of prestalk cells and that it is not a master control morphogen for stalk cell differentiation" (Jang and Gomer 2011).

The second major cell fate hypothesis was inspired by a series of experiments performed by Yasuo Maeda and colleagues. Their studies suggested that an amoeba's fate was determined by where it was in its cell cycle at the onset of starvation. In these experiments, cells were synchronized so that most of the cells in the population were at the same phase of the cell cycle. The cells were then labeled with dyes and mixed with unlabeled unsynchronized cells and allowed to develop. Cells in S and early G_2 phase at the time of starvation sort out to the anterior regions of developing *Dictyostelium* slugs and become predominantly prestalk cells, whereas cells in late G_2 phase at the time of starvation sort out to posterior regions and become predominantly prespore cells. (Jang and Gomer 2011)

These results were confirmed by video-microscopy experiments by Gomer and Firtel (1987), who postulated a "mechanism [that] could conceivably involve a substance that is synthesized at one phase of the cell cycle" (Jang and Gomer 2011). Gomer and Firtel also reported that up to 41% of cells in a slug remained in a null state for an extended period, uncommitted to either developmental pathway even while their neighbors were becoming prestalk or prespore cells; these cells were sometimes found to differentiate into prespore cells when they came into contact with other cells. Moreover, when one cell in a just-divided pair became a prespore or prestalk cell, its sister became a null cell (Gomer and Firtel 1987; Jang and Gomer 2011).

Other factors have been identified that may affect cell fate. For example, altering intracellular pH or calcium concentration can shift cells from a prestalk to a prespore fate, or vice versa. These results align well with the cell-cycle hypothesis because amoebae in the M phase, S, phase and early G_2 phase tend to exhibit low pH and low internal calcium concentration, while amoebae in the later G_2 phase exhibit higher pH and internal calcium (Jang and Gomer 2011).

Studies of mutant *Dictyostelium* strains have strengthened the link between the cell cycle and cell fate. In 1996, Gomer's laboratory isolated a gene, *rtoA*, whose mutant form disrupts the cell cycle in *D. discoideum* and produces slugs with an atypically high percentage of prestalk cells. They also found that, in *rtoA* mutants, cells at any point in the cell cycle gave rise, randomly and without preference, to prestalk and prespore cells (Wood et al. 1996). This contrasts starkly with the preferential differentiation of wild type cells at different points in the cell cycle, observed by Maeda and others. A later study showed that *rtoA* plays a role in vesicle fusion, and linked mutation in *rtoA* to disruption in pH levels during the cell cycle (Brazill et al. 2000). These results suggest

> that during evolution, an ancestor of *Dictyostelium* may have originally used a stochastic mechanism, possibly based on stochastic variations in cytosolic pH, to choose the initial cell type and that RtoA evolved to connect this mechanism to the cell cycle, so that the larger cells with more nutrient reserves (cells in late G_2) would become prespore and the smaller cells with fewer nutrient reserves (cells that had just emerged from a cell division) would tend to be come prestalk. (Jang and Gomer 2011)

This idea is echoed in more recent studies of other factors that may influence cell fate choice. A recent study suggests that lineage bias may be determined by a cell's nutritional history (Chattwood et al. 2013; Morgani and Brickman 2013). Using immunostaining and live cell imaging, Goury-Sistla et al. (2012) showed that the speed distribution of *D. discoideum* amoebae in starvation medium was bimodal, with 20% of the cells moving significantly faster than the others. This population showed changes in mobility-related molecules such as F-actin, had higher internal calcium, and appeared to coincide with the prestalk cells. The bimodality of speed distributions was reversible, disappearing when the cells were returned to a nutrient-rich medium, and was not observed at all in a mutant (triA⁻) with unstable prespore and prestalk populations.

Jaiswal et al. (2003) showed that the first cells to starve are the ones that initiate the aggregation process. Using distributions of pre-starved and newly starved clones of *D. discoideum* strains, Kuzdzal-Fick and colleagues showed that – contrary to expectation – early starvers are more likely to follow the prespore pathway.

> The first cells to starve have lower energy reserves than those that starve later, and previous studies have shown that the better-fed cells in a mix tend to form disproportionately more reproductive spores. Therefore, one might expect that the first cells to starve and initiate the social stage should act altruistically and form disproportionately more of the sterile stalk, thereby enticing other better-fed cells into joining the aggregate. This would resemble caste determination in social insects, where altruistic workers are typically fed less than repro-ductive queens. However, we show that the opposite result holds: the first cells to starve become reproductive spores, presumably by gearing up for competition and outcompeting late starvers to become prespore first. These findings pose the interesting question of why others would join selfish organizers. (Kuzdzal-Fick et al. 2010)

The authors noted that, while this contradicts the typical caste determination mecha-nism in social insects, there are species of paper wasps in which older and smaller workers ascend to queenship. Indeed, Kuzdzal-Fick et al. observed that "early starvers cheat by forming more than their fair share of the spores, while forcing late starvers to produce disproportionately more sterile stalk cells." Early starvers are likely produc-ing DIF, inducing the late starvers to become prestalk cells. But why haven't cells evolved a resistance to DIF, in order to avoid the genetic cul-de-sac of stalkhood? Kuzdzal-Fick et al. suggest that individual cells make a Pascal's wager on their future chances of passing on their genes, operating "under a veil of ignorance, having little information on how many surrounding cells precede or follow them in starvation and their likelihood of becoming a spore" (Kuzdzal-Fick et al. 2010). How such a bet-hedging strategy might interact with the timing of starvation onset, with the cell cycle, or with a cell's ability to produce or respond to DIF, is not yet understood.

The idea of a "veil of ignorance" favoring cooperation has been proposed as a more general principle by Queller and Strassmann (2013), who suggest that ignorance of payoff may act to suppress meiotic drive. They note that in meiosis in maize, tetrads of cells are formed; one of the end cells will become the egg. Certain chromosomes with extra spindle attachment sites jockey for position in one of the end cells, but seek out either end with equal probability, suggesting that they have no means of identifying which end cell will become the egg. As another example of ignorance of possible payoffs, they cite the example of ant colonies founded by small groups of unrelated queens, only one of which will survive. While they sug-gest that ignorance "may be an overlooked device supporting cooperation", Queller and Strassmann do note that in some situations ignorance can be detrimental to cooperation: cheaters can't be punished if they can't be identified.

*** *** ***

The formation of multicellular aggregates in *Dictyostelium* provides a rich and endlessly fascinating example of collective behavior. But there is a crucial aspect of this behavior that we have neglected until now: they dynamics of aggregation itself. In fact, individual amoebae stream toward each other to form a slug in an example of collective swarming far more complicated than Reynolds's boids model.

The complexity of the aggregation process begins with the dynamics of cAMP signaling, which provides a classic example of nonlinear biochemical oscillations. Albert Goldbeter and colleagues pioneered the development of nonlinear dynamical models for cAMP oscillations in *Dictyostelium*. Many of these models, and the biochemistry behind them, are described in Goldbeter's classic book *Biochemical Oscillations and Cellular Rhythms: the Molecular Bases of Periodic and Chaotic Behavior* (Goldbeter 1996).

The first observations of oscillating behavior in *Dictyostelium* aggregation were made by Gerisch and colleagues in the 1960s, and by Durston and others in the 1970s (Goldbeter 1996, p. 165). Time-lapse films of *Dictyostelium* aggregation, using dark-field and phase contrast microscopy, showed concentric rings – and sometimes spirals – of alternating light and dark bands in the zone where amoebae were aggregating. Using conventional microscopy to investigate the individual amoebae in different bands, Alcantara and Monk (1974) found that the "light bands consisted of elongated moving cells, while the darker interband areas contained rounded cells exhibiting randomly-oriented pseudopodia". Shortly before these studies were performed, cAMP had been identified as the "acrasin" of *D. discoideum* (Konijn et al. 1968). The natural conclusion was that pulsatile signals of cAMP might be driving the formation of wave fronts of moving and motionless amoebae (Fig. 10.9). As Alcantara and Monk put it,

> [t]he concentric wave pattern may easily be understood if each band of moving cells corresponds to the zone of influence, at a given time, of a wave of stimulation propagating outward form the centre of the aggregation field. The darker interband areas would then contain cells which have ceased to move in response to a signal that has just passed them and which have not yet been stimulated by the next signal. (Alcantara and Monk 1974)

The role of cAMP in driving this concentric wave pattern was confirmed when Tomchik and Devreotes (1981) imaged aggregating *D. discoideum* using a fluorescent antibody specific for cAMP. Oscillations in cAMP had previously been observed (Goldbeter 1996, p. 168), but Tomchik and Devreotes showed that the cAMP oscillations overlaid the pattern of cell morphology. High concentrations of cAMP were present in the dark bands of fast-moving cells, and cAMP concentration dropped significantly in the lighter bands.

The beautiful concentric waves and spiral patterns observed in the aggregating amoebae and in the cAMP concentration caught the attention of scientists working at the interface between physics and biology. The wave patterns were strikingly similar to spiral waves in the oscillating Belousov-Zhabotinsky chemical reaction (Winfree 1972, 1980) and to spiral waves observed in the heart during cardiac arrhythmias. These comparisons suggested that the amoebae were behaving like an excitable medium (Durston 1973).

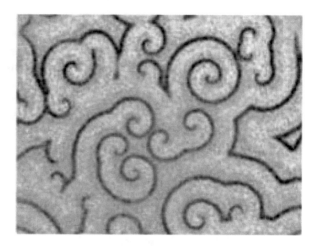

Excitable media result when a spatially extended group of coupled elements –
such as cells in the heart and brain, or localized chemical reactions in the Belousov-
Zhabotinsky reaction – exhibit nonlinear oscillations. For a physicist, this is easy to
visualize as a network of weakly coupled springs or pendulums. For a biologist, a
classical example of a nonlinear oscillator is an excitable cell like a neuron. When
the cell's transmembrane potential passes a threshold, voltage-gated sodium channels open, allowing influx of sodium. This is followed shortly after by the opening
of voltage-gated potassium channels, through which potassium ions flow into the
extracellular space. The channels then close, and cannot open again for a refractory
period, during which the channels relax back to their original configuration. These
concerted openings and closings cause a sharp spike, known as the action potential,
in the transmembrane voltage. The signal spreads within the axon, but not very far:
the axon is an extremely poor conductor. But the spread of the signal is just enough
to push nearby channels above their threshold voltage, triggering them to open.
Signals spread from one neuron to another via chemical synapses, and thus a group
of neurons can act as an excitable medium.

In a 1952 work lauded by generations of neuroscientists (and neurodynamicists),
Alan Hodgkin and Andrew Huxley (grandson of Thomas Huxley and half-brother
of Julian and Aldous Huxley) developed a set of nonlinear differential equations
describing the dynamics of voltage oscillations across the axonal membrane. Denis
Noble later extended the Hodgkin-Huxley equations to describe cardiac action
potentials, adding a term for the flow of calcium ions into the cell, which gives the
cardiac action potential a long plateau (and a duration of hundreds of milliseconds,
in contrast to the 1–2 ms duration of a neural action potential). Cardiac cells are
connected by gap junctions, which allow direct electrical communication between
the cells, and cause signals to spread faster through cardiac tissue than between
neurons. In the excitable medium of cardiac muscle, spiral waves have a clear pathological significance: they result when normal electrical flow through the atria or
ventricles is impeded, and can lead to potentially fatal arrhythmias. The study of
spiral wave breakup in cardiac tissue has been a highly active area of research for

decades, and detailed experimental images of spiral waves have been obtained using grids of recording electrodes and voltage-sensitive fluorescent dyes.

Together, the evidence that aggregating *Dictyostelium* amoebae behaved as an excitable medium and the co-localized cAMP oscillations observed by Tomchik and Devreotes suggested that cAMP signaling might play a causal role in driving the oscillatory, excitatory behavior of the aggregating cells. This was confirmed in Günther Gerisch's laboratory, using *D. discoideum* grown in liquid suspension rather than on a flat medium. Gerisch and Hess (1974) recorded oscillations in light scattering – reflective of changes in cell morphology – in these suspensions after the amoebae had been subjected to starvation conditions. They also showed that adding cAMP to the medium triggered phase shifts in the oscillatory period, with the clear implication that cAMP plays a causal role in driving the oscillations.[10] Testing periodic samples of both cells and supernatant for cAMP using a binding assay, Gerisch and Wick (1975) confirmed the existence oscillations in both intracellular and extracellular cAMP. The oscillation period (5–10 min) correlated with the oscillations in light scattering, and hence with the changes in cell morphology.

What were the mechanisms underlying these cAMP oscillations, and mediating their role in aggregation? Roos and Gerisch (1976) showed that adenylate cyclase, an enzyme that catalyzes the production of cAMP from ATP, was activated by the binding of extracellular cAMP to a receptor in the cell membrane, as illustrated schematically in Fig. 10.10. This means that

> [s]ynthesis of cAMP in *D. discoideum* therefore is a self-amplifying process...the more cAMP accumulates in the external medium, the more it binds to the receptor and activates the intracellular production of cAMP; transport of the latter into the extracellular medium results in a positive feedback loop that renders the whole process autocatalytic. (Goldbeter 1996, p. 176)

This process is analogous to the flow of sodium ions into a neuron pushing the membrane potential higher and higher above the threshold potential. In a form of positive feedback characteristic of nonlinear systems, in a single amoeba, the cAMP signal is amplified by positive feedback as the amoeba self-stimulates with its own secreted cAMP. The extracellular cAMP it produces, however, can also stimulate neighboring cells; it is in this manner that the ensemble of cells becomes mutually coupled, forming an excitable medium.

Goldbeter and Segel (1977) proposed a dynamical model for the cAMP oscillations. The model consisted of three coupled nonlinear differential equations, describing the time-variation of intracellular ATP, intracellular cAMP, and extracellular cAMP. The model incorporated the allosteric binding of extracellular cAMP to its receptor, cAMP transport from extracellular to intracellular space, and hydrolysis of extracellular cAMP by phosphodiesterase. The model generated cAMP oscillations that matched experimental observations with great accuracy.

---

[10] Similar dynamical phenomena are observed in other excitable biological systems. For example, action potentials in cardiac cells can be advanced by electrical stimulation in the so-called "early beat" phenomenon (Guevara et al. 1986). The electrical stimulus plays a role analogous to the addition of cAMP to the *Dictyostelium* suspension.

Fig. 10.10 cAMP
regulation in *D.
discoideum*, as modeled in
the simpler early versions
of Goldbeter's models

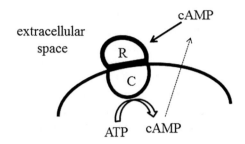

intracellular space

Despite its success, there remained a major discrepancy between the growing body of experimental evidence and the predictions of the Goldbeter-Segel model. The problem lay in the term describing the downstroke in the intracellular cAMP concentration. In the model, this was driven by a decline in the concentration of the substrate (ATP), as well as by the transport of cAMP from the intracellular to the extracellular space.[11] The downstroke term is critically important for the system dynamics, since it put the brakes on the autocatalytic, positive feedback process by which cAMP stimulates its own synthesis.[12] Yet experiments showed no appreciable drop in intracellular ATP concentration! The value remained constant around a value of 1.2 mM (Goldbeter 1996, p. 189). Might there be two "compartments" of ATP, one near the cell membrane, which became depleted, and another intracellular compartment that remained comparatively untouched?

Further experimental studies suggested that the intracellular cAMP downstroke oscillations were not due to substrate depletion, but to cAMP receptor desensitization. As the receptor's response to cAMP is diminished, the activation of adenylate cyclase drops off. Indeed, it was shown in the 1980s that repeated stimulation of the receptor by cAMP led to its covalent modification: it became phosphorylated, and in that condition was unable to stimulate adenylate cyclase activation at all. Once adenylate cyclase ceases making cAMP, the extracellular cAMP concentration (exported from the intracellular space) begins to decrease. This allows the receptor a chance to recover, become dephosphorylated, and ready itself once again to respond to extracellular cAMP. This produces the downstroke in intracellular cAMP. The time required for the receptor to dephosphorylate determines the system's refractory period.

In order to address these new experimental results, Martiel and Goldbeter (1987) proposed a new model, now containing 11 variables, that incorporated the phosphorylation of the receptor, cAMP binding to the receptor in its phosphorylated and dephosphorylated states, the rate of transition between the dephosphorylated

---

[11] The downstroke in the concomitant oscillations of extracellular cAMP was controlled by a term representing the activity of phosphodiesterase.

[12] In neurons, the positive feedback process of increasing voltage is halted by the closure of sodium channels at a sufficiently high transmembrane potential, and the opening of potassium channels at approximately the same voltage.

and phosphorylated states, and so on. By making various assumptions, such as a constant intracellular ATP concentration, they reduced the model to a more tractable three variables. A later version of the model, proposed in the 1990s, incorporated evidence that the activation of adenylate cyclase was mediated by G-proteins (Goldbeter 1996, p. 227).

Can these oscillatory models predict the spatial patterns seen in *Dictyostelium* aggregation? As early as 1970, Keller and Segel had proposed a generic slime mold aggregation model in which formation of discrete patches of amoebae are generated by a Turing instability in the diffusion of acrasin as well as other factors such as acrasin sensitivity and production rate. Other early spatiotemporal models, such as that of MacKay (1978), used a partial differential equation model to simulate the interaction and aggregation of 1000 cells generating and responding to acrasin; this model produced aggregates showing some of the streaming and spiral structures observed experimentally. Tyson et al. (1989) and Tyson and Murray (1989) used the Martiel and Goldbeter model, combined with the diffusion of cAMP between the simulated cells, in order to generate spiral wave structures. Also using the Martiel and Goldbeter model, Levine and Reynolds (1991) showed that the concentric target patterns observed experimentally could become unstable, resulting in a switch to the streaming patterns seen in the later stages of aggregation. Even more sophisticated models combined the Martiel-Goldbeter model of receptor desensitization with chemotactic movement of the amoebae themselves; such models also generate spiral waves and streaming patterns (Höfer et al. 1995).

Concentric and spiral wave patterns were also observed in an alternative to the Martiel-Goldbeter desensitization model. Monk and Othmer (1990), using a model in which influx of intracellular calcium, rather than cAMP receptor phosphorylation, caused desensitization,[13] showed wave patterns similar to those observed experimentally. In this case, however, the spiral waves were not generated by a Turing instability. In fact, various types of heterogeneity can be added to *Dictyostelium* models in order to generate spirals. One approach has been to introduce heterogeneous initial conditions such as random distributions of cAMP concentration, though this results more often in target patterns rather than spirals (Goldbeter 1996, p. 233). Halloy, Pontes and Goldbeter suggested that spiral waves could be induced by the intersection of orthogonal gradients of cells with activated receptors. Such a scenario could occur if nearby groups of amoebae entered the aggregation phase at approximately the same time, and their expanding aggregates grew to intersect with one another (Goldbeter 1996, p. 233). Monk and Othmer (1990) suggested that cell movement itself may be important in the formation of spiral waves. Experimental results suggest that aspects of this pattern-forming behavior are under genetic control. Pálsson et al. (1997) identified a mutant *D. dis-*

---

[13] Monk and Othmer questioned the role of receptor phosphorylation on the basis of experiments which, for example, showed that pertussis toxin blocks adaptation to increasing cAMP concentration, but does not affect phosphorylation of cAMP receptors. In recent years the relation of calcium and cAMP signaling in *Dictyostelium* has been shown to be ever more complex (see, for example, Malchow et al. 2004; Lusche et al. 2005).

coideum strain unable to produce spiral waves. This strain exhibited a mutation in the gene for an inhibitor of phosophodiesterase. This protein, secreted early in the starvation phase in normal cells, increases extracellular cAMP concentration by inhibiting its removal by phosphodiesterase.

Waves of activity are not confined to the aggregation process. Using time-lapse films after staining with neutral red dye, Durston and Vork (1979) observed wave-like motions of cells in the slug as it progressed toward culmination. Siegert and Weijer (1992) observed rotational cell movements in the tips of *D. discoideum* cells, and proposed that a twisted scroll wave is involved in organizing the developing slug; similar dynamics were observed in *D. mucoroides* (Dormann et al. 1997). Durston (2013) described spiral scroll waves of cAMP in *D. discoideum* slugs, and suggested that, rather than becoming twisted, the waves undergo a process of "dislocation", breaking into disconnected segments. He suggested that scroll wave dislocation may occur in metazoan embryonic development as well as in *Dictyostelium* morphogenesis.

How do these patterns initiate? Studies have suggested that the earliest cells to starve are the ones to trigger the aggregation response, as discussed above. But is one cell sufficient? And given the right conditions, can any amoeba take on this role? Despite experimental (Durston 1974a) and computational (DeYoung et al. 1988) investigations, the answers remained unclear until recently. A 2010 study by Gregor et al. showed that any amoeba can act as a nucleation site for aggregation (see also Prindle and Hasty 2010). Using fluorescence resonance energy transfer (FRET) to perform live-cell imaging of cAMP signalling, Gregor et al. recorded intracellular cAMP oscillations in starved *D. discoideum* populations. They found that cells exhibit sporadic synchronous pulses beginning about 5 h after nutrient deprivation.

> Over the next 2 h, the period of firing shortened to 8 min and thereafter to 6 min when cells began to aggregate. The entire population participated in the firing of the first pulse... Differences in the phase of the oscillations depending on the regions indicate[d] that pulses propagated in space as waves. Different spatial locations competed for wave initiation... However, a single region eventually dominated and determined the aggregation center. (Gregor et al. 2010)

Gregor et al. investigated the cAMP oscillatory response of groups of amoebae at different densities, and in media of different flow rates. As expected, higher densities led to increased extracellular cAMP production, while higher flow rates decreased the local concentration. For low flow rate and high density, periodic intracellular cAMP oscillations were observed; for lower flow rates, oscillations occurred only sporadically. When Gregor et al. plotted the rate of cAMP oscillations against the ratio of density to flow rate, they observed a sharp increase in the pulse rate at a critical ratio (see Fig. 2C in Gregor et al. 2010), which then leveled off at 0.167 pulses/min, or one pulse every 6 min. Studies performed on individual cells showed an optimal response when extracellular cAMP was applied every 6 min. A faster application of cAMP depressed the response, presumably due to the amoeba's refractory period, during which phosphodiesterase is needed to degrade extracellular cAMP.

Taken together, the results suggested that the oscillation response was a form of quorum sensing, with no individual cell taking the lead. Increased density led to increased extracellular cAMP, which triggered all cells to begin oscillating; at a critical density, when the local concentration of cAMP had built up sufficiently, these oscillations occurred at the "resonant" frequency of one cycle every 6 min. Whichever cell or group of cells reached this frequency first became the aggregation center. These cells could then entrain other nearby cells, stimulating them with extracellular cAMP, recruiting their neighbors into the growing synchronous cluster. "The initiation process," they concluded, "is inherently collective and stochastic."

There are multiple sources of nonlinearity in *Dictyostelium* cAMP dynamics. In addition to the nonlinear threshold response to cAMP concentration, the activation of adenylate cyclase is a far more complex process than assumed by the early models of Goldbeter and colleagues. Several G-proteins are involved, and, in contrast to the simple schematic diagram in Fig. 10.10, multiple cAMP-activated receptors are needed to activate a single adenylate cyclase enzyme. Given this complexity, it is not surprising that complex dynamical patterns such as bursting (a number of rapid oscillations followed by a quiescent interval) or even chaotic behavior can arise in *Dictyostelium* cAMP dynamics. Even a three-variable version of the Martiel-Goldbeter model can produce bursting and chaos. As the phosphodiesterase reaction rate constant is continuously varied, the system undergoes a series of period-adding bifurcations, moving from single spikes to pairs of spikes to up to seven spikes in a burst. For other ranges of the reaction rate constant, the system exhibits bistable behavior, hysteresis, and a classic period-doubling route to chaos. The system could even be nudged from one bistable state to another[14] by the precisely timed addition of extracellular cAMP (Goldbeter 1996, pp. 243 ff).

Could such complex dynamics occur in a natural *Dictyostelium* population? Gottmann and Weijer (1986) observed "doublet or triplet" waves in a mutant *Dictyostelium* strain, consistent with bursting cAMP oscillations produced by the aggregation center (Goldbeter 1996, pp. 262–263). Another mutant strain, *Fr17*, identified by Durston, was also reexamined as a possible source of chaotic dynamics. Durston (1974b) had measured the time interval between cAMP waves in *Fr17* and its parent wild-type strain, NC-4. In the wild type, a histogram of inter-wave intervals showed a well-defined peak, indicating clear rhythmicity. In contrast, the corresponding histograms for *Fr17* were extremely broad, indicating arhythmicity in the pacemaker cells' signaling, or in the relay process conducted by the responding cells, or both. While not definitively indicative of chaotic behavior, this observation intrigued Goldbeter and others (Goldbeter 1996, pp. 262–267).

Kessin (1977) had isolated a temperature-sensitive variant of *Fr17*, called HH201. Studies with these two strains showed that the normal developmental pattern was accelerated by about a factor of two in *Fr17* cells. For example, *Fr17* cells began to aggregate 4–5 h after starvation rather than the normal 7–8 h. When

---

[14] Such transitions can also be induced between bistable 1:1 and 2:1 states in cardiac dynamics (Yehia et al. 1999).

these cells formed fruiting bodies, they were disorganized in structure and often exhibited multiple short stalks with few spore cells (Kessin 1977; Couckell and Chan 1980). Kessin's studies suggested that *Fr17* and its derivative HH201 resulted from a single mutation. Due to the effect on developmental timing and aggregation, he suggested that the mutation related to some derangement of cAMP metabolism. The studies of Couckell and Chan (1980) strongly supported this idea. They found a much higher cAMP concentration in HH201 cells. Furthermore, while normal NC-4 cells responded to external cAMP by increasing their internal cAMP, the mutant cells did not. They also found a much higher rate of production of cAMP in the mutant cells, suggesting an increase in adenylate cyclase activity. Finally, they observed large, irregular oscillations in intracellular cAMP in HH201 cells after only 2 h of starvation, while NC-4 cells barely showed any variation in intracellular cAMP this early in the starvation process. The mutation clearly seemed to be affecting adenylate cyclase, and therefore cAMP metabolism.

Intrigued, Goldbeter and colleagues wondered whether chaotic oscillations might be obtainable in the *Fr17* strain, and, if so, whether these chaotic patterns might be reproduced by adjusting adenylate cyclase-related parameters in a computational model. They had high hopes for identifying "the first example of autonomous chaos at the cellular level", and also for finding an example of a "dynamical disease", a term coined by Mackey and Glass, in a unicellular organism.

> While the transition to chaos in *Dictyostelium* would result from some genetic mutation, the addition of drugs [had] been reported to elicit the transition from periodic to chaotic oscillations in a molluscan neuron... The model for cAMP signalling suggests that the addition of an appropriate amount of exogenous phosphodiesterase should transform chaos into periodic behaviour. This prediction could be tested during aggregation of the mutants *Fr17* or *HH201* on agar, in order to determine whether the broad histogram of intervals between successive waves becomes narrower in the presence of the enzyme... (Goldbeter 1996, pp. 263–264)

Goldbeter and Wurster (1989) investigated oscillations of light scattering in a cell suspension of HH201. To their surprise, they observed oscillations quite similar to those of the wild-type cells. They speculated that there might be a difference in the precise parameter values in their experiments in contrast to those of Durston (1974b) or Couckell and Chan (1980); those experiments were performed with *Dictyostelium* grown on agar plates, while Goldbeter and Wurster studied cell suspensions. Another possibility was that chaotic oscillations occurred in some cells, but were suppressed by the regular oscillations of neighboring amoebae (Li et al. 1992). Thus

> part of the cells within a continuously stirred suspension would oscillate in a chaotic manner if left on their own, while the remaining cells would oscillate periodically in the absence of the chaotic population. The coupling of the two populations within the same suspension could well suppress any manifestation of chaos by conferring upon the coupled system a global, regular behavior. (Goldbeter 1996, p. 267)

Whether such dynamical interactions play a role in a natural biological context, however, remains unclear.

Dictyostelium discoideum and its relatives provide a dramatic example of the transition to multicellularity within a single life cycle, as well as a host of other complex dynamical behaviors. Dictyostelium has a broad range of other behaviors and lifestyles that we have not even touched upon. It has a sexual life cycle (O'Day and Keszei 2012). It also practices a primitive form of agriculture (Brock et al. 2011; Boomsma 2011). Yet even its most complex behaviors may be traced to well-defined underlying causes. Genetic studies suggest that simple changes in regulatory pathways may be responsible for the evolution of cell differentiation in *Dictyostelium*. These pathways are fraught with the tension between competition and cooperation. *Dictyostelium* is particularly fascinating for the reversible nature of its multicellularity. Many other species, however, do not find their way back so easily from multicellularity, as we will explore in the following chapter.

References

Alcantara F, Monk M (1974) Signal propagation during aggregation in the slime mold *Dictyostelium discoideum*. J Gen Microbiol 85:321–334

Alvarez-Curto E, Rozen DE, Ritchie AV, Fouquet C, Baldauf SL, Schaap P (2005) Evolutionary origin of cAMP-based chemoattraction in the social amoebae. Proc Natl Acad Sci U S A 102(18):6385–6390

Bonner JT (1957) A theory of the control of differentiation in the cellular slime molds. Q Rev Biol 32:232–246. (Reprinted in 51: 296–312, 1976)

Boomsma JJ (2011) Farming writ small. Nature 469:308–309

Brazill DT, Caprette DR, Myler HA, Hatton RD, Ammann RR, Lindsey DF, Brock DA, Gomer RH (2000) A protein containing a serine-rich domain with vesicle fusing properties mediates cell cycle-dependent cytosolic pH regulation. J Biol Chem 275(25):19231–19240

Brock DA, Douglas TE, Queller DC, Strassmann JE (2011) Primitive agriculture in a social amoeba. Nature 469:393–396

Brookman JJ, Jermyn KA, Kay RR (1987) Nature and distribution of the morphogen DIF in the *Dictyostelium* slug. Development 100:119–124

Chattwood A, Nagayama K, Bolourani P, Harkin L, Kamjoo M, Weeks G, Thompson CR (2013) Developmental lineage priming in *Dictyostelium* by heterogeneous Ras activation. eLIFE 2:e01067

Clarke M, Dominguez N, Yuen IS, Gomer RH (1992) Growing and starving *Dictyostelium* cells produce distinct density-sensing factors. Dev Biol 152(2):403–406

Couckell MB, Chan FK (1980) The precocious appearance and activation of an adenylate cyclase in a rapid developing mutant of *Dictyostelium discoideum*. FEBS Lett 110(1):39–42

De Wit RJ, Konijn TM (1983) Identification of the acrasin of *Dictyostelium minutum* as a derivative of folic acid. Cell Differ 12(4):205–210

DeYoung G, Monk PB, Othmer HG (1988) Pacemakers in aggregation fields of *Dictyostelium discoideum*: does a single cell suffice? J Math Biol 26:487–517

Dormann D, Weijer C, Siegert F (1997) Twisted scroll waves organize *Dictyostelium mucoroides* slugs. J Cell Sci 110:1831–1837

Du Q, Kawabe Y, Schilde C, Chen Z, Schaap P (2015) The evolution of aggregative multicellularity and cell-cell communication in the Dictyostelia. J Mol Biol 427(23):3722–3733

Durston AJ (1973) *Dictyostelium discoideum* aggregation fields as excitable media. J Theor Biol 42(3):483–504

Durston AJ (1974a) Pacemaker activity during aggregation in *Dictyostelium discoideum*. Dev Biol 37:225–235

Durston AJ (1974b) Pacemaker mutants of *Dictyostelium discoideum*. Dev Biol 38:308–319

Durston AJ (2013) Dislocation is a developmental mechanism in *Dictyostelium* and vertebrates. Proc Natl Acad Sci 110(49):19826–19831

Durston AJ, Vork F (1979) A cinematographical study of the development of vitally stained *Dictyostelium discoideum*. J Cell Sci 36:261–279

Eichinger L et al (2005) The genome of the social amoeba *Dictyostelium discoideum*. Nature 435:43–57

Ennis HL, Dao DN, Pukatzki SU, Kessin RH (2000) *Dictyostelium* amoebae lacking an F-box protein form spores rather than stalk in chimeras with wild type. Proc Natl Acad Sci U S A 97(7):3292–3297

Fortunato A, Queller DC, Strassmann JE (2003a) A linear dominance hierarchy among clones in chimeras of the social amoeba *Dictyostelium discoideum*. J Evol Biol 16:438–445

Fortunato A, Strassmann JE, Santorelli L, Queller DC (2003b) Co-occurrence in nature of different clones of the social amoeba, *Dictyostelium discoideum*. Mol Ecol 12:1031–1038

Foster KR, Shaulsky G, Strassmann JE, Queller DC, Thompson CRL (2004) Pleiotropy as a mechanism to stabilize cooperation. Nature 431:693–696

Gerisch G, Hess B (1974) Cyclic-AMP-controlled oscillations in suspended *Dictyostelium* cells: their relation to morphogenetic cell interactions. Proc Natl Acad Sci U S A 71(5):2118–2122

Gerisch G, Wick U (1975) Intracellular oscillations and release of cyclic AMP from *Dictyostelium* cells. Biochem Biophys Res Commun 65(1):364–370

Goldbeter A (1996) Biochemical Oscillations and Cellular Rhythms: The Molecular Bases of Periodic and Chaotic Behaviour. Cambridge University Press, Cambridge

Goldbeter A, Segel LA (1977) Unified mechanism for relay and oscillation of cyclic AMP in *Dictyostelium discoideum*. Proc Natl Acad Sci U S A 74(4):1543–1547

Goldbeter A, Wurster B (1989) Regular oscillations in suspensions of a putatively chaotic mutant of *Dictyostelium discoideum*. Experientia 45:363–365

Gomer RH, Firtel RA (1987) Cell-autonomous determination of cell-type choice in *Dictyostelium* development by cell cycle phase. Science 237:785–762

Gottmann K, Weijer CJ (1986) In situ measurements of external pH and optical density oscillations in *Dictyostelium discoideum* aggregates. J Cell Biol 102:1623–1629

Goury-Sistla P, Nanjundiah V, Pande G (2012) Bimodal distribution of motility and cell fate in *Dictyostelium discoideum*. Intl J Dev Biol 56:263–272

Gregg JH, Hackney AL, Krivanek JO (1954) Nitrogen metabolism of the slime mold *Dictyostelium discoideum* during growth and morphogenesis. Biol Bull Woods Hole 107:226–235

Gregor T, Fujimoto K, Masaki N, Sawai S (2010) The onset of collective behavior in social amoebae. Science 328:1021–1025

Guevara MR, Shrier A, Glass L (1986) Phase resetting of spontaneously beating embryonic ventricular heart cell aggregates. Am J Phys 251(Heart. Circ. Physiol. 20):H1298–H1305

Harper RA (1932) Organization and light relations in *Polysphondelium*. Bull Torrey Bot Club 59(2):49–84

Höfer T, Sheratt JA, Maini P (1995) *Dicstyostelium discoideum*: cellular-self-organization in an excitable biological medium. Proc R Soc Lond B 259:249–257

Jaiswal JK, Mattoussi H, Mauro JM, Simon SM (2003) Long-term multiple color imaging of live cells using quantum dot bioconjugates. Nat Biotechnol 21(1):47–51

Jang W, Gomer RH (2011) Initial cell type choice in *Dictyostelium*. Eukaryot Cell 10(2):150–155

Kay RR, Jermyn KA (1983) A possible morphogen controlling differentiation in *Dictyostelium*. Nature 303:242–244

Kay RR, Thompson CRL (2001) Cross-induction of cell types in *Dictyostelium*: evidence that DIF-1 is made by prespore cells. Development 128:4959–4966

Keller EF, Segel LA (1970) Initiation of slime mold aggregation viewed as an instability. J Theor Biol 26:399–415

Kessin RH (1977) Mutations causing rapid development of *Dictyostelium discoideum*. Cell 10:703–708

Kessin RH (2000) Cooperation can be dangerous. Nature 408:917–918

Kessin RH (2001) Dictyostelium: Evolution, Cell Biology, and the Development of Multicellularity. Cambridge University Press, Cambridge

Konijn TM, Barkley DS, Chang YY, Bonner JT (1968) Cyclic AMP: a naturally occurring acrasin in the cellular slime molds. Am Nat 102(925):225–233

Konijn TM, Van de Meene JGC, Chang YY, Barkley DS, Bonner JT (1969) Identification of adenosine-3',5'-monophosphate as the bacterial attractant for myxamoebae of *Dictyostelium discoideum*. J Bacteriol 99(2):510–512

Kopachik W, Oohata A, Dhokia B, Brookman JJ, Kay RR (1983) *Dictyostelium* mutants lacking DIF, a putative morphogen. Cell 33:397–403

Kuzdzal-Fick JJ, Queller DC, Strassmann JE (2010) An invitation to die: initiators of sociality in a soeical amoeba become selfish spores. Biol Lett 6:800–802

Levine H, Reynolds W (1991) Streaming instability of aggregating slime mold amoebae. Phys Rev Lett 66(18):2400–2403

Li YX, Halloy J, Martiel JL, Wurster B, Goldbeter A (1992) Suppression of chaos by periodic oscillations in a model for cyclic AMP signalling in *Dictyostelium* cells. Experientia 48:603–606

Loomis WF (2006) The *Dictyostelium* genome. Curr Issues Mol Biol 8(2):63–74

Louis JM, Saxe CL III, Kimmel AR (1993) Two transmembrane signaling mechanisms control expression of the cAMP receptor gene *CAR1* during *Dictyostelium* development. Proc Natl Acad Sci U S A 90:5959–5973

Lusche DF, Bezares-Roder K, Happle K, Schlatterer C (2005) cAMP controls cytosolic Ca^{2+} levels in *Dictyostelium discoideum*. BMC Cell Biol 6:12

MacKay SA (1978) Computer simulation of aggregation in *Dictyostelium discoideum*. J Cell Sci 33:1–16

Malchow D, Lusche DF, Schlatterer C (2004) A link of Ca^{2+} to cAMP oscillations in *Dictyostelium*: the calmodulin antagonist W-7 potentiates cAMP relay and transiently inhibits the acidic Ca^{2+} store. BMC Dev Biol 4:7

Martiel J-L, Goldbeter A (1987) A model based on receptor desensitization for cyclic AMP signaling in *Dictyostelium* cells. Biophys J 52:807–828

Monk PB, Othmer HG (1990) Wave propagation in aggregation fields of the cellular slime mould *Dictyostelium discoideum*. Proc R Soc Lond B 240:555–589

Morgani SM, Brickman JM (2013) Survival of the fattest. eLIFE 2:e01760

O'Day DH, Keszei A (2012) Signalling and sex in the social amoebozoans. Biol Rev 87:313–329

Olive EW (1902) Monograph of the *Acrasiae*. Proc Boston Soc Natl Hist 30(6):451–513

Pálsson E, Lee KJ, Goldstein RE, Franke J, Kessin RH, Cox EC (1997) Selection of spiral waves in the social amoebae *Dictyostelium*. Proc Natl Acad Sci U S A 94:13719–13723

Ponte E, Bracco E, Faix J, Bozzaro S (1998) Detection of subtle phenotypes: the case of the cell adhesion molecule *csaA* in *Dictyostelium*. Proc Natl Acad Sci U S A 95:9360–9635

Prindle A, Hasty J (2010) Stochastic emergence of groupthink. Science 328(5981):987–988

Queller DC, Strassmann JE (2013) The veil of ignorance can favour biological cooperation. Biol Lett 9:2013065

Raper KB (1935) *Dictyostelium discoideum*, a new species of slime mold from decaying forest leaves. J Agric Res 50:135–147

Raper KB (1940) Pseudoplasmodium formation and organization in *Dictyostelium discoideum*. J Elisha Mitchell Sci Soc 56:241–282

Roos W, Gerisch G (1976) Receptor-mediated adenylate cyclase activation in *Dictyostelium discoideum*. FEBS Lett 68(2):170–172

Santorelli LA, Thompson CRL, Villegas E, Svetz J, Dinh PA, Sucgang R, Kuspa A, Strassmann JE, Queller DC, Shaulsky G (2008) Facultative cheater mutants reveal the genetic complexity of cooperation in social amoebae. Nature 451:1107–1110

Santorelli LA, Kuspa A, Shaulsky G, Queller DC, Strassmann JE (2013) A new social gene in *Dictyostelium discoideum*, *chtB*. BMC Evol Biol 13:4

Sawai S, Guan XJ, Kuspa A, Cox EC (2007) High-throughput analysis of spatio-temporal dynamics in Dictyostelium. Genome Biol 8(7):R144

Schaap P (2011a) Evolutionary crossroads in developmental biology: *Dictyostelium discoideum*. Development 138(3):387–396

Schaap P (2011b) Evolution of developmental cyclic adenosine monophosphate signaling in the *Dictyostelia* from an amoebozoan stress response. Develop Growth Differ 53(4):452–462

Shaulsky G, Kessin RH (2007) The cold war of the social amoebae. Curr Biol 17:R684–R692

Shimomura O, Suthers HLB, Bonner JT (1982) Chemical identity of the cellular slime mold *Polysphondylium violaceum*. Proc Natl Acad Sci U S A 79:7376–7379

Siegert F, Weijer CJ (1992) Three-dimensional scroll waves organize *Dictyostelium* slugs. Proc Natl Acad Sci U S A 89:6433–6437

Strassmann JE, Zhu Y, Queller DC (2000) Altruism and social cheating in the social amoeba *Dictyostelium discoideum*. Nature 408:965–967

Tomchik KT, Devreotes PN (1981) Adenosine 3′,5′-monophosphate waves in *Dictyostelium discoideum*: a demonstration by isotope dilution-fluorography. Science 212:443–446

Town C, Stanford E (1979) An oligosaccharide-containing factor that induces cell differentiation in *Dictyostelium discoideum*. Proc Natl Acad Sci U S A 76(1):308–312

Tyson JJ, Murray JD (1989) Cyclic AMP waves during aggregation of *Dictyostelium* amoebae. Development 106:421–426

Tyson JJ, Alexander KA, Manoranjan VS, Murray JD (1989) Spiral waves of cyclic AMP in a model of slime mold aggregation. Physica D 34(1–2):193–207

Van Haastert PJM, De Wit RJW, Grijmpa Y, Konijn TM (1982) Identification of a pterin as the acrasin of the cellular slime mold *Dictyostelium lacteum*. Proc Natl Acad Sci U S A 79:6270–6274

Winfree AT (1972) Spiral waves of chemical activity. Science 175:634–636

Winfree AT (1980) The Geometry of Biological Time. Springer, New York

Wood SA, Ammann RR, Brock DA, Li L, Spann T, Gomer RH (1996) RtoA links initial cell type choice to the cell cycle in *Dictyostelium*. Development 122:3677–3685

Yehia AR, Jeandupeux D, Alonso F, Guevara MR (1999) Hysteresis and bistability in the direct transition from 1:1 to 2:1 rhythm in periodically driven single ventricular cells. Chaos 9(4):916–931

Chapter 11
Multicellularity: *Volvox*

On phone.

Bernard Black

VOLVOX MEANS "fierce roller". These little algae rolled fiercely into human consciousness in an encounter with the Dutch microscopist Anton van Leeuwenhoek. "I saw," he wrote,

> that they were not simply round, but that their outermost membrane was everywhere beset with many little projecting particles...the little bodies...never lay still...their progression was brought about by a rolling motion...Each of these little round bodies had enclosed within it 5, 6, 7, nay, some even 12, very little round globules, in structure like to the body itself wherein they were contained... While I was keeping watch, for a good time, on one of the biggest round bodies...I noticed that in its outermost part an opening appeared, out of which one of the inclosed round globules, having a fine green colour, dropt out, and took on the same motion in the water as the body out of which it came...soon after a second globule, and presently a third, dropt out of it; and so one after another till they were all out, and each took on its proper motion. (Dobell 1960, pp. 257–258; also quoted by Kirk 1998, p. 1)

In the years following Leeuwenhoek's observations (Fig. 11.1), *Volvox* was viewed as a prime example of Bonnet's "emboîtement". But what ultimately came to intrigue the scientific community most was the diverse range of forms taken on by the various species of *Volvox* and its more distant cousins. Some are single cells, while others form colonies of identical cells. Still others, like those observed by Leeuwenhoek, differentiate into germ and somatic cells.

August Weismann keenly recognized the importance of these varied forms in explaining the transition to multicellularity, describing them as "simpler forms in which phyletic transitions are represented" (Kirk 1998, pp. 4–5). In *The Evolution Theory*, Weismann wrote that *Volvox* encapsulated the core question of cell differentiation (Fig. 11.2). "How can a cell by division give rise not only to others like itself, but also to the body-cells, which are of quite different structure? This is, in its simplest form, the fundamental problem of all reproduction through germ-cells" (Weismann 1904, Vol. I, p. 259).

Volvox is a genus within the class *Chlorophyceae*, a type of green algae. *Volvox* belongs to the order *Volvocales* and the family *Volvocaceae*. We will encounter various genera within the *Volvocaceae* in this chapter: *Gonium, Pandorina, Eudorina,*

© Springer Science+Business Media B.V. 2018
S. Bahar, *The Essential Tension*, The Frontiers Collection,
DOI 10.1007/978-94-024-1054-9_11

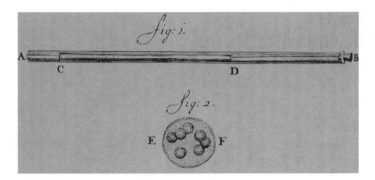

Fig. 11.1   An illustration from Leeuwenhoek's 1700 letter on *Volvox*. *Figure i* shows the glass tube stopped with a cork within which Leeuwenhoek collected water containing *Volvox* (*Figure 2*)

Pleodorina, Volvulina, and *Volvox* itself. *Volvox* (Fig. 11.3) and its cousins are closely related to, and may have descended from, another well-known *Chlorophycean,* the single-celled, bi-flagellated green alga *Chlamydomonas reinhardtii* (Fig. 11.4).

The volvocaceans were initially thought to provide a prime example of Darwinian inference of history through a series. Just as *Dictyostelium* passes from single-celled to multicelled within a single life cycle, the volvocaceans were thought to do so within their evolutionary lineage (Fig. 11.5). As David Kirk writes,

> it would be quite satisfying if one could identify a group that…comprises an extant collection of closely related organisms that range in complexity from unicellular forms through homocytic colonial forms to heterocytic multicellular forms, with different cell types and a complete division of labor. (Kirk 1998, p. 13)

In order to shed light on the evolutionary origins of multicellularity, the divergence of organisms within such a group would need to be

> of such recent origin (in a geological frame of reference) that there is some hope that its various members may still retain within their genomes – unblurred by long eons of genetic drift – traces of the genetic changes that permitted transitions from one level of organizational complexity to the next. (Kirk 1998, p. 13)

Indeed, ribosomal RNA sequence comparisons in the late 1980s suggested that *Volvox carteri,* one of the best-studied *Volvox* species, and *Chlamydomonas reinhardtii* shared a common ancestor only 35 million years ago (Kirk 1998, p. 43). A more recent study from Michod's group pushes the divergence point back significantly further, into the Triassic, about 200 million years ago (Herron et al. 2009; Fig. 11.6). Umen (2014) throws some shade on this result, however, noting that, of all the multiple fossil calibrations used by Herron and colleagues, the only green alga they used, *Proterocladus,* "has produced incongruences in other analyses and may not be part of the taxonomic group it was originally identified with". Nonetheless, if correct, the divergence time proposed by Herron et al. is still far more recent than the appearance of the first filamentous prokaryotes (3.5 billion years ago) and the first multicelled prokaryotes with distinct cell types (2 billion

Fig. 11.2 Illustrations of
Volvox aureus from
Weismann's 1904 *The
Evolution Theory*. From
author's collection

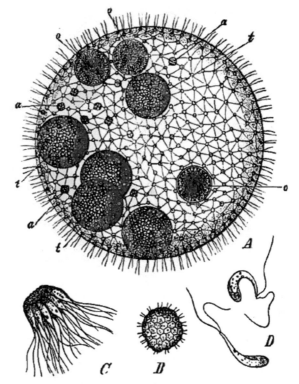

FIG. 63. *Volvox aureus*, after Klein and Schenck.
A, besides the small flagellate somatic cells of the colony
there are five large egg-cells (*t*) which are capable of
parthenogenetic development, three recently fertilized egg-
cells (*o*) and a number of male germ-cells (*a*) in process of
multiplication. From each of these, by continued division,
a bundle of spermatozoa arises. *B*, a bundle of thirty-two
sperm-cells in process of development, seen from above.
C, the same seen from the side. Magnified 687 times.
D, individual spermatozoa, magnified 824 times.

years ago) (Kirk 1998, pp. 8–9), or *Grypania spiralis*, a putative eukaryotic alga
dating back 2.1 billion years (Han and Runnegar 1992).

The initial picture of the development of multicellularity in the *Volvocaceae*,
known as the volvocine lineage hypothesis, envisioned a linear, monophyletic
development. The purported earliest ancestor was a *Chlorophycean* alga like
Chlamydomonas reinhardtii, which typically lives a unicellular lifestyle, propelling
itself with flagella and reproducing asexually. *C. reinhardtii,* however, are able to
resorb their flagella and form a multicellular conglomerate called a palmella (after
a relative, *Palmella*, that typically forms such structures as part of its normal life
cycle). This conglomerate is held together by a mixture of secreted glycoproteins
called mucilage.

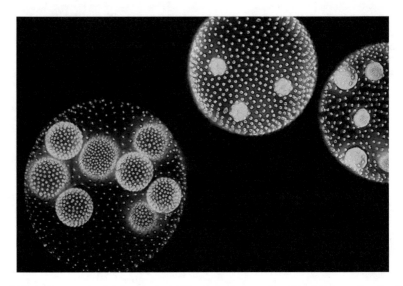

Fig. 11.3 A photograph of Volvox by Frank Fox (http://www.mikro-foto.de). Reproduced under a CC BY-SA 3.0 DE (http://creativecommons.org/licenses/by-sa/3.0/de/deed.en) license via Wikimedia Commons

In order of increasing complexity, the next member of the putative volvocine lineage is the genus *Gonium*. Like all multicellular members of this lineage, *Gonium* cells reproduce by *multiple fission*. After enlarging to 2^n times its original volume, a cell rapidly undergoes n fission events, producing 2^n daughter cells. In unicellular forms, each of these daughter cells detaches and swims off. In *Gonium* and other multicellular species, "the 2^n cells produced by a round of cell division become cemented to one another by extracellular materials before hatching from their mother [cell] wall" (Kirk 1998, p. 25). This sort of cluster is called a coenobium. *Gonium* typically forms very small irregularly shaped coenobia, of perhaps eight cells. In the late 1980s, Batko and Jakubiec identified a new *Gonium* species, *G. dispersum*, which could be considered a "missing link" between single-celled and multicelled forms. This species undergoes multiple fission as do other *Gonium* species, but then some – though not all – newly divided cells disperse to live a free-swimming individual lifestyle reminiscent of *Chlamydomonas* (Kirk 1998, pp. 25–26).

Next up in the hypothetical volvocine lineage is *Pandorina*, which, forms tightly packed spherical colonies of about 16 cells. These cells are so tightly packed that they take on a keystone-like configuration. *Pandorina* colonies show some anterior-posterior polarity, such as a gradient in eyespot size (Kirk 1998, 2005).[1] (Even *Gonium* exhibits some differences between cells as result of their location in the coenobium. In the case of *Gonium*, this presents in the form of differences in the

---

[1] The eyespot gradient becomes more dramatic in other related species. Indeed, the posterior cells in *V. rousseletti* lack eyespots entirely, appearing to be "blind" (Ueki et al. 2010).

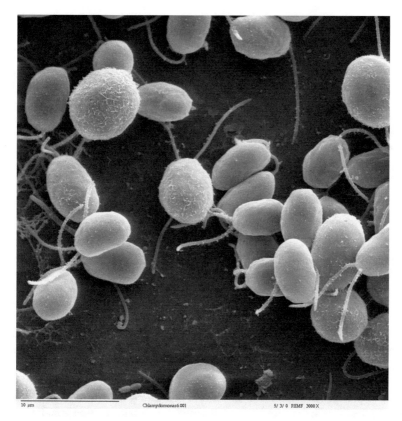

Fig. 11.4 A scanning electron microscope (SEM) image of *Chlamydomonas reinhardtii*. Note their pairs of flagella. *Scale bar* shows 10 μm. Image released to the public domain by the Dartmouth Electron Microscope Facility, Dartmouth College

rotation of the basal bodies. This, however, is a polarity between the center and the periphery of the structure, rather than between the anterior and posterior regions.)

Like *Pandorina*, *Eudorina* is composed of a collective of cells. However, *Eudorina* forms larger colonies than *Pandorina*, and with less tightly packed cells. Surrounded by extracellular matrix, they retain a spherical shape. As Kirk explains, "[i]n some larger colonies, the anteriormost four cells sometimes fail to enlarge; they continue to provide flagellar motility while the other cells enlarge, redifferentiate, and divide" (Kirk 1998, p. 28).

Eudorina is closely related to *Pleodorina*, which *always* has an anterior-posterior cell-size gradient. *Pleodorina*'s larger anterior cells become germ cells, called *gonidia*, and divide, while the smaller cells maintain a somatic role, providing flagellar motility. Motility is essential for volvocaceans, which must execute diurnal movements in still waters. During the day, they stay near the water surface, undergoing photosynthesis; at night, they move deep below the water surface in order to sequester phosphorous (Kirk 1998, p. 51). Here at last, it seems, is a permanent and

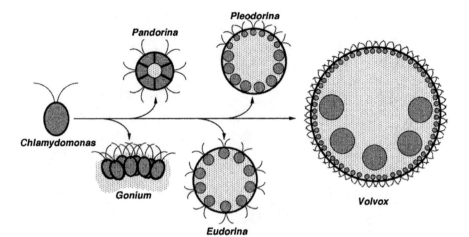

Fig. 11.5 Would it were so simple. The initial hypothesis for the evolution of the volvocine lineage. Reprinted from D.L. Kirk, Seeking the ultimate and proximate causes of volvox multicellularity and cellular differentiation. *Integr. Comp. Biol.* 43(2): 247–253, 2003, by permission of Oxford University Press

full-fledged differentiation. However, in the 1960s, M. E. Goldstein showed that some species of *Eudorina* and *Pleodorina* could interbreed. More recent studies of these hybrids have been carried out by Coleman (2002).

Last in the hypothesized volvocine lineage comes the genus *Volvox* itself. In its asexual reproductive phase, *Volvox* lacks a cell-size gradient. Instead, it has two distinct populations: thousands of small somatic cells and 16 large gonidia.

The elegant logical progression from *Chlamydomonas* through *Gonium*, *Pandorina* and *Eudorina*, to *Pleodorina* to *Volvox*, however, is quite different from the messy reality. Even though they can successfully hybridize, ribosomal RNA sequences show that *Eudorina* and *Pleodorina* are comparatively distantly related (Kirk 1998, p. 31). There are also substantial differences between strains even within a *Eudorina* species. Goldstein had hybridized a strain of *Eudorina elegans* with *Pleodorina illinoisensis*. Kirk and colleagues compared the RNA sequences of *Eudorina elegans* and *P. californica*. But the two strains of *E. elegans* were subsequently shown to have significant genetic differences. Moreover, when Kirk's group examined the ribosomal RNA sequence of Goldstein's *E. elegans* strain, they found it "nearly identical" to *P. illinoisensis* (Kirk 1998, pp. 31–32)! While this makes sense of the hybridization results, it also indicates the extraordinary complexity of the lineages of these related strains and species. Other studies have identified mutants of *Volvox powersii* phenotypically similar to *Pleodorina*, and mutants of *Volvox carteri* that, "if found in nature, might be classified as *Eudorina*" (Kirk 1998, pp. 28–29). Complicating matters still further, a number of isolated mating groups can be identified within single species, and the

> genetic distance [between mating groups]…is not related to the extent of geographical separation in any simple way. A *P[andorina] morum* individual isolated from a small pond

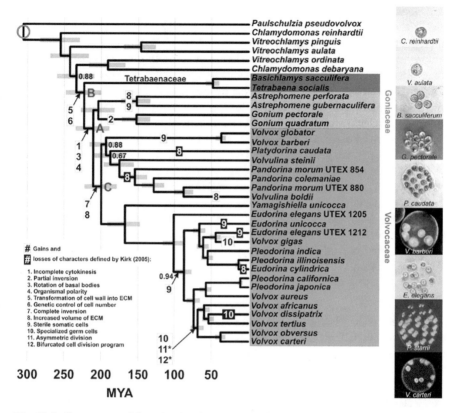

Fig. 11.6 Chronogram of the volvocine lineage, with reference to the twelve steps proposed by Kirk. Reprinted with permission from M.D. Herron, J. D. Hackett, F. O. Aylward, R. E. Michod, Triassic origin and early radiation of multicellular volvocine algae, *Proc. Natl. Acad. Sci. U.S.A.* 106(9): 3254–3258, 2009

in the American Midwest can be chromosomally indistinguishable from, and interfertile with, a strain isolated from Europe or Asia, while being reproductively isolated and chromosomally very different from a second *P. morum* individual taken from the same pond! (Kirk 1998, p. 31)

The geographical distribution of these so-called "syngens" is less mysterious when one considers that these algae can be transported across vast distances on – and in – the bodies of birds. In the spore stage, they can even be transported on air currents (Kirk 1998, p. 46).

While *Volvox* and its relatives may line up in an organized morphologic sequence, their phylogeny is clearly not so obliging. Kirk and his colleagues used ribosomal RNA sequences to derive a genealogical tree that placed some organisms in close proximity to others with quite different morphologies. Larson et al. (1992) found, for example, that *Volvox aureus* is genealogically closer to *Pleodorina californica* than to *Volvox carteri*, while *Volvox powersii* is a close relative of *Pleodorina illi-*

noisensis and one strain of *Eudorina elegans*. Another study using 18S and 2S rRNA sequences, by Buchheim and Chapman (1991), found similarly complex – though not identical – lineages, as did Nozaki et al. (1995), comparing sequences of chloroplast genes. Even the study by Herron et al. (2009), which pushed the estimate for divergence between *Volvox* and its unicellular ancestors back into the Triassic, agreed with Kirk and others that the genealogical tree presents a "complex picture in which phylogeny does not strictly mirror ontogeny; some traits have multiple independent origins and reversals from derived to ancestral states". The lineage proposed by Herron et al. (2009) is shown in Fig. 11.6.

More recently, the genomes of both *Chlamydomonas reinhardtii* and *Volvox carteri* have been sequenced (by Merchant et al. in 2007 and Prochnik et al. in 2010, respectively), providing new data and insights, including striking similarities between the genomes of *C. reinhardtii* and *V. carteri*. This new data, however, has not resolved fundamental complexity of the volvocine lineage (Umen and Olson 2012). As recently as 2014, Nozaki et al. identified an entirely new genus in Lake Isanuma. They named it *Colemanosphaera*, after pioneering *Volvox* researcher Annette Coleman.

*** *** ***

There are clear advantages – predator avoidance, improved ability to store nutrients – to a multicellular lifestyle. But as the volvocine lineage shows, this lifestyle can be adopted in a variety of ways. A comparison between two canonically different *Volvox* species, *V. rousseletti* and *V. carteri*, helps illuminate the molecular origins of the transition to multicellularity. As cells in *V. rousseletti* grow, they remain connected by cytoplasmic bridges. This gives them a "bell-like" shape when viewed from the side (Kirk 1998, p. 35). All developing cells in this species have a pair of flagella, but a few cells grow slightly larger than their neighbors; work in 1933 by Pocock suggests that these cells stop dividing one cycle before their fellows (Kirk 1998, p. 34). These cells – the gonidia – then reabsorb their flagella and divide, while continuing to grow, without breaking their cytoplasmic bridges to their neighbors. After multiple divisions, the gonidia have created an embryonic *V. rousseletti* spheroid. In contrast, *V. carteri* cells break their cytoplasmic bridges at the conclusion of embryogenesis; their gonidia, which never even have flagella or eyespots, "put all of their energy into growth, eventually becoming 500–1,000 times larger (in volume) than somatic cells before they begin to divide" (Kirk 1998, p. 34). They then stop growing and undergo a rapid series of cleavages in order to generate an embryonic spheroid.

Size is not always advantageous in the "time-share" habitat enjoyed by volvocaceans, where seasonally changing conditions of turbidity, nutrient abundance, and water temperature gradient favor different species in different seasons. These environmental variations help to answer the questions posed by Kirk (among many others): "Given that *Chlamydomonas* is still found in such environments, why is there also *Volvox*? Alternatively, however, we might also ask this: Given that there is now *Volvox* in such environments, why is there also still *Chlamydomonas*?" (Kirk 1998,

p. 52). Nonetheless, since our focus is on the evolution of multicelluarity, factors favoring an increase in size are of particular interest.

Large size is well known to help organisms avoid predation; the "predation threshold" is considered to be around an organismal volume of 8000 μm^3; *Chlamydomonas* is much smaller than this, and indeed often lives in moist soil rather than in aquatic environments to avoid predation by rotifers and copepods (Kirk 1998, pp. 55–56). Why would organisms grow larger than the predation limit? Here, the ability to store nutrients during times of scarcity plays a decisive factor. Graham Bell hypothesized that phosphate storage in the extracellular membrane would enable cells to sequester this essential (and often limiting) nutrient, while preventing the phosphates from competitively inhibiting their own influx. With more overall extracellular membrane, multicellular organisms would have a clear advantage. Sequestration of nutrients during times of abundance would enable these larger organisms to subsist from their internal stores during times of scarcity, providing a further selective advantage. Various experimental studies in natural pond settings support this hypothesis (Kirk 1998, pp. 60–61).

Bell's most important insight into the relation between organism size and nutrient storage, however, came with his realization that storage can be optimized even further when multicellular organisms exhibit differentiation. Bell and Koufoupanou (1991) proposed the *source-sink hypothesis*. In the case of volocaceans, Kirk explains,

> the extracellular matrix in the center of a volvocacean colony can be thought of as a 'sink' into which cells at the periphery of the colony (the 'source') can pump acquired nutrients so as to maintain a high rate of nutrient uptake and thereby outcompete unicells for this resource… An organism in which one set of cells harvest essential resources while the cells of a second set use them to produce progeny should be successful only if the organism can thereby produce a greater mass of progeny than it would have if all cells had participated in both processes. Because the effect of a nutrient sink on rates of uptake is realized only when external concentrations of those nutrients are high [since at lower concentrations uptake will not lead to competitive inhibition, and thus sequestration in a "sink" is unnecessary], the Bell source-sink model predicts that a germ-soma division of labor should be advantageous only in a nutrient-rich environment. (Kirk 1998, pp. 61–62)

Koufopanou and Bell (1993) verified this hypothesis in *Volvox carteri*. They measured the growth rate of gonidia as a function of nutrient concentration under three different conditions: gonidia in intact spheroids[2] with somatic cells and extracellular matrix present, gonidia in broken spheroids (but with some somatic cells still attached) and isolated gonidia. At every nutrient concentration, the gonidia produced a greater mass of progeny in the intact spheroids. Moreover, the growth rate increased more steeply as a function of nutrient concentration in the intact spheroids than in the other two cases. Broken spheroids performed similarly to the isolated gonidia, highlighting the critical role of the intact extracellular matrix. While Koufopanou and Bell did not present growth curves for a unicellular volvocacean for comparison, they calculated a growth curve for a hypothetical spheroid

---

[2] A general term for a colonial mass of algae.

consisting only of gonidia growing at the rate measured for isolated gonidia. They found that, under certain assumptions, this "compensation curve" was exceeded by the growth rate of the intact spheroid only at high nutrient concentration. These results are consistent with the source-sink hypothesis, and also with the emergence of selective pressure acting on the differentiated spheroid as a whole.

Mary Agard Pocock's early studies on *V. rousseletti* were consistent with the idea that somatic cells provide nutrients to the gonidia. However, in contrast to *V. carteri*, the gonidia of *V. rousseletti* are directly fed by somatic cells, without the intermediary use of the extracellular matrix. Recall that in *V. rousseletti* the cytoplasmic bridges between somatic cells and nascent gonidia are not broken, as they are in *V. carteri*. In the 1930s, Pocock observed that more cytoplasmic bridges were retained between somatic cells and gonidia than between adjacent somatic cells. She also found that

> that once *V. rousseletti* gonidia begin to grow and divide, somatic cells cease growing altogether, and later actually decrease in size, particularly in the posterior of the spheroid – in the vicinity of the embryos... Pocock observed that a somatic cell that apparently had become isolated from all others because of an accidental severance of all of its cytoplasmic bridges was larger than all of its neighbors, and she suggested that the increased size may have resulted from its lost ability to export materials through the cytoplasmic-bridge system. (Kirk 1998, p. 65)

The cytoplasmic bridge feeding system is so efficient that *V. rousseletti* gonidia grow an astounding 5600-fold during development (Kirk 1998, p. 66).

As important as the source-sink hypothesis may be in providing an advantage for the division of labor in *Volvox*, it only holds under high-nutrient conditions. Fluctuating environmental conditions could conceivably favor such a source-sink process, since volvocaceans with germ-soma differentiation would have an advantage at least some of the time. But metabolism is not the only source of evolutionary pressure toward differentiation. Flagella provide a *structural* constraint favoring division of labor between germ and soma. As Koufopanou and Bell (1993) explain, "cells of *Volvox* and related genera do not maintain functional flagella during embryonic cleavage, apparently because their basal bodies move to the mitotic poles and away from the flagellar bases... Sterile tissue [i.e., somatic cells incapable of reproduction] may have evolved to provide functional flagella during embryogenesis." As Kirk emphasizes, the source-sink model represents a solution to an ecological constraint, while use of somatic cells for motility represents the solution to a cytological constraint. This is a vivid example of what Stephen Jay Gould would have referred to as historical "channeling", an evolutionary lineage whose innovations are driven by a set of inescapable limitations, as the roots of city trees grow between slabs of pavement.

Green flagellates must be able to execute phototaxis, moving toward the light when needed, but not steering too close, in order to avoid the production of toxic photooxidants. A precise relation has developed between the location of the eyespot[3] and the flagella. This relationship is maintained during reproduction by "using

---

[3] The role of the eyespot here provides an easy answer to the standard "creationist" question, "what use is 1% of an eye?" Clearly, it is of great use for survival in cases where an organism needs to

the very same cytoskeletal components to link critical parts of the dividing cell in an equally stereotyped manner" (Kirk 1998, p. 91). In *Chlamydomonas*, for example, the eyespot is asymmetrically placed, allowing the cell to perceive the direction of incoming light.[4] The flagella are located near the eyespot, but they too are asymmetrical. The flagellum closer to the eyespot is known as the *cis* flagellum, and the one farther from the eyespot is known as the *trans* flagellum. The two flagella respond differently to incoming light. As a result, when swimming, *Chlamydomonas* performs an asymmetrical breaststroke, so that its eyespot moves in a helical path, surveying all 360° of the environment as it turns (Kirk 1998, p. 71).

Positive and negative phototaxis are accomplished by differential responses of the *cis* and *trans* flagella to light exposure. For positive phototaxis, the flagellum on the darker side beats more strongly; for negative phototaxis, which occurs when the organism is exposed to light beyond a threshold intensity, the flagellum on the light side beats more strongly. These responses are regulated by calcium channels in the eyespot region. *In vitro* studies suggest that positive phototaxis occurs at calcium concentrations of $\sim 10^{-7}$ M and negative phototaxis at concentrations of $\sim 10^{-9}$ M (Kirk 1998, p. 75).

The conflict between motility and reproduction occurs because the basal bodies that anchor the flagella also function as centrioles during cell division. But, because of the cell wall structure of the green flagellates, these two functions are physically incompatible. If the organism continues to swim, it cannot reproduce (Koufopanou 1994). In many eukaryotic cells, the centrioles are passed down semi-conservatively, with one daughter cell retaining the parental centriole and the other receiving the newly synthesized one. Experiments on mutant *Chlamydomonas* in the early 1980s showed that the *cis* flagellum is always associated with the younger basal body (centriole). As a result,

> green flagellates like *Chlamydomonas* must have a way to assure that the eyespot and the younger BB [basal body] will end up on the same side of the cell after each round of cell division. A cell that divided in a way that the parental and daughter BBs were located at random with respect to the eyespot clearly would produce daughter cells in which the response to light would be nonadaptive; such cells would be very likely to disappear from the population without leaving progeny. (Kirk 1998, p. 78)

A precise structured relation must be maintained between the *cis* and *trans* flagella in order for their effective strokes to be oppositely directed. This is controlled by a 180° rotational symmetry between the basal bodies, and this symmetry, as well as the proximity of the *cis* flagellum to the eyespot, must be maintained (Kirk 1998, pp. 78–80). The eyespot is anchored to a microtubule rootlet projecting from the daughter basal body (Kirk 1998, p. 82). But the basal bodies are also coupled to the

---

respond to light, but not to actually form an image. For a brilliant model of the simplicity of the pathway by which an image-forming eye might evolve from a directionally-sensitive eyespot, see the classic paper by Nilsson and Pelger (1994), and also Richard Dawkins's discussion of the topic in *Climbing Mount Improbable* (1996).

[4] The structure of the eyespot itself provides added directionality (see Kirk 1998, p. 72, for more details).

mitotic spindle (Kirk 1998, p. 93). This is essential for proper segregation of new and old basal bodies in each daughter cell. One mother-daughter basal body pair must also be aligned with each mitotic spindle, in such a way that the *cis* basal body is on the side of the eyespot in each nascent cell. Since these connections must be made and preserved during the movement of organelles and other intracellular structures, and since the flagella project from the basal bodies through channels in the plasmalemma and cell wall, the ancestors of *Chlamydomonas* faced a stunningly complex engineering problem.

In simpler "naked" and scale-covered flagellate species, which lack a cell wall,

> the BBs simply move apart in the plane of the plasmalemma – with flagella attached and actively beating. Then, at telophase, a cytokinetic furrow forms between the BBs and the cell divides, having remained fully motile through mitosis and cytokinesis. (Kirk 1998, p. 98)

Some algal species with cell walls have successfully resolved their engineering dilemma by jettisoning their flagella during the asexual phase of their life cycle to enable mitosis, and sloughing off their cell walls during their sexual phase to enable motility. However, this strategy is not optimal in all environments. Single-celled green flagellates like *Chlamydomonas reinhardtii* take the approach of resorbing their flagella during mitosis. Living in moist soil, they simply "resorb and hang on", developing a sticky coating as the flagella are resorbed, and using this to cling to a nearby substrate while dividing (Kirk 1998, p. 101).

Water-dwelling flagellates cannot simply resorb and hang on. Not only is there no substrate to attach to, but, without flagella, the organisms could not execute phototaxis and would be unable to maintain their position within the water column, potentially depriving themselves of key nutrients during cell division (Kirk 1998, p. 101). Pond-dwelling *Chlamydomonas* species and multicelled volvocaceans up to the size of *Eudorina* solve this problem by decoupling their flagella and basal bodies. The flagella continue to beat, maintaining cell motility, while the intracellular components execute their well-choreographed cytokinetic dance beneath them (Kirk 1998, p. 101). But flagella can remain decoupled only so long before they begin to beat in an uncoordinated fashion and eventually tear themselves loose from the cell (Kirk 1998, p. 104). This can be avoided by one last creative solution to the flagellation constraint: division of labor.

Division of labor in volvocaceans is facilitated by the fact that green flagellates with cell walls undergo multiple, rather than binary, fission. Indeed, multiple fission appears to be closely coupled to the existence of a cell wall: descendants of walled flagellates that lose their cell walls revert to binary fission.

> Why these two phenomena are so tightly coupled is not certain. However, a credible working hypothesis is that the flagellation constraint was first resolved (as it is today in *C. reinhardtii*) with loss of motility during division, and thus variants that combined several divisions in one immotile period per day would have survival advantage over ancestral forms that had to become immotile several times per day in order to complete the same number of divisions. In this regard, it may well be significant that most species that lose their flagella while dividing do all of their dividing at night and are motile while the sun is shining... Whether or not this working hypothesis about its origin has any validity, multiple

fission is currently a fact of life for all walled chlamydomonads and all volvocaceans. (Kirk 1998, p. 102)

For cells held closely together during multiple fission, a next step toward permanent multicellularity could be facilitated by incomplete cytokinesis. Recall that *V. rousseletti* maintains cytoplasmic bridges between cells; such bridges are a natural part of cytokinesis, used to "hold sister cells in a fixed spatial relationship until cell wall deposition has begun" (Kirk 1998, p. 102). Such incomplete division was observed in some of the *Gonium dispersum* colonies studied by Batko and Jakubiec.

> It is noteworthy that the walls of adjacent cells within a *G. dispersum* colony are so loosely joined that individual cells frequently break free after hatching. A second step toward multicellularity very likely involved coalescence of walls between adjacent cells, as is now seen in all other species of *Gonium*... From there it would have been a small step to the formation of wall elements that joined the outer ends of adjacent cell walls into a single "colony boundary," as occurs today as a second, discrete step during the development of each juvenile colony of *Pandorina*... At that point, variants of increasing size and cell number could have begun to be selected for if (but only if) those larger variants could find some new way to remain motile while their cells divided. (Kirk 1998, p. 103)

This could be accomplished by maintaining sterile somatic cells. Never reproducing, they would never need to shed their flagella. Koufopanou (1994) has suggested that is the origin of somatic cells in the Volvocales. This is supported by the observations that, in *Pleodorina* and *Eudorina*, the ratio of somatic cells to gonidia increases with the total cell number per spheroid, as does the relative size of the gonidia (Koufopanou 1994; Kirk 1998, p. 105). With a decreasing proportion of gonidia, the evolutionary bottleneck becomes tighter, a hallmark of the onset of multicellular individuality (Buss 1987).

Division of labor appears to be reversible in *Pleodorina californica*. Gerisch showed in 1959 that isolated *P. californica* somatic cells can grow and reproduce; this ability is lost in wild-type *Volvox* somatic cells, though restored in some mutant strains (Kirk 1998, p. 103). This is consistent with the idea that signaling between cells in a colony suppresses the reproductive potential of somatic cells, and that removal of the presence of the other cells in the spheroid allows this potential to be re-expressed. This suggests that the development of multicellularity should not be reduced to an argument over voluntary altruism in *Volvox* any more than it can be in *Dictyostelium*. "Why would a somatic cell sacrifice its autonomy and reproductive potential for the good of the group?" may be the wrong question to ask. A more appropriate question might be: "How does the group act to suppress the autonomy and reproductive potential of a somatic cell in order to exploit its resources?" We will return to such speculations below.

In a 2005 paper with the tongue-in-cheek title "A twelve-step program for evolving multicellularity and a division of labor", Kirk placed the source-sink hypothesis and the flagellation constraint in the broad context of evolutionary history.[5] Here,

---

[5] The steps are: (1) Incomplete cytokinesis; (2) Incomplete inversion of the embryo; (3) Rotation of the basal bodies; (4) Establishment of organismic polarity; (5) Transformation of cell walls into an extracellular matrix; (6) Genetic modulation of cell number; (7) Complete inversion of the embryo;

mutants of a strain of *Volvox carteri* are particularly important, since they show how single genetic changes can disrupt the multicellular, differentiated lifestyle. This presents in stark relief the ease with which the "forward direction" toward increasing complexity may have proceeded in evolutionary history.

In order to understand how *Volvox* mutations can reveal the possible evolutionary steps to the development of cytodifferentiation, we need to take some time to survey the normal biology of *Volvox carteri*, which Kirk refers to as "a Rosetta stone for deciphering the origins of cytodifferentiation". Readers who wish a more detailed (and endlessly fascinating) tour through *Volvox* biology are referred to Chap. 5 of Kirk's (1998) text.

Different strains of *Volvox carteri* exhibit radically different spectra of mutations. The strain used in most developmental studies is *Volvox carteri* forma *nagariensis*, obtained from a pond in Japan. This strain exhibits key spontaneous mutants that shed important light on the origins of multicellularity and the division of labor, as we shall see. As Kirk writes,

> these mutants exhibit interesting aberrations of asymmetric division, germ-soma differentiation, the switch from asexual to sexual reproduction, or other fundamental developmental processes. Such mutants not only indicate that all of these processes are under rather direct genetic control, but also provide tools for analyzing the nature of such controls. (Kirk 1998, p. 112)

In contrast to *nagariensis*, *Volvox carteri* forma *weismannia* is not only reproductively isolated from *nagariensis* in its sexual stage, but produces none of the mutants found in *nagariensis*! It has been suggested that the difference may lie in the DNA repair systems of the two strains.

Volvox carteri f. *nagariensis* is typically studied in descendants of the original strains, HK10, which is female in the sexual stage, and HK9, which is male. As adult spheroids, they contain 2,000–4,000 somatic cells and up to 16 gonidia. Their growth cycle occurs over 48 h and is closely regulated by the light-dark cycle, a fact exploited by researchers who wish to synchronize their spheroids to the same life cycle stage. Somatic cells have a polarized structure similar in many respects to *Chlamydomonas*, with basal bodies at the apical end. A striking difference from *Chlamydomonas*, however, is the orientation of the flagella.

In *V. carteri*, somatic cells are arrayed along the outer surface of the spheroid. If each cell performed the breaststroke, they would simply cause the water to swirl around. All flagella on one side of the spheroid must beat in the same direction, while those on the other side beat in the opposite direction, in order for the entire spheroid, rather than a single cell, to do the breaststroke.[6] The arrangement of the

---

(8) Increased volume of extracellular matrix; (9) Partial germ-soma division of labor; (10) Complete germ-soma division of labor; (11) Asymmetric division; (12) Bifurcation of the cell division program. More recently, Herron and Michod (2008) have suggested that the order of the first six steps may need to be revised, and that embryo inversion may have evolved in a single step (see also Nishii and Miller 2010). See Fig. 11.6 for how Herron et al. (2009) fit their proposed volvocine lineage together with Kirk's twelve steps.

[6] In fact, *Volvox* does list toward one side as it swims, since all the flagella beat with a slight bias to the right, so that the spheroid rotates counterclockwise when viewed from behind, giving rise to the name "fierce roller" (Kirk 1998, p. 180).

basal bodies in *Chlamydomonas* is not compatible with this. In *Volvox*, instead of being mutually oriented at 180°, the basal bodies are aligned parallel to one another. This may a physical constraint on the somatic cells that acts as a form of conflict suppression, in the sense of Michod (see Chap. 13). With this basal body alignment, somatic cells are incapable of "making it on their own" since, separated from the rest of their galley crew, they could not swim, let alone execute phototaxis. Isolated, they would simply tumble in place.

Gonidia, unsurprisingly, are structurally quite different from somatic cells. They lack an eyespot and have rudimentary flagella only at an early stage of development; these are eventually completely resorbed. Gonidia have a radial configuration, with a central nucleus surrounded by large vacuoles. The extracellular matrix that fills the spheroid interior has a crystalline structure similar to the crystalline layers of *C. reinhardtii*, suggesting a common evolutionary origin. Indeed, Kirk notes, "*C. reinhardtii* and *V. carteri*, when stripped of their own crystalline layers, can nucleate the assembly of each other's crystalline-layer glycoproteins in a normal crystal lattice, whereas *C. reinhardtii* and *C. eugametos* are incapable of such cross-nucleation" (Kirk 1998, p. 123).

During the asexual[7] life cycle, the gonidia within an adult spheroid undergo a series of rapid (approximately one every 35 min) cleavage events, generating all the smaller somatic cells and the next generation of gonidia that will be present when they hatch from the parent spheroid (see Figs. 11.7 and 11.8A). Cleavages are stereotyped and asymmetrical. The cell fate determination of whether a cell will follow a somatic or gonidial path appears to depend on size alone, with cells 8 µm or larger in diameter becoming gonidia (Kirk 1998, pp. 137–138). The mechanism by which cells "sense" their own size still appears to be unclear, but it is known that asymmetrical division is regulated in *Volvox* by a gene called *glsA*, to which we will return below.

During the rounds of asymmetric cleavage in the gonidia that produce the embryonic new spheroids, incomplete cytokinesis results in cytoplasmic bridges between nearby cells. These bridges form a network that plays a critical role in the gonidial maturation process, because during the cleavage stage, new gonidia form on the outside of the growing embryonic spheroid, and new somatic cells on the inside. An

---

[7] The sexual life cycle is induced when *Volvox* is exposed to a sex-inducing pheromone. As Kirk notes, "the sexual pheromone of *V. carteri* is one of the most potent bioactive molecules known, capable of inducing full sexual development in wild-type spheroids at a concentration of less than 10^{-16} M" (Kirk 1998, p. 244). Once a spheroid is exposed to this pheromone, its gonidia produce more new gonidia than in the asexual cycle; in females, these develop into eggs, and in males they undergo a further round of cleavage and differentiation to produce a packet of sperm. Once fertilized, eggs can live under extremely harsh conditions as zygospores. In some species and strains, eggs and sperm can be produced by the descendants of a single clone. Many other sexual variants exist, including strains of *V. aureus* in which gonidia are directly induced to become zygospores by the presence of the sex-inducing pheromone. See Kirk (1998), pp. 126–127, for more detailed discussion and references.

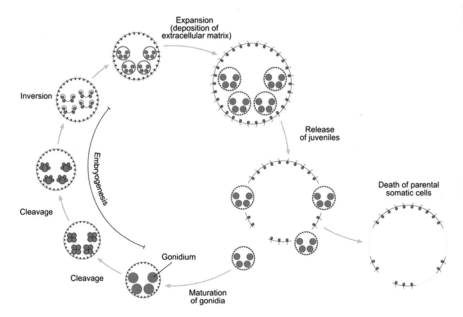

Fig. 11.7   Schematic representation of *V. carteri*'s asexual life cycle (from Hallmann, 2011)

inversion process must occur in order for the juvenile[8] spheroids to be able to swim free from their parents after hatching. This inversion process (see Fig. 11.8B) is accomplished by successive waves of cell shape change; somatic cells initially have a spindle shape, but those closest to an open region called the phialopore take on a flask-like configuration,[9] elongating at one end. As this happens, the cells make use of the cytoplasmic bridges, which initially link the cells like a belt that "constitutes the only structural element against which the cells can exert force to effect inversion" (Kirk 1998, p. 156). Cells adjacent to the open region slide perpendicular to the cytoplasmic bridge axis, until they are attached to the bridge at one end rather than along their midsection. This causes the edge of the phialopore to fold over like a lip. As a result, the region of maximum curvature moves back toward the posterior of the spheroid, and the cells at the maximum curvature undergo the spindle-to-flask shape change, while cells over the edge of the lip relax to their spindle configuration. Finally, the lip regions

> nearly surround the uninverted posterior hemisphere… At that point, the posterior hemisphere 'snaps' through the opening at the equator, and eventually the phialopore lips come together once again at what will be the posterior pole of the adult… the flagellar ends of all

---

[8] Spheroids are considered to have undergone a transition from their embryonic to their juvenile stage after they have completed the inversion process (Kirk 1998, p. 153).

[9] In the 1960s and 1970s, Kelland showed that any slit made in an embryonic spheroid would induce changes in cell shape, followed by the initiation of the inversion process (Kirk 1998, p. 155).

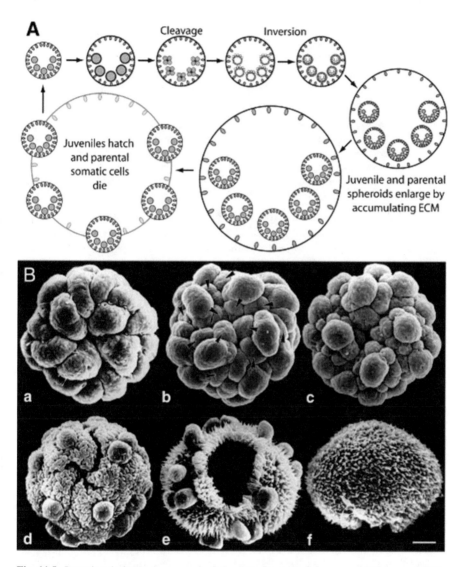

Fig. 11.8 Inversion during embryogenesis. Schematic diagram of the asexual *Volvox carteri* life cycle (A). Embryogenesis (B), showing the 32-cell stage where all cells are similar in size (**a**), the first set of asymmetric cells after the sixth division (**b**), the phialopore (**d**) and embryo inversion (**e**, **f**). Reproduced with permission from D.L. Kirk, A twelve-step program for evolving multicellularity and a division of labor. BioEssays 27(3): 299-310, 2005. © 2005 Wiley Periodicals, Inc

somatic cells are on the exterior, and the gonidia, which protruded from the surface of the pre-inversion embryo, have been moved to the interior. (Kirk 1998, pp. 152–153)

The anterior-posterior axis is thus reversed during the inversion process.

Embryogenesis typically takes place in the dark (though Kirk and colleagues observed that the *V. carteri* life cycle can be completed in 48-h light conditions, albeit with the cells drifting out of phase with each other). When exposed to light during a normal 48-h light-dark cycle, the cells in the embryo, until then visually distinguishable only by size, begin the differentiation process in earnest. Kirk and colleagues showed that the translation of mRNA begins almost immediately, suggesting that transcription of somatic-specific genes may already have begun during embryogenesis. Kirk notes that during early embryogenesis

> [t]he specificity of accumulation of early transcripts is not symmetrical, however: At very early stages, presumptive somatic cells accumulate 'gonidia-specific' transcripts at nearly the same low rate as do the young gonidia; it is only just after cytodifferentiation has begun that these transcripts disappear from somatic cells and earn the term 'gonidia-specific.' (This may speak to the ability of the presumptive somatic cells to differentiate as gonidia under some conditions....) (Kirk 1998, pp. 159–160)

The genes expressed during this early period primarily code for extracellular matrix proteins. The extracellular matrix of a juvenile spheroid is richly complex, with various layers and zones, such as the crystalline layer mentioned above. The matrix holds somatic cells and gonidia in precise locations within the spheroid as the cytoplasmic bridges break down; gonidia are located near the somatic cells to which they are most closely related. The positioning is critical for the future survival of the spheroid: it anchors somatic cells in the proper orientation with respect to the spheroid's anterior-posterior axis so that their flagella can beat in a coordinated manner. The detailed mechanism behind this coordinated assembly does not appear to be fully understood, though a temperature-sensitive mutant, *flgC11*, has been identified that disrupts somatic cell placement, leading to uncoordinated swimming (Kirk 1998, pp. 175–176).

The "lush extracellular tapestry" of the extracellular matrix, as Kirk describes it, is deposited during differentiation. The somatic cells also grow their flagella during this period. Curiously, the *cis* and *trans* flagella have different rates of growth, though they eventually reach a similar length in the mature somatic cell.

After maturation, the juvenile spheroids hatch from their parent. I cannot match Kirk for his dramatic description of this process.

> The first sign that hatching is imminent comes when the parental somatic cells located closest to the center of an underlying juvenile spheroid begin to rock within their cellular compartments. These oscillations, which are feeble at first, become more vigorous until, one after another, the cells wrench free of the spheroid and tumble away. This process spreads laterally until all of the cells lying above the juvenile have tumbled off – but it spreads no farther. At about that time, the juvenile, which has been rotating restlessly beneath the somatic cell layer, typically becomes frozen in place momentarily (presumably as it gets caught in an opening whose diameter is slightly smaller than its own); then after rolling outward ever so slowly, it suddenly bursts free of its invisible restraints and spins rapidly out of view. Within an hour or two after the process begins, the parent has become a mere "ghost": a spheroid with gaping holes equal in number and diameter to the juveniles that were within it earlier, and with the rest of its surface dotted with somatic cells that are already beginning to fade as their chlorophyll content diminishes. (Kirk 1998, pp. 178–179)

The opening of the pores through which the juveniles hatch is mediated by H-lysin, a serine protease similar to the V-lysin secreted by juvenile *Chlamydomonas* before they hatch from their parent. It appears to be still unknown, however, which cells secrete this protease. A range of questions remain unanswered, such as how the juvenile spheroids protect themselves from the protease, and why the protease has an effect only on the somatic cells of the adult spheroid that lie above the juveniles, so that these hatch through the "ceiling" while leaving the "floor" and "walls" around them intact (see Kirk 1998, pp. 179–180).

After hatching, somatic cells in the parental "ghost" undergo programmed cell death, releasing nutrients into the water. Meanwhile, the new spheroids execute phototaxis in a manner generally similar to that used by *Chlamydomonas*, including threshold effects in order to remain in optimal light intensity. Now, however, it is a question of ensembles of flagellar pairs changing their coordinated beat frequency due to light exposure, rather than differential responses of two flagella within a single pair. The role of calcium channels in transducing the phototactic response in *V. carteri*, compared to *Chlamydomonas* and other flagellates, has not yet been fully elucidated. Most recently, *V. carteri* and *Chlamydomonas* (Drescher et al. 2010) and *V. rousseletti* (Ueki et al. 2010) have been the subject of extensive studies as models of biological fluid dynamics; Drescher and colleagues identified a "hydrodynamic bound state" in which nearby *Volvox* spheroids are drawn toward each other. A recent experimental study suggests that hydrodynamic interactions aid the synchronous beating of flagellar ensembles in *V. carteri* (Brumley et al. 2014). In a recent review, Goldstein (2015) summarizes some of this recent work and places it within the broad context of biological "microswimmers".

∗∗∗ ∗∗∗ ∗∗∗

As described earlier, *V. carteri* exhibits mutant forms with distinct developmental phenotypes. Studies of *Volvox* mutants are primarily due to the work of three laboratories – those of Richard Starr, first at Indiana University and then at the University of Texas at Austin, Robert Huskey at the University of Virginia, and David Kirk of Washington University in St. Louis, whose seminal *Volvox* text has been our guide for much of the present chapter. These laboratories identified a panoply of mutants – MulB, MulC, MulD, MulX, MulA, MegA, R-1, R-2, RadA, doughnut, double-posterior, S16, pld, Inv, Dis, Exp, Rel, Flg, Eye, Rot, and many others. As one would expect, each of these mutants results in identifiable pathologies in structure and development.

The Mul mutants are known as *pattern-switching* mutants, and cause asexual spheroids to develop in a pattern similar to male or female spheroids. MulB leads to asexual spheroids with gonidia distributed similarly to the eggs in a female spheroid; MulC and MulD lead to distribution of gonidia similar to that of sperm packets in male spheroids. In all cases, the mutants disrupt the cleavage pattern in the developing gonidia, which become asymmetrical at different division cycles in the normal asexual, male and female developmental pathways. These mutants are particularly interesting since they show a decoupling between patterns of cell division and germ cell differentiation. In normal cells, exposure to the sexual pheromone will trigger a male or female pattern of asymmetric cell division as well as the production of sperm packets or eggs. In Mul mutants, the male or female cell division pattern is followed, but "normal" gonidia are produced.

In the doughnut mutant, the anterior-posterior polarity is disrupted, so that a phialopore is formed, and gonidia are produced, at *both* ends of the embryonic spheroid. The spheroid therefore attempts inversion from both ends, and eventually forms a toroidal shape. Somatic cells appear normal, and exhibit a normal gradient of eyespot size over the abnormal toroidal topology. The double-posterior mutant forms gonidia at both ends, and has no gradient of eyespot size in its somatic cells, but only attempts inversion from one end. A wide range of Inv mutants (of which Kirk's lab alone isolated 60 separate strains!) exhibit abnormalities in the inversion process, ranging from incomplete inversion to failing to invert at all. These mutants may have defective cytoplasmic bridges that are unable to withstand the physical strain of the inversion process (Kirk 1998, p. 233). Remarkably, some of these partially inverted spheroids are viable.

Other mutants are characterized by damage to various components of the extracellular matrix. In the Dis mutant, juvenile spheroids dissociate into individual somatic cells that flail about ineffectually, since their flagella beat in the same direction, and gonidia that, remarkably, are capable of undergoing cleavage and producing embryos of their own (Kirk 1998, p. 235). The delayed dissolver (d-Dis) mutant releases its juvenile spheroids early; evidence suggests this results from defective structural integrity of the extracellular matrix rather than early secretion of H-lysin (Kirk 1998, p. 235). Other mutants lack key components of the extracellular matrix and, while they do not dissociate, have a "deflated" appearance. Rel mutants exhibit delayed release of juvenile spheroids, and "sometimes as many as three or four generations of juveniles will develop within a parental Rel spheroid. Needless to say, within such cramped quarters, development tends to become progressively more abnormal with each generation" (Kirk 1998, p. 236). This defect appears to derive from insufficient H-lysin, rather than matrix abnormality.

Flg, Eye and Rot mutants all exhibit pathological cell orientation within the spheroid. The somatic cells of Flg (flagellaless) mutants either lack flagella entirely (though they retain the flagellar channels), or have truncated flagella. Somatic cells lacking flagella are randomly oriented, both with respect to the other cells and with respect to the anterior-posterior axis of the entire spheroid. This, as well as the behavior of a temperature-sensitive mutant flgC11, suggested that the presence of functional flagella during the time of extracellular membrane deposition is essential for proper somatic cell orientation (Kirk 1998, p. 237).

Eye mutants have normal flagella, but their somatic cells are oriented incorrectly within the spheroid with respect to their eyespots. These mutants are unable to navigate properly. The Eye mutant phenotype shows that functional flagella are necessary, but not sufficient, for proper somatic cell orientation (Kirk 1998, p. 238). One of the most striking orientation mutants is Rot, which has its cells oriented backwards. These mutant spheroids swim with a clockwise rotation when viewed from behind, and swim "backwards", with their gonidia end forward.

A variety of mutations affect sexual reproduction. Some result in female or male cells that generate eggs or sperm packets without pheromone activation. Eggs in females with one of these mutations (at the *sex^cA* locus) are unable to redifferentiate as gonidia if unfertilized. Such redifferentiation is typical in wild type *V. carteri*. The "70–3" strain studied by Starr and colleagues has greatly reduced fertility: sexual spheroids are rarely produced even when exposed to four orders of magnitude more pheromone than typically needed. Exposure to UV radiation induces gender reversal in a wild-type male strain of *V. carteri*, producing spheroids that, when exposed to pheromone, exhibit the typical male pattern of cleavage, but produce eggs rather than sperm packets. As Kirk explains, "[t]hese eggs are then capable of being fertilized by sperm produced by the unmutagenized brothers of the mutant individual. When the progeny of this incestuous coupling are exposed to pheromone, half of them develop as normal males, and half as egg-producing males," suggesting that the gender reversal is the result of a single mutation within the mating-type locus of the *V. carteri* genome (Kirk 1998, p. 244). This also implies that male *V. carteri*, which are haploid, contain the genetic information needed to follow at least some of the female developmental program.

Perhaps the most fascinating mutants are those that affect the normal germ-soma dichotomy. These affect the genes that are essential for the division of labor. The first of these, at a locus known as *regA* (for "regenerator" phenotype) was studied first by Starr's laboratory, and then by Huskey's and Kirk's. After behaving like normal somatic cells for a time, *regA* somatic cells retract their flagella and redifferentiate as gonidia. These gonidia develop normally, except that, upon maturity, somatic cells in their daughter spheroids express the regenerator phenotype themselves. Several *regA* variants have been identified, varying in the number and location of somatic cells that redifferentiate as gonidia. Experiments with a temperature-sensitive mutant show that all these different variants result from a mutation at the same locus; the different spatial patterns of regenerating somatic cells (throughout the spheroid, just at one of the poles, etc.) depend on the time during a critical window of embryogenesis when the mutant *regA* is expressed.

The protein coded for by the gene at the *regA* locus is a transcriptional repressor that targets genes for chloroplast proteins (Meissner et al. 1999); there is no orthologous gene in *C. reinhardtii* (Umen and Olson 2012). There is a homologous[10] gene, however, in *C. reinhardtii*, called *rls1* (Nedelcu 2009), orthologous to a different gene in *V. carteri*, *rlsD* (Hallmann 2011). These results suggest that *regA* might have arisen in a *V. carteri* ancestor by duplication of *rlsD* followed by mutation (Hallmann 2011). Nedelcu and Michod have suggested that

> the evolution of soma in multicellular lineages involved the co-option of life-history genes whose expression in their unicellular ancestors was conditioned on environmental cues (as an adaptive strategy to enhance survival at an immediate cost to reproduction), through

---

[10] *Orthologous* genes share a common descent by speciation from a common ancestor and typically retain a similar function in the extant species. *Homologous* genes constitute a more broadly defined group, covering any set of genes that share a common ancestor; orthologous genes are a type of homologous genes, but so are *paralogous* genes, which are related by gene duplication.

shifting their expression from a temporal (environmentally induced) into a spatial (developmental) context. (Nedelcu 2009)

Nedelcu hypothesized that *rls1* is such an environmentally-cued gene, "generally induced under conditions when the temporary down-regulation of photosynthesis is beneficial in terms of survival, though costly in terms of immediate reproduction" (Nedelcu 2009). A distant relative of *rls1*, *regA* has been co-opted into a spatial context. By preventing the transcription of chloroplast proteins, *regA* prevents somatic cells from growing large enough to reproduce. It thus acts as a master switch to shunt cells away from a gonidial fate. RegA mutants revert to a *Chlamydomonas*-like developmental program, in which every cell passes through both vegetative and reproductive stages.

The *lag* (late-gonidia) mutants follow the "first vegetative, then reproductive" ancestral pathway, though in a quite different manner. Gonidia expressing this mutation (which can arise from one of at least four loci) undergo apparently normal asymmetric division during embryogenesis. The larger cells formed during the division process, however, do not enter the gonidial developmental pathway on schedule. Rather, they temporarily become abnormally large somatic cells. They spend an extra day in this state before resorbing their flagella, losing their eyespots, and redifferentiating as gonidia. Somatic *lag* mutants develop norally. From these observations, Kirk concluded that, in the wild-type, the *lag* gene(s) suppress the somatic cell differentiation pathway in large cells destined to become gonidia. In *lag* mutants, the suppression of the somatic pathway is delayed.

A third mutant revealing key steps in the transition to multicellularity is the *gls* (gonidialess) mutant. Here, asymmetrical cell division never takes place in the embryonic gonidia, and spheroids contain exclusively somatic cells. Since they clearly cannot reproduce, the *gls* mutant is typically studied against the "background" of the *regA* mutation. In *gls/reg* double mutants, only somatic cells are produced, but they can reproduce, since they have the additional mutation that allows them to differentiate as gonidia. The life cycle of such double mutants is reminiscent of *Eudorina*, in which a colony of biflagellate cells later redifferentiate as reproductive cells (Kirk 1998, p. 253). *Gls* mutant cells are incapable of initiating asymmetric division. Kirk suggested that *gls* loci "encode products that are required for shifting the division plane from the center of the cell to one side" (Kirk 1998, p. 253).

Using a technique called transposon tagging, Miller and Kirk (1999) cloned *glsA*, and found that it coded for a protein associated with the mitotic spindle. The protein had sequence similarities with the human protein MPP11 (mitotic phase phosphoprotein 11), which is phosphorylated during mitosis and associated with the mitotic spindle; homologues were also found with proteins in *C. elegans*, *S. cerevisiae*, and mice. The GlsA protein was found to contain a domain found in Hsp40 molecular chaperones, providing a further clue to its mechanism of action. Miller and Kirk suggested that GlsA might bind to and activate another Hsp protein, Hsp70; this was confirmed by Miller's laboratory (Cheng et al. 2005, 2006), which also showed that GlsA can co-localize with histones. Pappas and Miller (2009) found that some fraction of GlsA is also associated with the ribosome, and concluded that the protein likely plays a role in regulating both transcription and translation.

Based on these experiments, Kirk proposed a simple developmental programme leading to germ-soma specification (Fig. 11.9). After the gonidia undergo five symmetrical divisions, leading to a 32-cell embryo, the *gls* loci are activated, which leads to asymmetrical division. In any large cell generated by the asymmetrical divisions that follow, the *lag* loci are activated, which represses somatic genes, causing the cell to follow the gonidial developmental pathway. In small cells, *regA* is expressed, suppressing gonidial genes, and the cell becomes a somatic cell. One remaining chicken-or-the-egg question (among many) is how small cell size triggers the activation of *regA*, which itself inhibits somatic cell growth by repressing the production of chloroplast proteins.

Kirk's hypothesis leads to a number of predictions that can be tested when the genomes of other species in the volvocine lineage are sequenced.

> First of all, we can predict…that the closest "homocytic" relative of *V. carteri*, such as some isolate of *Eudorina*, will be found to possess genes homologous to many of the early-somatic and gonidial genes of *V. carteri* and will express these genes simultaneously, in a pattern similar to that seen in the Gls/Reg mutant of *V. carteri*… Second, we can predict that the isolate of *Pleodorina* that appears to be the closest living relative of *V. carteri* will possess a homologue of the *V. carteri regA* gene and will express this gene selectively in those cells that are destined to remain small, flagellated, and incapable of reproduction… But we predict that *Pleodorina* will not have functional equivalents of the *gls* or *lag* genes, because *Pleodorina* species neither divide asymmetrically nor have cells that fail to differentiate as biflagellate motile cells before redifferentiating as gonidia. The most distant relative in which we would expect to find a functional equivalent of the *lag* genes of *V. carteri* is some species of *Volvox*, such as *V. tertius*, in which there are no asymmetric cleavage divisions, but in which nevertheless a discrete set of cells begin to enlarge and differentiate as gonidia immediately after the end of embryogenesis, without first passing through the biflagellate, quasi-somatic condition. Correspondingly, the most distant relative in which we would expect to see functional equivalents of the *gls* loci of *V. carteri* would be *V. obversus*, the only other *Volvox* species that exhibits asymmetric division (albeit in a very different spatial pattern). In short, a specific set of predictions to be tested in such an analysis would be that the three types of genetic functions that are now believed to play central roles in germ-soma specification in *V. carteri* were added to the volvocacean genetic repertoire in the sequence first *regA*, then *lag*, and then *gls*. (Kirk 1998, pp. 326–327)

Furthermore, since *V. rousseletti* does not undergo asymmetric cell division, one would expect it to lack genes homologous to the *gls* gene(s) in *V. carteri*. These predictions can be compared with the phylogenetic lineage postulated by Herron et al. (2009) and the twelve-step program suggested by Kirk (see Fig. 11.6).

While the developmental programme proposed by Kirk is elegant and eminently logical, it is undeniable that changes at these three groups of loci alone are not sufficient to account for volvocine complexity. For example, as Domozych and Domozych (2014) remark, the *Volvox carteri* genome was shown by Prochnik et al. (2010) to be 17% larger than that of *Chlamydomonas*. Many of these additional genes are likely to code for proteins involved in the elaborate, multi-compartmental extracellular matrix of *V. carteri*. For example, the genome of *Chlamydomonas*

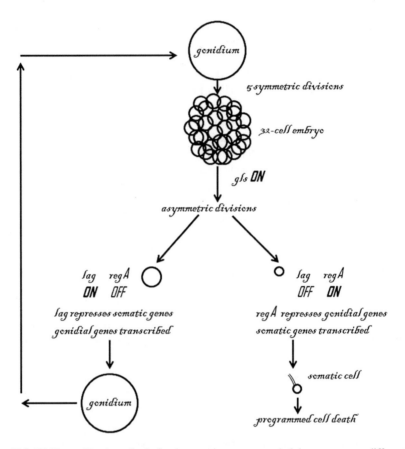

Fig. 11.9 Kirk's working hypothesis for the genetic program underlying germ-soma differentiation in *Volvox carteri*. Kirk notes that "[u]nderlying this hypothesis is the assumption that the ancestral program of volvocacean differentitation ("first vegetative and then reproductive") is also the default program in *V. carteri*" (Kirk, 1998, p. 256). Figure adapted by the author from Fig. 6.7, Kirk, 1998)

codes for 22 pherophorin proteins (extracellular matrix proteins that also have homology to sex-inducing pheromones), while the *V. carteri* genome codes for 49 (Nishii and Miller 2010).

The role of the extracellular matrix in allowing the assembly of a 4,000-cell structure cannot be underestimated. Domozych and Domozych write that

> it is estimated that each *Volvox* cell produces an ECM [extracellular matrix] that is 10,000 times larger than the ECM/cell wall of a *Chlamydomonas* cell… This strongly suggests that major elaborations of the ECM/cell wall were critical in the evolution of the multicellular habit in volvocine algae. ECM/wall components form the structural framework that provides the resistive force that counterbalances turgor pressure which would otherwise make formation/maintenance of the cytoplasmic bridges impossible. (Domozych and Domozych 2014)

Recall that the cytoplasmic bridges are critical to maintenance of structure during inversion. The complex ensemble of extracellular matrix proteins in *Volvox* may have had another major advantage. Hydroxyproline-rich glycoproteins coating the exterior of *Volvox* have evolved to play a role in sexual signaling, thus filling a developmental, rather than purely structural, function (Prochnik et al. 2010; Domozych and Domozych 2014).

What does the remarkable life cycle of *V. carteri* tell us about the tension between competition and cooperation in the emergence of biological collectives? First, the initial diversion of a lineage into various branches is not a particularly difficult task for evolution to accomplish. With the action of a very few genes, a temporal vegetative-then-reproductive lineage can be diverted into a spatially distributed lineage, with coexisting vegetative and reproductive cells. This remarkable passage from time into space allows us to infer history from a series in true Darwinian fashion, and provides a window into the moments in evolutionary history when a group becomes an individual. The great complexity of the volvocine lineage, however, reminds us that this transition is no simple linear march toward greater complexity. In order to to explore the transition to multicellularity with more precision, one would have to reproduce such evolutionary transitions in the laboratory. And that is precisely what researchers in the cutting-edge field of experimental evolution are doing.

References

Bell G, Koufopanou V (1991) The architecture of the life cycle in small organisms. Philos Trans R Soc Lond B 332:81–89

Brumley DR, Wan KY, Polin M, Goldstein RE (2014) Flagellar synchronization through direct hydrodynamic interactions. eLIFE 3:e02750

Buchheim MA, Chapman RL (1991) Phylogeny of the colonial green flagellates: a study of 18S and 26S rRNA sequence data. BioSystems 25(1–2):85–100

Buss LW (1987) The Evolution of Individuality. Princeton University Press, Princeton

Cheng Q, Pappas V, Hallmann A, Miller SM (2005) Hsp70A and GlsA interact as partner chaperones to regulate asymmetric division in *Volvox*. Dev Biol 286(2):537–548

Cheng Q, Hallmann A, Edwards L, Miller SM (2006) Characterization of a heat-shock-inducible *hsp70* gene of the green alga *Volvox carteri*. Gene 371(1):112–120

Coleman AW (2002) Comparison of *Eudorina*/*Pleodorina* ITS sequences of isolates from nature with those from experimental hybrids. Am J Bot 89(9):1523–1530

Dawkins R (1996) Climbing Mount Improbable. W. W. Norton & Company, New York

Dobell C (1960) Antony van Leeuwenhoek and His "Little Animals". Dover Publications, New York

Domozych DS, Domozych CE (2014) Multicellularity in green algae: upsizing a walled complex. Front Plant Sci 5:649

Drescher K, Goldstein RE, Michel N, Polin M, Tuval I (2010) Direct measurement of the flow field around swimming microorganisms. Phys Rev Lett 105(16):168101

Goldstein RE (2015) Green algae as model organisms for biological fluid dynamics. Annu Rev Fluid Mech 47:343–375

Hallmann A (2011) Evolution of reproductive development in the volvocine algae. Sex Plant Reprod 24:97–112

Han TM, Runnegar B (1992) Megascopic eukaryotic algae from the 2.1-billion-year-old Negaunee iron-formation, Michigan. Science 257(5067):232–235

Herron MD, Michod RE (2008) Evolution of complexity in the volvocine algae: transitions in individuality through Darwin's eye. Evolution 62:436–451

Herron MD, Hackett JD, Aylward FO, Michod RE (2009) Triassic origin and early radiation of multicellular volvocine algae. Proc Natl Acad Sci U S A 106(9):3254–3258

Kirk DL (1998) Volvox: Molecular-Genetic Origins of Multicellularity and Cellular Differentiation. Cambridge University Press, Cambridge

Kirk DL (2003) Seeking the ultimate and proximate causes of *Volvox* multicellularity and cellular differentiation. Integr Comp Biol 43(2):247–253

Kirk DL (2005) A twelve-step program for evolving multicellularity and a division of labor. BioEssays 27(3):299–310

Koufopanou V (1994) The evolution of soma in the Volvocales. Am Nat 143:907–931

Koufopanou V, Bell G (1993) Soma and germ: an experimental approach using *Volvox*. Proc R Soc Lond B 254:107–113

Larson A, Kirk MM, Kirk DL (1992) Molecular phylogeny of the volvocine flagellates. Mol Biol Evol 9(1):85–105

Meissner M, Stark K, Cresnar B, Kirk DL, Schmitt R (1999) *Volvox* germline-specific genes that are putative targets of RegA repression encode chloroplast proteins. Curr Genet 36(6):363–370

Merchant SS, Prochnik SE, Vallon O, Harris EH, Karpowicz SJ, Witman GB, Terry A, Salamov A, Fritz-Laylin LK, Maréchal-Drouard L, Marshall WF, Qu LH, Nelson DR, Sanderfoot AA, Spalding MH, Kapitonov VV, Ren Q, Ferris P, Lindquist E, Shapiro H, Lucas SM, Grimwood J, Schmutz J, Cardol P, Cerutti H, Chanfreau G, Chen CL, Cognat V, Croft MT, Dent R, Dutcher S, Fernández E, Fukuzawa H, González-Ballester D, González-Halphen D, Hallmann A, Hanikenne M, Hippler M, Inwood W, Jabbari K, Kalanon M, Kuras R, Lefebvre PA, Lemaire SD, Lobanov AV, Lohr M, Manuell A, Meier I, Mets L, Mittag M, Mittelmeier T, Moroney JV, Moseley J, Napoli C, Nedelcu AM, Niyogi K, Novoselov SV, Paulsen IT, Pazour G, Purton S, Ral JP, Riaño-Pachón DM, Riekhof W, Rymarquis L, Schroda M, Stern D, Umen J, Willows R, Wilson N, Zimmer SL, Allmer J, Balk J, Bisova K, Chen CJ, Elias M, Gendler K, Hauser C, Lamb MR, Ledford H, Long JC, Minagawa J, Page MD, Pan J, Pootakham W, Roje S, Rose A, Stahlberg E, Terauchi AM, Yang P, Ball S, Bowler C, Dieckmann CL, Gladyshev VN, Green P, Jorgensen R, Mayfield S, Mueller-Roeber B, Rajamani S, Sayre RT, Brokstein P, Dubchak I, Goodstein D, Hornick L, Huang YW, Jhaveri J, Luo Y, Martínez D, Ngau WC, Otillar B, Poliakov A, Porter A, Szajkowski L, Werner G, Zhou K, Grigoriev IV, Rokhsar DS, Grossman AR (2007) The *Chlamydomonas* genome reveals the evolution of key animal and plant functions. Science 318(5848):245–250

Miller SM, Kirk DL (1999) *glsA*, a *Volvox* gene required for asymmetric division and germ cell specification, encodes a chaperone-like protein. Development 126(4):649–658

Nedelcu AM (2009) Environmentally induced responses co-opted for reproductive altruism. Biol Lett 5:805–808

Nilsson DE, Pelger S (1994) A pessimistic estimate of the time required for an eye to evolve. Proc Biol Sci 256(1345):53–58

Nishii I, Miller SM (2010) *Volvox*: simple steps to developmental complexity? Curr Opin Plant Biol 13:646–653

Nozaki H, Itoh M, Sano R, Uchida H, Watanabe MM, Kuroiwa T (1995) Phylogenetic relationships within the colonial Volvocales (*Chlorophyta*) inferred from *rbc*L gene sequence data. J Phycol 31(6):970–979

Nozaki H, Yamada TK, Takahashi F, Matsuzaki R, Nakada T (2014) New "missing link" genus of the colonial volvocine green algae gives insights into the evolution of oogamy. BMC Evol Biol 14(1):37

Pappas V, Miller SM (2009) Functional analysis of the *Volvox carteri* asymmetric division protein GlsA. Mech Dev 126(10):842–851

Prochnik SE, Umen J, Nedelcu AM, Hallmann A, Miller SM, Nishii I, Ferris P, Kuo A, Mitros T, Fritz-Laylin LK, Hellsten U, Chapman J, Simakov O, Rensing SA, Terry A, Pangilinan J, Kapitonov V, Jurka J, Salamov A, Shapiro H, Schmutz J, Grimwood J, Lindquist E, Lucas S, Grigoriev IV, Schmitt R, Kirk D, Rokhsar DS (2010) Genomic analysis of organismal complexity in the multicellular green alga *Volvox carteri*. Science 329(5988):223–226

Ueki N, Matsuyaga S, Inouye I, Hallmann A (2010) How 5000 independent rowers coordinate their strokes in order to row into the sunlight: Phototaxis in the multicellular green alga *Volvox*. BMC Biol 8:103

Umen JG (2014) Green algae and the origins of multicellularity in the plant kingdom. Cold Spring Harb Perspect Biol 6(11):a016170

Umen JG, Olson BJ (2012) Genomics of volvocine algae. Adv Bot Res 64:185–243

Weismann A (1904) The Evolution Theory (trans. JA Thomson & MR Thomson). Edward Arnold, London

Chapter 12
Experimental Evolution

Keep your eyes peeled for a small black iron door.

David Mitchell

LONG BEFORE it was given such a name, experimental evolution took place during the serial passages of pathogenic viruses and bacteria necessary to develop live attenuated vaccines (Kawecki et al. 2012). The field of experimental evolution as we know it now, however, has developed rapidly and dramatically during the past three decades. Probably the best known experimental evolution study is Richard Lenski's long-term evolution experiment (LTEE) at Michigan State University. Starting with twelve replicate *E. coli* cultures in February 1988, the experiment reached an astonishing 66,000 generations in November 2016, and (of course) continues. Using a strain that lacks a mechanism for genetic exchange, any changes seen in the twelve lines are a result of mutations alone, under the action of selection.

When the experiment was first started, Lenski envisioned it lasting only 2000 generations. "I'd already had success with some shorter duration, more traditionally designed experiments," Lenski explained in a recent interview with Jeremy Fox for *PLoS Biology*.

> So it wasn't a total shot in the dark – I knew the LTEE would yield data. I also knew, though, it was an unusually abstract, open-ended, and nontraditional experiment... Maybe I was overly confident, but I was pretty sure the outcomes – whatever they might be – would be cool. I knew enough about what would happen – based on the experiments I had already done – that I was confident the data and analyses would be informative with respect to at least some of my questions. (Fox and Lenski 2015)

The outcome was far more than he could have dreamed of in 1988. The study has continued for decades, resulting in a wealth of publications on the role of historical contingency in the development of novel phenotypes and a host of other striking results regarding the tempo and mode of evolutionary change. One particular advantage of the LTEE (and other experimental evolution studies inspired by it) is that samples from any generation can be frozen and stored,[1] creating a living fossil

---

[1] Lenski's lab typically freezes samples of every 500th generation at −80 °C. Unfrozen "fossil" samples are allowed one day to thaw and acclimate to experimental conditions before being used for comparisons of fitness, cell size, or other assays.

© Springer Science+Business Media B.V. 2018
S. Bahar, *The Essential Tension*, The Frontiers Collection,
DOI 10.1007/978-94-024-1054-9_12

235

record. An organism can be thus set in competition with a resuscitated ancestor – allowing the experimenter to realize the evolutionary *gedanken* experiment of "replaying the tape", imagined by Gould (1989) in his narrative of the reassessment of the Burgess Shale fossils, *Wonderful Life*. The LTEE also allows, at least to some extent, investigation of macroevolutionary trends, and their relation to microevolutionary changes.

During their first 2000 generations, all *E. coli* strains in the LTEE exhibited rapid increases in average cell volume, and also in relative fitness with respect to their common ancestor (assayed as the ratio of growth rates between "competing" bacteria under identical experimental conditions). But by the time Lenski and his then graduate student Michael Travisano published their assessment of the LTEE at 10,000 generations, things had changed significantly. The growth rate and relative fitness of all the strains appeared to have plateaued over generations 5000–10,000.

A plateau in cell size was observed for all twelve replicate populations, suggesting that initial selection for larger cell size had been succeeded by period of stasis driven by "a genetic/developmental constraint (such that no new mutations could increase cell size further) or … stabilizing selection (such that both smaller and larger variants continued to appear but were purged by natural selection)" (Lenski and Travisano 1994). Of course, they noted, "whether or not cell size was a target for selection, the population may have continued to adapt to the environment (after size was static) by changing other traits." It was not possible to determine with certainty "whether cell size was the actual target of selection (or merely a correlated response to selection on other traits)". The cell size plateau, however, was not the only surprise that the LTEE provided at its landmark 10,000th generation. Continuing the trend observed in the first studies of the LTEE up to 2000 generations (Lenski et al. 1991), the populations continued to diverge *from each other* in cell size, as well as from their ancestor (Fig. 12.1). The among-population standard deviation of cell volume showed an increasing trend that could be fit, like the cell size data, by a hyperbolic curve.

Though it fluctuated more than the cell size data, fitness increased rapidly during the initial 2000 generations, and then seemed to plateau. As with the cell size data, the fitness trajectories of the different populations diverged, though there was no clear trend in the standard deviation data for fitness as there had been for cell size. Either the populations had drifted apart and eventually come to occupy different adaptive fitness peaks or were "climbing slowly along different ridges of unequal elevation toward the same peak" (Lenski and Travisano 1994). The results were indicative of "macroevolutionary" trends toward both divergence and stasis.

Lenski and Travisano next investigated the relationship between fitness and cell size. Both measures showed initial increases followed by plateaus, and a plot of cell size against relative fitness could be roughly fit by a line with positive slope. However, the relation between fitness and size within a single population over time (longitudinal regression), and among all populations at any given time point (cross-sectional regression), told a more complex story. Within any given lineage, cell size and fitness increased together, though the slope of this increase differed between lineages. The cross-sectional regressions, however, had slopes insignificantly differ-

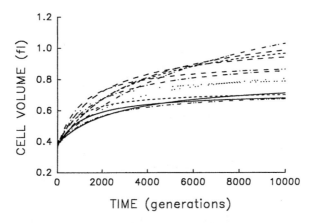

TIME (generations)

Fig. 12.1 Divergence of cell volume in 12 replicate *E. coli* populations in the first 10,000 generations of the LTEE. The *curves* show the best fit of a hyperbolic model from each population.
Reprinted with permission from R. E. Lenski and M. Travisano, Dynamics of adaptation and diversification: a 10,000-generation experiment with bacterial populations. *Proc. Natl. Acad. Sci. U.S.A.*
91: 6808–6814, 1994. Copyright (1994) National Academy of Sciences, USA

ent from zero. These results "do not support the hypothesis that the functional relationship between size and fitness is causal and rigidly fixed", the authors concluded.
In fact, the results suggest *a divergence in the relation between fitness and cell size
among the populations*.

Despite the divergent behavior of the populations, their evolutionary trajectories
for both size and fitness showed clearly parallel trends among all twelve populations. Lenski and Travisano suggest that this parallelism might be a result of the
identical experimental environments, as well as inevitable mutational redundancies,
given the number of population replications (~7.5×10^{11}) and the baseline mutation
rate (~2.5×10^{-3} mutations per genome replication). They noted that a researcher
stumbling upon the observed variability in their data but unfamiliar with the experimental conditions (essentially, the situation in which a paleontologist might find
herself when comparing two fossil beds)

> might attribute this diversity to environmental heterogeneity or phylogenetic constraints,
> but any such 'just-so story' would clearly be misguided in this case. Instead, our experiment
> demonstrates the crucial role of chance events (historical accidents) in adaptive evolution.
> (Lenski and Travisano 1994)

Broadly, they interpreted their results as indicative of a strong role for historical
contingency.

> Sustained divergence in mean fitness supports a Wrightian model of evolution in which
> replicate populations found their way onto different fitness peaks. Although the experimen-
> tal populations were so large that the same mutations occurred in all of them, the order in
> which various mutations arose would have been different… As a consequence, some
> populations may have incorporated mutations that were beneficial over the short-term but
> led to evolutionary dead-ends. (Lenski and Travisano 1994)

They noted the difficulty of moving to a higher adaptive peak in the traditional Wrightian fitness landscape model, since this would be hindered by selection against mutations that led to transient drops in fitness. "In this respect," they wrote, "it is important that, in our experiment, populations were not on one adaptive peak and asked to 'jump' to another; instead, they were thrown into an arbitrary environment and asked to climb an accessible peak."

The behavior of the LTEE exhibited rapid diversification into a new environment.

> The initial rapid evolution was presumably due to intense selection triggered by the sudden environmental changes imposed at the start of our experiment. Although the ancestors of the founding bacterium used in this study had been "in captivity" for several decades, they were not systematically propagated under the experimental conditions we imposed (serial dilution in glucose-minimal medium). The experimental regime was therefore an unusual environment... [T]he most reasonable interpretation for the eventual stasis in our experimental populations is that the organisms have "run out of ways" to become much better adapted to their environment. (Lenski and Travisano 1994)

There were some hints of "quasi-punctuated dynamics" as well. During the first 2000 generations, relative fitness exhibited transient plateaus followed by step-like increases. Lenski and Travisano attributed this to the fact that a favorable allele appears at an initially low frequency and does not have an appreciable effect until it has spread through a large portion of the population. "However," they noted,

> we saw no compelling evidence for any more radical punctuation, such as when one adaptive change sets off a cascade of further changes.... Such an effect might have been manifest by a period of renewed, rapid evolutionary change in a population that had previously been at or near stasis. Perhaps 12 populations and 10,000 generations were too few to see such rare events. (Lenski and Travisano 1994)

The punctuated pattern was observed only when sampling every 100 generations, not at the coarser intervals of every 500 generations. A study by Elena et al. (1996) revealed even clearer "step-like" changes in cell volume when sampling every 100 generations (Fig. 12.2). Importantly, this showed an instance in which a punctuated pattern could arise as a result of only "the two most elementary population genetic processes: mutation and natural selection" (Elena et al. 1996). Of course, a *punctuated pattern* is not *punctuated equilibrium* per se; that term refers to macroevolutionary dynamics at the species level or higher, above over much longer time scales (Gould 2002, pp. 931–934). Punctuated patterns have been observed in many other biological growth processes, including human growth (Gould 2002, pp. 934–935).

A follow-up study at 50,000 generations (Wiser et al. 2013) provides an interesting epilogue to this aspect of the LTEE story. Measurements of relative fitness over an additional 40,000 generations revealed a surprising trend. Instead of leveling off, fitness continued to increase, though at a far slower rate than in the early years of the experiment. Moreover, 6 of the 12 populations had acquired a "hypermutator" phenotype – their point-mutation rates had increased by a factor of about 100. Comparison of the hypermutators to the other populations showed that the former had a faster rate of fitness increase (Fig. 12.3).

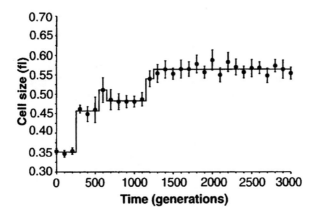

Fig. 12.2 A punctuated pattern in *E. coli* cell size in the LTEE. Change in average cell size (1 fl = 10^{-15} L) over 3000 generations. Data points show the mean of 10 replicate assays, and error bars show 95% confidence intervals. The data is fit with a step function model. From S. F. Elena, V. S. Cooper, and R. E. Lenski, Punctuated evolution caused by selection of rare beneficial mutations, *Science* 272: 1802–1804, 1996. Reprinted with permission from AAAS

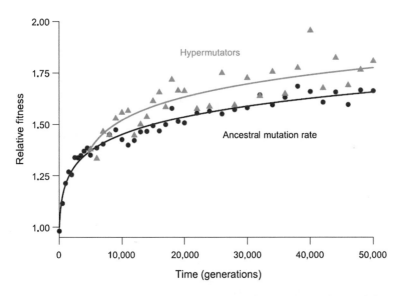

Fig. 12.3 Fitness continues to increase. Wiser et al. (2013) found that mean fitness of the twelve ancestral *E. coli* populations in the LTEE continued to increase over 50,000 generations. Six of the populations (*black circles*) retained their ancestral mutation rate, while the other six (*green triangles*) showed a hypermutator phenotype. *Curves* show theoretical predictions of dynamical models; the model shown with the *green curve* has a mutation rate 100 times that of the one with the *black curve*. From M.J. Wiser, N. Ribeck, and R.E. Lenski, Long-term dynamics of adaptation in asexual populations, *Science* 352: 1364–1367, 2013. Reprinted with permission from AAAS

Consistent with the lack of a plateau, hyperbolic functions did not fit the data well over this extended period of time. Not only was the data much better fit by a power law, but a power-law fit to the first 10,000 generations predicted the future growth curve far better than did the original hyperbolic fit used in 1994. A theoretical model accounting for clonal interference[2] and diminishing-returns epistasis[3] provided a similar result. Looking back at 25 years of data, Lenski and his colleagues could now conclude that "[b]oth our empirical and theoretical analyses imply that adaptation can continue for a long time for asexual organisms, even in a constant environment" (Wiser et al. 2013).

Among the many fascinating results of the LTEE, one of the most dramatic was the evolution of a novel phenotype that enabled the bacteria to grow aerobically using citrate as a nutrient source (Blount et al. 2008, 2012). Under normal conditions, *E. coli* cannot metabolize citrate. Indeed, the "inability to use citrate as an energy source under oxic conditions has long been a defining characteristic of *E. coli* as a species"; it lacks a citrate transporter that can import citrate under oxic conditions (Blount et al. 2008). However, one case of spontaneous mutation to a Cit⁺ phenotype was observed in the 1980s, and the *E. coli* genome does encode a citrate-succinate exchanger, though this appears to only be induced under anoxic conditions. *E. coli* do need citrate, since they use a ferric dicitrate transport system in order to facilitate influx of iron. Since the standard DM25 medium used for the LTEE contained 1700 μM citrate as a chelating agent, cells did have the opportunity to develop mutations that would enable them to utilize citrate as a carbon source[4]; after all, by the 30,000th generation of the LTEE, each of the dozen populations had undergone billions of mutations (Blount et al. 2008).

At generation 33,127, an increase in the turbidity of samples from population Ara-3[5] was observed, indicating a sharp rise in population size (Fig. 12.4). Sudden turbidity increases had been observed before, but had previously been traced to contaminants. Here, however, careful analysis showed that the increased population size originated from the Ara-3 strain, not from a contaminant. After thousands of generations of comparative stasis, a key innovation had suddenly appeared, metabolizing the citrate present in the medium.

---

[2] In clonal interference, competition occurs among organisms with different beneficial mutations, which makes it more difficult for the beneficial mutations to spread throughout the population; this is a particular issue in asexual populations, which do not undergo genetic recombination, rendering the beneficial mutations incapable of "joining forces".

[3] In diminishing-returns epistasis, "the marginal improvement from a beneficial mutation declines with increasing fitness" (Wiser et al. 2013).

[4] The DM25 medium contained 139 μM glucose, serving as a carbon source for non-Cit⁺ *E. coli*.

[5] When establishing the LTEE in 1988, Lenski and colleagues used twelve *E. coli* strains derived from a single clone, identical except for their ability to utilize arabinose. Six strains had a mutation that allowed them to utilize arabinose (strains Ara+1, Ara+2, etc.…), and six did not (Ara-1, Ara-2, etc.…) This mutation had no effect on the strains under the conditions of the LTEE, which did not include arabinose in the DM25 medium, but did allow the strains to be distinguished from each other when grown on tetrazolium-arabinose plates, where the Ara⁻ cells make red colonies and the Ara⁺ cells make colonies that appear white.

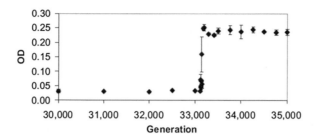

Fig. 12.4 Emergence of the Cit[+] phenotype. A sudden increase in optical density (OD) at 420 nm indicated a dramatic increase in the *E. coli* population. *Error bars* indicate the range of three measured OD readings for each generation. Reprinted with permission from Z.D. Blount, C.Z. Borland, and R.E. Lenski, Historical contingency and the evolution of a key innovation in an experimental population of *Escherichia coli*. *Proc. Natl. Acad. Sci. U.S.A.* 105(23): 7899–7906, 2008. Copyright (2008) National Academy of Sciences, USA

Lenski and colleagues investigated "frozen fossil" samples from earlier generations in order to determine precisely when the Cit[+] phenotype had arisen. They found that the phenotype had increased in the population for some time, and then decreased in prevalence for a while, before finally reaching a significant[6] and stable level.

Cit[+] clones could be readily isolated from the frozen sample of population Ara-3 taken at generation 33,000. To estimate the time of origin of the Cit[+] trait, we screened 1,280 clones randomly chosen from generations 30,000, 30,500, 31,000, 31,500, 32,000, 32,500, and 33,000 for the capacity to produce a positive reaction on Christensen's citrate agar, which provides a sensitive means to detect even weakly citrate-using cells. No Cit[+] cells were found in the samples taken at 30,000, 30,500, or 31,000 generations. Cit[+] cells constituted ≈0.5% of the population at generation 31,500, then 15% and 19% in the next two samples, but only ≈1.1% at generation 33,000. It appears that the first Cit[+] variant emerged between 31,000 and 31,500 generations, although we cannot exclude an earlier origin. The precipitous decline in the frequency of Cit[+] cells just before the massive population expansion suggests clonal interference..., whereby the Cit[-] subpopulation produced a beneficial mutant that outcompeted the emerging Cit[+] subpopulation until the latter evolved some other beneficial mutation that finally ensured its persistence. The hypothesis of clonal interference implies that the early Cit[+] cells were very poor at using citrate, such that a mutation that improved competition for glucose could have provided a greater advantage than did marginal exploitation of the unused citrate. (Blount et al. 2008)

In the context of the LTEE's potential to shed light on macroevolutionary trends, the new Cit[+] variant was potentially revelatory. Was it simply the result of a rare mutation[7] finally making its appearance? Or was facilitated by the background of

[6] Cit[+] cells never completely dominated the population; about 1% of the population retained the Cit[-] phenotype when Cit[+] and Cit[-] cells were grown together. As Blount et al. (2008) observed, "[a]lthough the Cit[+] cells continued to use glucose, they did not drive the Cit[-] subpopulation extinct because the Cit[-] cells were superior competitors for glucose. Thus, the overall diversity increased as one population gave rise evolutionarily to an ecological community with two members, one a resource specialist and the other a generalist."

[7] Ara-3 was *not* one of the populations that had evolved a hypermutator phenotype!

prior chance mutation(s), and thus "contingent on the particular history" of the Ara-3 population? If so, it would provide a singular experimental example of how "random and deterministic processes become intertwined over time such that future alternatives may be contingent on the prior history of an evolving population" (Blount et al. 2008).

In order to test whether the Cit[+] phenotype had arisen in a historically contingent manner, Lenski and his colleagues resuscitated an array of cells ancestral to Ara-3 in order to replay the tape of evolution. If the innovative ability to metabolize citrate was contingent on the presence of a particular background of genetic changes, then one would expect the Cit[+] phenotype to arise with higher probability in populations older[8] than some threshold ancestral generation, when the necessary genetic background presumably arose. In contrast, if the Cit[+] phenotype was the result of a single mutation independent of the genetic background, it should arise with equal probability when the tape was replayed for any Ara-3 fossil population, regardless of their age. A range of experiments, involving hundreds of replicates, supported the former hypothesis. Cit[+] never arose in ancestral populations that had been frozen before generation 20,000, and were most likely to arise after around generation 30,000. A further series of "fluctuation experiments", using seven Ara-3 clones that had produced Cit[+] cells in at least one of the replay experiments, showed that this "potentiated" genetic background increased the likelihood of the evolution of the Cit[+] strain by at least a factor of two compared to ancestral Ara-3. This "increased" rate, however, was startlingly low (indeed, three orders of magnitude lower!) compared to the baseline mutation rate in the LTEE, suggesting that the Cit[+] phenotype arose by a more complicated mechanism than simply a single point mutation.

Were the mutational changes identical each time the Cit[+] phenotype re-evolved? In order to answer this question, and to investigate the specific genetic changes driving the transition to citrate metabolism, Lenski's lab undertook the genomic analysis of the Cit[+] populations. They expected to identify a number of genetic changes in each variant, noting that

> emergence of the Cit[+] phenotype in population Ara-3 indicates at least two important genetic events: the origin of the function in its weak form, and its subsequent refinement for efficient use of citrate. The replay experiments indicate an even more complex picture that must involve, at a minimum, three important genetic events. At least one mutation in the LTEE was necessary to produce a genetic background with the potential to generate Cit[+] variants, while the distribution and dynamics of Cit[+] mutants in fluctuation tests indicate at least two additional mutations are involved. (Blount et al. 2008)

When the genomic analysis was published four years later, however, it yielded an even more complex picture than could have been imagined in 2008. Blount et al. (2012) sequenced 29 clones taken from various generations of the Ara-3 lineage, including nine Cit[+] clones and three Cit[−] clones from generations after the potentiating mutations had presumably occurred. They looked for various mutations such as single-nucleotide polymorphisms (SNPs) as well as insertions, deletions, and chromosomal rearrangements, and reconstructed the entire phylogenetic history of

---

[8] Note that "older" means later in the lineage, not "more ancestral".

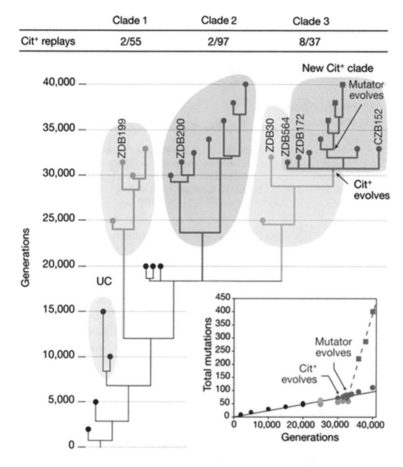

Fig. 12.5   Ara-3 phylogeny (Figure 1, Blount et al., 2012). Symbols at the branch tips correspond to the 29 sequenced clones. Reprinted by permission from Macmillan Publishers Ltd., Z.D. Blount, J.E. Barrick, C.J. Davidson, and R.E. Lenski, Genomic analysis of a key innovation in an experimental *Escherichia coli* population. *Nature* 489: 513–518, 2012, copyright 2012

Ara-3 (Fig. 12.5). They identified multiple clades that had arisen before 20,000 generations. One of these appeared to become extinct by about 15,000 generations; they called this "UC", for "unsuccessful clade". Three others (C1, C2 and C3) persisted through the emergence of the Cit⁺ phenotype. A molecular clock estimate showed that C1 had diverged from the common ancestor of C2 and C3 comparatively early in the LTEE, before 15,000 generations. C2 and C3 diverged from each other by generation 20,000, and the Cit⁺ phenotype arose in clade C3.

One particularly interesting result from the genetic analysis was that Cit⁺ clones after generation 36,000 had an SNP in the *mutS* gene, which codes for a protein involved in DNA mismatch repair. No evidence of this SNP was found in generations before 36,000, indicating that the mutation had arisen *after* the first cells began to express the Cit⁺ phenotype. Once it arose, this SNP facilitated further mutations, so

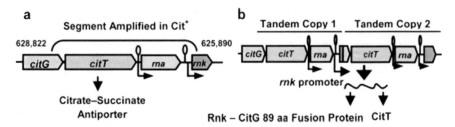

Fig. 12.6 Gene amplification actualizes the Cit⁺ phenotype. Panel **a** shows the ancestral arrangement of citT, rna, and the rnk promoter region. Panel **b** shows the regulatory changes that result from tandem amplification. Reprinted by permission from Macmillan Publishers Ltd., Z.D. Blount, J.E. Barrick, C.J. Davidson, and R.E. Lenski, Genomic analysis of a key innovation in an experimental *Escherichia coli* population. *Nature* 489: 513–518, 2012, copyright 2012

that Cit⁺ cells after generation 36,000 accumulated mutations much faster than their Cit⁻ cousins and their Cit⁺ ancestors (see inset in Fig. 12.5). Recall that Ara-3 was *not* one of the lineages that had previously developed a hypermutator phenotype!

What mutations caused the "actualization" of the Cit⁺ trait, enabling the cells to metabolize citrate under aerobic conditions? A prime suspect was identified when all nine sequenced Cit⁺ genomes were found to contain tandem copies of part of the *cit* operon, which controls citrate fermentation. The amplified segment included *citT*, which encodes the citrate-succinate exchanger, a protein typically inactive except under anaerobic conditions. The segment was

> not present in the ancestor or any of the sequenced Cit⁻ genomes and it is only found in population samples after the evolution of the Cit⁺ lineage… Polymerase chain reaction (PCR) screens also failed to detect this segment in 27 Cit⁻ clones from generations 33,000 through 40,000, whereas it was found in all 33 Cit⁺ clones from the same generations. (Blount et al. 2012)

Southern blot and PCR studies of 13 isolates from a generation 33,000 clone that had reverted to Cit⁻ showed no evidence of the segment.

The repeated segment contained *citT* and another gene, *rna* (which codes for RNase I), as well as fragments of genes from either end of the *cit* operon. When the segment was repeated, this produced a hybrid sequence containing a promoter region for *citT*, aligned adjacent to *citT* in the next tandem copy (Fig. 12.6). If the mutated promoter[9] region functioned under aerobic conditions, Blount et al. hypothesized, "the new *rnk-citT* regulatory module might allow CitT expression during aerobic metabolism, and thereby confer a Cit⁺ phenotype".

To test this prediction, they expressed the altered *rnk-citT* module, via a plasmid, in the ancestral LTEE cell line, as well as in two clones from clade C3. One of these, ZDB30, was a potentiated Cit⁻ clone from generation 32,000; the other, ZDB172,

---

[9] Blount et al. note that "[c]omparative studies have shown that gene duplications have an important creative role in evolution by generating redundancies that allow neo-functionalization… Our findings highlight the less-appreciated capacity of duplications to produce new functions by promoter capture events that change gene regulatory networks."

was a "weakly Cit⁺ clone", also from generation 32,000. They found expression of the *rnk-citT* module in the latter two clones, but not in the ancestral population. Thus, in the genetic background that allowed for Cit⁺ to evolve, the mutant regulatory region was indeed capable of expressing a citrate transporter under aerobic conditions. A version of ZDB30 with *rnk* promoter inserted immediately upstream of *citT* exhibited a weak Cit⁺ phenotype. While these cells were barely able to utilize citrate, they were able to out-compete ZBD30 cells lacking the mutant promoter.

Evidence suggested that actualization and potentiation act epistatically, since expression of *rnk-citT* in the ancestor and in clones from clades C1 and C2 exhibited much weaker Cit⁺ phenotypes than expression in potentiated clones from C3. Potentiation of Cit⁻ lineages that exhibit an increased propensity to evolve into Cit⁺ may be due to a mutation in *arcB*, which codes for a histidine kinase whose disruption upregulates the tricarboxylic acid cycle.

Expression of the *rnk-citT* segment from early Cit⁺ clones produced very weak Cit⁺ phenotypes, able to metabolize citrate, but at low efficiency. Blount et al. next set about investigating how this phenotype was "refined", growing stronger and more competitive in subsequent generations. Later generations had increased copy numbers of the *rnk-citT* segment, up to nine copies around generation 33,000. Later clones went down to four copies, however, without a drop in the ability to metabolize citrate, suggesting that other "stabilizing" mutations might have occurred in in the interim, and that refinement itself is a multi-step process.

The *citT* region appears to be quite labile in Ara-3. Nineteen of the "re-evolved" Cit⁺ strains from the "replaying the tape" experiments showed duplicated of portions of the *citT* region, though none of these duplications are spliced at exactly the same sites. In one strain, a large portion of the *cit* operon moved to a location downstream of the promoter for a different gene entirely. The mutations that might have potentiated this lability, and other factors controlling the evolution of the Cit⁺ phenotype, are currently being explored by a number of research groups (Leiby et al. 2012; Leiby and Marx 2014; Quandt et al. 2015; Turner et al. 2015).

*** *** ***

One of the most exciting areas of experimental evolution – and the area most relevant for the theme of this book – is the evolution of multicellularity. These studies have consistently shown how easy it is for multicellularity to arise, under straightforward selective conditions and often with a minimum of genetic change. One striking example of this is provided by the bacterial swarming study published by Velicer and Yu (2003). Wild-type, *Myxococcus xanthus* can form swarms and fruiting bodies in a manner reminiscent of *Dictyostelium*. In order to swarm on soft agar, *M. xanthus* uses a mechanism known as S-motility, which is mediated by extracellular fibrous extensions called type IV pili. Pilin, a major structural component of these fibers, is coded for by the *pilA* gene. Velicer and Yu studied eight populations of a lineage in which they had deleted a large portion of *pilA*, so that the bacteria were unable to swarm effectively. They performed a series of transfer experiments for each lineage; in each transfer cycle (performed every 2 weeks), they

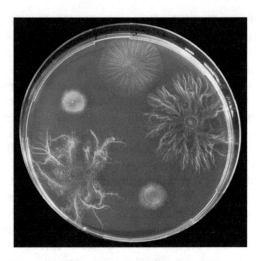

Fig. 12.7 Colony swarming in Velicer and Yu's experiment. Colonies counterclockwise from top are wild-type, A1, E7, A2, and E8. S-motility was inhibited in A1, A2, E7, and E8. New swarming phenotype E7 evolved from A1, and E8 evolved from A2. Reprinted by permission from Macmillan Publishers Ltd., G.J. Velicer and Y.T. Yu, Evolution of novel cooperative swarming in the bacterium *Myxococcus xanthus. Nature* 425(6953): 75–78, 2003, copyright 2003

took a sample of cells that had reached the edge of the agar plate, and plated them in fresh agar.

After about 30 transfers, each lineage had increased its swarming rate relative to the ancestral strain, though all strains exhibited smaller swarming rates than wild type (WT). Two lineages in particular, E7 and E8, "exhibited particularly dramatic improvements". They also showed significantly different patterns of swarming compared to wild type (and compared to one another), as shown in Fig. 12.7. "Whereas WT radiates evenly outward in a circular swarm," the authors wrote, "E7 extends outward in a pattern of branching tentacles and E8 expands in broad fans." Moreover, the *mechanism* of swarming in the evolved strains was completely different from that of wild type. Velicer and Yu verified that the *pilA* deletion was still present in E7 and E8, since antibodies to pilin failed to detect the protein, and electron microscopy showed no pili on the cells. Rather than using pili, swarming in E7 and E8 appeared to be mediated "by enhanced production of an extracellular fibril matrix that binds cells – and their evolutionary interests – together", Velicer and Yu reported.

The bacteria had co-opted a completely different cellular apparatus in order to swarm, taking advantage of A-motility, which is typically deployed for individual cell movement on hard agar rather than swarming on soft agar. "These results," Velicer and Yu concluded, "show that fundamental transitions to primitive cooperation can readily occur in bacteria." This cooperation did not come without a cost, however. *Inhibition* of the production of the fibrils necessary for this new mechanism of swarming increased the growth rate of strains E7 and E8 in liquid culture (by 5.3% and 4.6%, respectively; these growth increases were statistically significant). "Thus," Velicer and Yu concluded, evolved fibril production slows individual reproductive rate under unstructured, asocial conditions, but appears to be favoured by selection at the kin-group level in a structured habitat by its mediation of enhanced group migration. Superior migration allowed clustered groups of related cooperative genotypes to move into territory and consume resources that were unavailable to non-cooperative, poorly swarming competitors. (Velicer and Yu 2003)

In the same September 2003 issue of *Nature* in which Velicer and Yu published their *M. xanthus* study, Rainey and Rainey demonstrated the evolution of multicellularity using *Pseudomonas fluorescens*. *P. fluorescens* is a rod-shaped bacterium that grows under aerobic conditions. Rainey and Rainey observed the evolution of *P. fluorescens* as the bacteria adapted to take advantage of various niches in a heterogeneous broth environment. Most interesting was the "wrinkly spreader" (WS) phenotype, so named in contrast to its smooth (SM) ancestor. The wrinkly spreaders formed a biofilm at the air-broth interface. "Colonization of this niche," wrote Rainey and Rainey, "enables cells to avoid the anoxic conditions that rapidly build up in unshaken broth culture."

Rainey and Rainey found that the cooperating groups of *P. fluorescens* were potentiated by mutations in a set of ten genes on the *wss* operon. These mutations result in overproduction of an adhesive, acetylated cellulose polymer that provides a structural matrix for the biofilm (Spiers et al. 2002, 2003). Group adaptation comes at an individual cost, however. Competition experiments pitting individual WS cells against their ancestral SM genotype under abundant resource conditions showed that the WS cells had only 80% the fitness of their ancestor.[10] "Despite a much reduced doubling time," the authors reported, "WS readily invades (from a single mutant cell) populations dominated by the ancestral genotype to reach population densities that exceed those of the originally dominant ancestral type" (Rainey and Rainey 2003).

A genotype useful for a group of related organisms, but costly for individuals on their own, provides a ripe breeding ground for cheaters looking for a free ride. In this case, cheating cells avoided the metabolic cost of cellulose production by reverting to the SM genotype. They spread within the *P. fluorescens* colony; after 3 days, 24% of cells in the biofilm mat were cheaters. Despite their early success, cheaters ultimately proved detrimental to the group, since the decreased amount of cellulose weakened the biofilm mat and caused it to collapse.[11]

The problem of cheating factored prominently in experiments performed by Velicer and colleagues[12] demonstrating the evolution of *asocial* behavior in *M. xanthus* (Velicer et al. 1998, 2002; see Velicer and Stredwick 2002 for review). Six populations were derived from each of two ancestral clones. The two ancestors dif-

---

[10] Relative fitness was measured as the ratio of doublings of the competing genotypes, using a method described by Lenski and colleagues in the context of the LTEE (Lenski et al. 1991).

[11] Even colonies without cheaters collapse eventually, when the mat becomes too heavy to remain at the air-broth interface and sinks below the surface, where all the cells perish due to lack of oxygen. Colonies with cheaters collapse faster than purely cooperative colonies.

[12] Gregory Velicer was a graduate student, and later a postdoc, in Lenski's laboratory before moving to Indiana University and later leading his own research laboratories at the Max Planck Institute in Tübingen and at the ETH in Zurich.

fered only in a single gene (for resistance to rifampicin), which allowed the investigators to distinguish between the two derived lineages when grown together. The 12 populations were grown under conditions that promoted asociality: a liquid medium rich in nutrients. After 1000 growth cycles (each a 100-fold dilution into new medium), the bacterial growth rate had increased by an average of 36%. The cells, however, had essentially lost S-motility, the ability to form fruiting bodies, and the ability to sporulate. All 12 lines were capable of A-motility on hard agar, but moved significantly slower than their ancestors. The three lineages that did attempt to aggregate formed defective fruiting bodies (Velicer et al. 1998).

While all the lineages gave up their capacity for social behavior in exchange for greater fitness in the nutrient-rich medium, they did so in markedly different ways. There was significant variability in the change in growth rate among the different populations. Among lineages that completely lost their social capabilities, one increased its growth rate by 53%, and another by only 21%. A third lineage increased its growth rate significantly, but retained motility, sporulation, and fruiting body formation, albeit in somewhat reduced form (Velicer et al. 1998; Velicer and Stredwick 2002). Given the rapidity of the development of the various asocial phenotypes, the investigators concluded that these changes were likely due to selection against "costly and unnecessary" social functions. Moreover, "the variation among the derived lines reflects the stochastic appearance of several different mutations that all improve fitness in the asocial regime while having heterogeneous effects on social functions (depending on the underlying genes)" (Velicer et al. 1998). Some of these evolved asocial strains behaved as cheaters when co-cultured with wild type *M. xanthus*, and their descendants were disproportionately represented in spores arising from a mixed population (Velicer et al. 2000).

Even asocial cheaters can be reformed. A few years later, Velicer's group used experimental evolution to turn a strain that had been evolved into an obligate cheater into a "superior cooperator". This "was caused by a single mutation of large effect that confers fitness superiority over both ancestral genotypes [original cooperator and evolved cheater], including immunity from exploitation by the ancestral cheater" (Fiegna et al. 2006).

Kim et al. (2016) recently demonstrated the ease with which division of labor evolves in a bacterial population. They studied a mucoid strain of *Pseudomonas fluorescens* referred to as the "M strain", which produces a glucose-based polymer. When grown on plates, M strain clones spread outward rapidly in a fan-like configuration. This in itself was not so surprising, as this sort of configuration is often observed in bacterial colonies. What struck the researchers, however, was that the spreading colonies contained not just the original M strain clone, but also a second strain that had evolved quite quickly from this first one. This strain, called "D" because of its dry, wrinkly appearance, allowed the colony to spread much more rapidly than the M cells could alone. In fact, D cells literally pushed the M cells outward toward the edge of the colony, leading to faster spreading (Fig. 12.8). The pushing behavior could be attributed to the relatively inflexible nature of the D cells, which have rigid extracellular polymers on their surfaces. The more lubricated M cells are easily pushed along by the D cells growing on top of them. This leaves

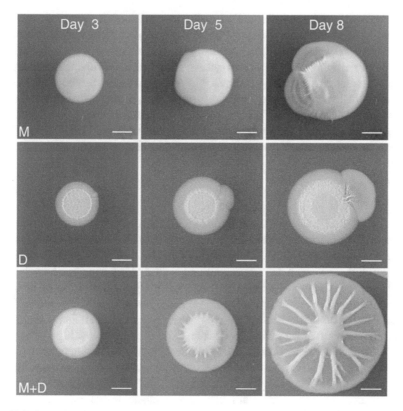

Fig. 12.8 Spreading fans of *P. fluorescens* strains. *Top row* shows the M-strain only, *middle row* the D-strain, and the *bottom row* a mixed population of M and D cells (initially at a 1:1 ratio). Scale bar shows 5 mm. Reproduced from Kim et al. (2016), under a Creative Commons Attribution 4.0 International License

"narrow tracks of stationary M cells underneath the moving D cells akin to scraped earth beneath the base of a moving glacier". Sitting on top of M cells, the D cells are able to access a large supply of oxygen, and thus their proportion increases rapidly within the colony. The two strains typically reached a balance of 90% D and 10% M, regardless of their initial proportions. Together, this combined phenotype was able to spread better than either cell type alone, and remained robust to perturbations.

The D phenotype resulted from a two-nucleotide deletion in a gene involved in the production of cyclic di-3′-5′-guanylate (c-di-GMP), a second-messenger protein that regulates motility and induces a wrinkly phenotype. Further experiments showed that the D phenotype could evolve against a background of M cells quite easily. Different instances involved different mutations, but all showed mutations in the locus regulating c-di-GMP production (Kim et al. 2016).

The evolution of D cells against an M cell background was mirrored by the evolution of M-type cells in a population of D cells. A proportion of M cells consis tently exhibited mutations disabling c-di-GMP production. When these newly evolved M cells were grown alone, they too allowed for the evolution of a new D morphotype in their midst. The D phenotype could arise, then, even in the absence of a functional c-di-GMP production system. Kim et al. concluded that "robust bidirectional evolution… reliably generates whichever partner is missing for collective spreading". The necessary mutations arose with ease, so that "[t]he robustness and organization of the collective phenotype contrasts with the simplicity of its origin"; all the mutations involved the activation or deactivation of a single intracellular messenger which controls the transition from a motile to a sessile state (Kim et al. 2016).

*** *** ***

Experimental studies of the evolution of multicellularity have been pioneered by Michael Travisano, who, after training at Michigan State with Richard Lenski, now runs his own laboratory at the University of Minnesota. Travisano and colleagues, working with both yeast (*S. cerevisiae*) and with *Chlamydomonas*, have laid out simple sets of conditions under which multicellularity and division of labor can arise under selection. In the first of these studies, Travisano and his group set out to investigate the conditions under which a transition to multicellularity might take place (Ratcliff et al. 2012). Does such a transition occur more readily by "aggregation of genetically distinct cells, as in biofilms, or by mother-daughter adhesion after division"? Under what conditions would selection on multicelled clusters come to dominate over selection on individual cells? Would the answers to these questions depend on the particular mechanism of cluster formation?

Previous studies showed how ecological conditions could favor the formation of multicellular clusters: the work of Boraas et al. (1998) had demonstrated that "predation by a small-mouthed ciliate results in the evolution of eight-celled clusters in the previously single-celled algae *Chorella*" (Ratcliff et al. 2012). Koschwanez et al. (2011) had shown that "metabolic cooperation among cluster-forming yeast allows them to grow at low densities prohibitive to growth of single-celled yeast" (Ratcliff et al. 2012). In contrast to these earlier works, Travisano's group set out to create simple experimental conditions that would select for multicellularity in a precisely controllable way. To do this, they chose gravity to select for multicellularity in *S. cerevisiae*. While perhaps not biologically realistic for yeast, this experimental scheme allowed precise measurement and tuning of conditions under which multicelled groups of yeast might be favored.

Ten isogenic populations of yeast were separated and grown in nutrient-rich medium, in shaking flasks. Then, for a 45 min period before transfer to a new flask, the shaking was stopped, and the yeast were allowed to settle. Only a sample of the yeast that had settled most (the bottom 100 µl of the flask) were then transferred to new medium.[13] This process was repeated for 60 transfer cycles; after the first week, the settling step was changed to centrifugation of each sample for 10 s at 100 g.

---

[13] Note how this essentially mimics a reverse of ZoBell's "bottle effect" (Chap. 9).

Fig. 12.9 Rapid, covergent evolution of the "snowflake" phenotype in the yeast *S. cerevisiae*. Replicate population number is shown in the lower right. In all ten populations, five of which are shown here, clusters developed which do not dissocate after budding. Reprinted with permission from W.C. Ratcliff, R.F. Denison, M. Borello, and M. Travisano, Experimental evolution of multicellularity. *Proc. Natl. Acad. Sci. U.S.A.* 109(5): 1595–1600, 2012

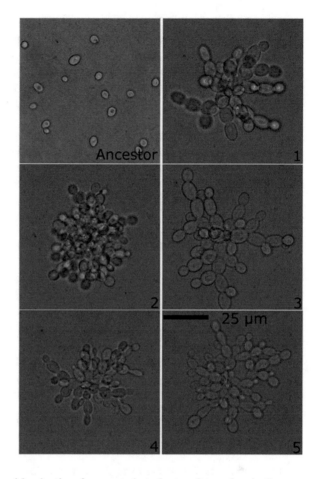

Ratcliff et al. observed rapid selection for yeast that clustered together in "snowflake" configurations (Fig. 12.9), and were thus better able to settle in their flasks. The snowflake clusters were not formed by aggregation, but by adhesion following cell division. In contrast to clusters of yeast (hyphae) that form under low nutrient conditions, the snowflake phenotype was maintained both in high and low nutrient concentrations. Competition experiments that pitted the snowflake clusters against normal unicellular *S. cerevisiae* showed that the snowflakes had a slightly lower fitness under normal conditions (i.e., without selection for settling). When they were allowed to continue growth, the snowflake clusters split into smaller clusters ("propagules"), after the parent cluster reached a certain size. In other words, this "the snowflake phenotype exhibits juvenile/adult life stage differentiation" (Ratcliff et al. 2012).

   In the next stage of their experiment, the investigators set up a protocol that allowed selection for rapidity of settling. They took nine replicates from a population in which the snowflake phenotype had already evolved, and exposed three replicates to a series of transfers with 5 min allowed for gravitational settling, another three to transfers with 15 min for settling, and the last three to transfers with 25 min for settling. As before, each transfer was performed with a sample from the bottom of each aliquot. "With mutation as the only source of within-cluster genetic variants," they wrote, "selection among clusters was expected to dominate within-cluster selection, leading to adaptation in multicellular traits" (Ratcliff et al. 2012). In other words, they had experimentally established the conditions for group selection.

As expected, the stronger selection conditions (reduced time to settle) selected for snowflakes that settled faster; these clusters were larger, and broke into larger propagule clusters, than the original snowflake ancestor population. Larger cluster size at the time of reproduction (i.e., when a snowflake split into propagule clusters) represented a "longer juvenile phase". Most importantly, this was an emergent multicellular trait. "Because the response to selection changed the multicellular phenotype," they wrote, "we conclude that selection was acting on the reproduction and survival of individual clusters rather than on that of their component cells" (Ratcliff et al. 2012).

The snowflake experiments showed the division of labor as well as the evolution of simple multiclelularity. Snowflake yeast experienced a tradeoff between settling rate and growth rate. Larger clusters settled faster and were thus selected for, but these clusters grew more slowly, likely because cells in the interior of the cluster were starved for access to resources. An adaptive strategy arose to compensate for this: the "self-sacrifice" of a percentage of cells by apoptosis.

> Snowflake yeast that produce smaller propagules can make more of them, increasing a cluster's fecundity, and smaller propagules will be relatively faster growing than larger propagules… To generate proportionally smaller propagules, each reproductive event must be asymmetric, with propagules having less than half the biomass of the parent. Apoptotic cells may generate 'weak links' that allow small branches to separate from larger clusters. (Ratcliff et al. 2012)

Observations bore out this interpretation. As cluster size increased, propagule size decreased: between 14 and 60 transfers, snowflake clusters doubled in size, but propagule size decreased from 40% of parental size to less than 20%. Dihydrorhodamine-123 staining showed that the proportion of apoptotic cells correlated with settling rate during later transfers, but not during the early ones. This suggested that apoptosis co-evolved with, but was not a direct result of, larger snowflake size. Video microscopy showed that apoptosis did indeed occur at the locations where the propagules broke off from the parent cluster; this tended to happen at the site of the oldest cells in the cluster. This function of apoptosis had "no obvious parallel in the unicellular ancestor" of their snowflakes, marking it clearly as an evolved cluster-level trait. Moreover, the number of apoptotic cells never exceeded 2% of the population, consistent with a hypothesis proposed by Willensdorfer (2009) that "the earliest somatic tissue should constitute only a small percentage of the multicelled organism's biomass; otherwise, the fitness cost of nonreproductive tissue would outweigh the benefit of divided labor".

In a second study of the snowflake yeast, published the following year, Ratcliff, Pentz and Travisano showed that faster settling conditions selected not only for snowflake clusters, but also for larger individual yeast cells (which could lead to up to a 45% increase in the settling rate). They also observed selection for more hydrodynamically streamlined clusters; shorter branches within the clusters reduced drag and increased the settling rate. Variations in the "tempo and mode" of the transition to multicellularity were also apparent. Initially, clusters increased in size; during a subsequent phase, cell size increased (by a staggering 216% cell mass); during a third phase, cell size dropped slightly, and cluster shape became less "branchy". The authors concluded "that costs associated with the first adaptations to faster settling, larger cluster size, impose secondary selection for novel, more complex adaptations" (Ratcliff et al. 2013a). This is consistent with observations of increased size followed by increased complexity in *Volvox* evolution. It is also reminiscent of the elaboration of internal tissue surface area in larger metazoans to offset the decreasing surface-area-to-volume ratio (a problem mentioned by Galileo himself in the opening pages of his *Dialogues Concerning Two New Sciences*).

In a remarkable follow-up study, Ratcliff et al. (2015) performed a genetic analysis of snowflake yeast. Comparing the ancestral strain to snowflake yeast after 7 days of transfers, they found that the expression level of over 1000 genes differed significantly between the two cases. Among the most downregulated genes were *CTS1*, which codes for an endochitinase required for separation of cells following mitosis; *DSE1*, which codes for a protein expressed in daughter cells and whose deletion disturbs natural cell separation following division; and *DSE2* and *DSE4*, which code for proteins secreted by daughter cells and which degrade the cell wall between mother and daughter, from the daughter side. Most strikingly, these and the three other most downregulated genes are *all regulated by the transcription factor ACE2*.

Two years before, a study by Oud et al. (2013) had (in their words) "inadvertently mimicked" the design of the original Ratcliff et al. experiment. Performing sequential bioreactor cultures of *S. cerevisiae*, they noticed that "[t]he vertical pipe used to empty the bioreactor after each cultivation cycle did not reach the bottom of the vessel. Consequently, fast-sedimenting cells were enriched in the small remaining volume used as inoculum for the next batch cultivation cycle" (Oud et al. 2013). Examining these fast-sedimenting cells, they found that they had taken on the snowflake phenotype described by Ratcliff and colleagues just the year before. After performing whole genome sequencing of the snowflake yeast, Oud et al. found gene duplications of, and frameshift mutations in, *ACE2*.

Ratcliff et al. sequenced ACE2 from each of their ten snowflake populations. Five populations had synonymous mutations in *ACE2* (i.e., a nucleotide was changed, but not in a way that would alter the amino acid sequence of the resulting protein). The other five populations either had mutations that caused amino acid substitutions at or near the *ACE2* protein's zinc finger-binding domain, or mutations that caused truncation of the protein. An *ACE2* knockout yeast strain was found to exhibit snowflake clustering. Ratcliff et al. concluded that *a mutation in a single gene could produce the snowflake phenotype*. The simplicity with which a transition to multicellularity can occur cannot possibly be underestimated. The machinery for the emergence of collective ensembles of cells does not arise by any mysterious means: it is right there in the zinc finger-binding motif of a single protein.

The genetic results just described clearly show that the yeast snowflake phenotype arose by incomplete division rather than by aggregation. Thus, barring mutations during mitosis, a snowflake cluster is formed of clones, as are its propagule clusters. This exemplifies the importance of genetic bottlenecks in suppressing competition between individual cells in a multicellular organism and enabling selection to "see" the collective as an individual in its own right. The idea of conflict suppression has been explored by Michod and colleagues (Michod 1996, 2003), and the the importance of a genetic bottleneck is discussed by Leo Buss in *The Evolution of Individuality* (Buss 1987). A bottleneck reduces genetic diversity at the lower (single cell) level, reducing competition. Moreover, as Ratcliff et al. (2015) point out, "[c]lonal collectives align the fitness interests of lower-level units, and as a result the primary way for a lower-level unit (for example, a cell) to increase its fitness is by enhancing the collective's fitness." It is the lack of such an alignment, Ratcliff et al. suggest, that has prevented multi-clonal collectives such as *Dictyostelium*, and multi-species collectives, such as biofilms, from making a full transition to "a higher level of individuality". These ideas could easily be recast from a gene's eye view, since all the lower-level units have similar (though after some generations and inevitable mutations, no longer identical) genomes. As Stephen Jay Gould would have hastened to point out, however, by this point the dominant locus of selection has shifted to the multicellular collective, not the cell, and even less the individual gene.

In addition to the yeast experiments, Travisano's laboratory investigated the evolution of multicellularity using *Chlamydomonas reinhardtii* (Ratcliff et al. 2013b). The investigators initiated their experiment with an outbred *C. reinhardtii* population, in order to start with a significant level of genetic variation. Using a protocol similar to that of the yeast experiments, twenty replicates from the initial population were spun down at 100 g between transfers in order to select for settling; only the bottom 100 µl was transferred to fresh medium. Ten similar populations were transferred without selection for settling, as a control. Transfers were performed every 3 days, and a total of 73 transfers were carried out.

In contrast to the yeast experiments, only 1 of the 20 test *C. reinhardtii* populations exhibited a rapid transition to multicellularity, finally acheiving a greater settling speed by the 46th transfer (Fig. 12.10). Its clusters contained hundreds of cells, and were "held together by a transparent extracellular matrix". As with the yeast, time-lapse microscopy showed that clusters formed by incomplete separation following mitosis rather than aggregation of independent cells.[14]

Did the *C. reinhardtii* clusters reproduced via a unicellular genetic bottleneck, like the yeast snowflakes? To test this, the investigators took ten isolates from the population that had evolved multicellularity. Portions of the genomes of these isolates were sequenced, and found to be identical at five unlinked loci, implying that

---

[14] Recall from Chap. 11 that *C. reinhardtii* reproduces by multiple fission, and offspring do not break out from the parent immediately.

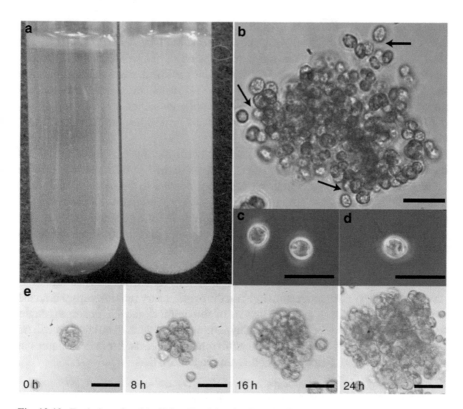

Fig. 12.10 Evolution of multicellular *C. reinhardtii*. Panel **a** shows tubes of *C. reinhardtii* after 20 minutes of settling; the algae in the tube on the *left* are multicellular. Panel **b** shows a multicellular cluster, with *arrows* indicating the extracellular matrix. Panel **c** shows the motile propagules released from the multicellular cluster, and **d** shows the ancestral form, identical in phenotype to the propagules. Panel **e** shows the formation of a cluster from a single parent cell. Scale bars are 25 μm. Reprinted by permsission from Macmillan Publishers Ltd., W.C. Ratcliff, M.D. Herron, K. Howell, J.T. Pentz, F. Rosenzweig, and M. Travisano, Experimental evolution of an alternating uni- and multicellular life cycle in *Chlamydomonas reinhardtii*. *Nat. Commun.* 4: 2742, 2013, copyright 2013

the isolates shared a common ancestor. The initial genetic variation in the strain that evolved multicellularity had thus been whittled down to a single lineage.

After 24 h in culture, investigators observed individual, actively swimming cells in each of the ten isolates. In order to observe the life cycle in more detail, they transferred the cultures to fresh medium and imaged them using time-lapse microscopy. For the first 4 h, no single cells detached from the parent clusters. Then, after about 4 h, something remarkable began to happen. "Cells in clusters began to activate their flagella, causing the cluster to convulse rapidly," they recounted. "This change in state, from stasis to full movement, took 3 min… Six minutes after the onset of flagellar activity, the first motile individual cells broke free from the cluster" (Ratcliff et al. 2013b). After about a day of observation, individually swimming cells were no longer observed, as all the individual cells had undergone fission and formed clusters. Two days after the culture was started, clusters had nearly tripled their volume. "Together," the investigators concluded, "these observations demonstrate that experimentally evolved *C. reinhardtii* possess a novel multicellular life cycle consisting of alternating phases: a dispersal phase, in which clusters reproduce via motile unicellular propagules and do not grow larger, and a growth phase, during which clusters produce few propagules and instead increase in cell number" (Ratcliff et al. 2013b).

Ratcliff and colleagues noted the contrast between the clusters they observed and the naturally forming palmella[15] clusters of *C. reinhardtii* that form when a single generation of new cells fails to break free from the mother cell wall. The clusters that resulted from settling selection were much larger, sometimes exceeding 100 cells. Moreover, these clusters "include[d] multiple mitotic generations of descendants of the founding cell" and "not held together by a parental cell wall but rather by a gelatinous extracellular matrix" (Ratcliff et al. 2013b). Analysis of the components of the gelatinous extracellular matrix, within which individual cells remain bounded by their own cell walls, and comparison to the extracellular matrix of *V. carteri*, will likely shed interesting light on the evolutionary relationships within the volvocine lineage. A full genetic analysis of the multicellular *C. reinhardtii* strain has not yet been published, but one can hypothesize that, when it is, changes will be observed in the genes coding for extracellular matrix proteins. From their studies in both yeast and algae, Ratcliff and colleagues concluded that

> [m]ost broadly, our finding that simple multicellularity can evolve in less than a year in both *Chlamydomonas* and *Saccharomyces* suggests that the genetic barriers (for example, few mutational paths to multicellularity) may be less restrictive than ecological barriers, namely a lack of persistent selective advantages for cellular clusters. (Ratcliff et al. 2013b)

Selective pressure for the evolution of multicellularity does not necessarily imply selection for a bottlenecked life cycle. This would be spectacularly unhelpful, for example, in an environment where predators preferred single-celled prey. The development of a unicellular bottleneck is clearly important for the suppression of intra-organismal conflict. However, Ratcliff and colleagues emphasize that "there is no direct evidence that it originally arose as an adaptation *for* this function" (Ratcliff et al. 2013b, my emphasis). They suggest that it is more likely that "the evolution of a single-cell genetic bottleneck by co-option of the ancestral [unicellular] phenotype may represent a rapid and general route by which this trait can arise". Such multicellular organisms would be "preadapted" for prevention of between-cell conflict. Another way of looking at this, which omits the false implication of causality, would be in Gould's sense of a spandrel or exaptation (see Chap. 15). To emphasize this point, Ratcliff and colleagues used probabilistic calculations of the number of surviving offspring per cluster as a function of propagule size, and the number of doublings undergone before selection for settling, in order to show that the single-

---

[15] See Chap. 11.

cell propagules provide a survival advantage in their own right, totally separate from any potential benefit in preventing between-cell conflict.

*** *** ***

Most experimental evolution studies to date have involved single species. Others, however, have sought to investigate the long-term evolution of multi-species eco-systems. Hasan Celiker and Jeff Gore at MIT recently studied an experimental eco-system populated with six different species of soil-dwelling bacteria. The experiment was performed in 96 identical replicates, in order to examine the role of historical contingency. In an initial short-term study, the investigators found that one of the species (*Pseudomonas putida*) did poorly, decreasing to only 1% of the total population, while other species claimed 15–35% of the total. In a longer-term version of the study, over 400 generations, the researchers found all species but *P. putida* happily coexisting in each of the 96 replicates. *P. putida* typically went extinct, except for 10 of the 96 replicates; in 9 of these, *P. putida* actually became the dominant species in the ecosystem. They speculated that this might the result of a rare mutational event that occurred before the species could become extinct (Celiker and Gore 2014).

Some species behaved quite differently in isolation than in the multi-species community. While isolated lines of *P. putida* quickly went extinct, another species, *Pseudomonas veronii*, which typically thrived in the multispecies environment, did not do well in isolation. Clearly, interactions among the species within the mixed community had an effect on survival.

Using a clustering algorithm to analyze relative species abundance, Celiker and Gore identified several distinct patterns of community structure. These patterns were largely dictated by "driver species"; typically, one species would begin to do better in the multispecies culture than it would in isolation, likely as a result of a rare mutational event, and this set the tone for the interactions within the entire community going forward. Was this the result of an evolutionary process, or just ecological sorting? The evolutionary nature of the change in community structure was confirmed by creating "reconstituted" communities from mixtures of the isolated lines. The behavior of these communities was significantly less complex than that of the multispecies cultures that had grown up together over 400 generations. Further studies in this system will surely include analysis of the mutational events leading to the dominance of particular strains.

*** *** ***

As organisms adapt to changing environments, changes in *gene regulatory networks* can have decisive and cascading effects. Recent studies have focused on the evolution of gene regulatory networks. The group headed by Gábor Balázsi (formerly at the M. D. Anderson Cancer Center in Houston, and now at the State University of New York at Stony Brook), has studied a synthetic gene circuit integrated into the genome of *S. cerevisiae* (González et al. 2015). Balázsi's group used

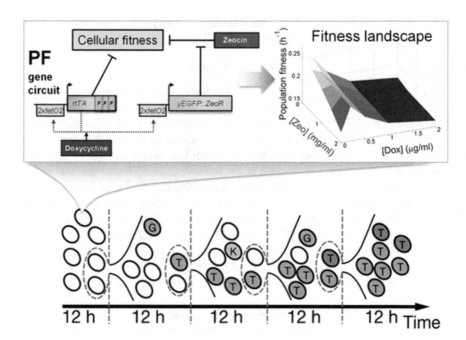

Fig. 12.11 Summary of results from Gonzalez et al. 2015, showing the evolution of a synthetic gene circuit inserted into the yeast geome. In *lower panel, empty circles* show cells with the intact PF circuit; *cirlces* labeled *G, K* and *T* show the generic, knockout and tweaking mutations, respectively. See text for details. Reprinted under a C.C. BY 3.0 Creative Commons License

a synthetic gene circuit consisting of two genes, a transcriptional regulator (*rtTA*) and an antibiotic-resistance gene fused to a fluorescence reporter (*yEGFP::zeoR*), shown schematically in Fig. 12.11. In the presence of as the inducer doxycycline (an antibacterial compound that does not harm yeast), the rtTA protein enhances transcription of its own gene in a positive feedback loop, and also enhances the expression of *yEGFP::zeoR*, which confers resistance to the antibiotic zeocin (Nevozhay et al. 2012). But rtTA is toxic under some conditions, and thus antibiotic resistance can come at a steep cost. Since expression levels vary, each cell will have a different balance between antibiotic resistance and rtTA toxicity.

S. cerevisiae is unaffected by doxycycline, which enabled the researchers to decouple the stress (zeocin, an antibiotic which causes DNA double-strand breaks, ultimately arresting the cell cycle and causing cell death) with the inducer of the stress response (doxycycline). Treatment with doxycycline alone, in the absence of zeocin, led the synthetic gene circuit to "respond gratuitously to a harmless environmental change", leading to toxicity without any accompanying benefit. In contrast, cells treated with zeocin alone were unable to respond to a harmful stress. A more complicated scenario resulted in complex cost-benefit tradeoffs: treatment with a combination of doxycycline and zeocin resulted enabled the cells to respond suboptimally to stress.

Using two computational models, the researchers plotted the fitness landscape (with fitness measured as the rate of cell division) as a function of the concentrations of inducer and antibiotic (Fig. 12.11). They predicted that, in the doxycycline-only condition, mutations that *decrease rtTA toxicity* would increase fitness, while in the zeocin-only condition mutations would be beneficial if they *increased zeocin resistance*, either enhancing *yEGFP::zeoR* expression or up-regulating native stress responses. When both doxycycline and zeocin were present, the situation became more complex, since no single mutation of the above types would be optimal (González et al. 2015).

Based on the computational models, Balázsi's team predicted that, when doxycycline and zeocin were both present in relatively high concentrations, "tweaking" ("T") mutations, which weakened rtTA toxicity, would predominate. In contrast, when the concentration of doxycycline was significantly lowered (from 2 to 0.2 µg/ml), both T mutations and mutations that affect the response to zeocin ("generic drug resistance", or "G", mutations) would occur. The model further suggested that only G mutations would be favored in the zeocin-only case, while in the doxycycline-only case, "K" or "knockout" mutations, which abolished rtTA toxicity, should be favored.

The team tested their predictions with yeast populations under the various conditions. In the doxycycline-only condition, with a high (2 µg/ml) concentration of doxycycline, fluorescence quickly increased and remained high during the first 40 generations, indicating (unnecessary) expression of *yEGFP::zeoR*, and then decreased to baseline, consistent with the appearance of "knockout" mutations abolishing rtTA expression (and therefore rtTA toxicity) and *yEGFP::zeoR* expression. "As fluorescence levels dropped," the authors wrote, "population growth rate increased significantly…, indicating that the initial cost of futile response disappeared" (González et al. 2015). The cells were no longer wasting resources expressing *yEGFP::zeoR*, nor were they experiencing the toxic effects of rtTA. Gene sequencing showed that multiple mutations occurred in the rtTA coding sequence, including the introduction of STOP codons and a 78-base-pair deletion.

When Balázsi's team performed experiments at a lower concentration of doxycycline (0.2 µg/ml), they found results that deviated from their prediction. Selection pressure was weaker in this case, due to reduced rtTA toxicity. However, the computational models had predicted that K mutants would still dominate the population. Instead, a T mutation occurred, and the fitness of the overall population did not change with statistical significance (González et al. 2015).

In the zeocin-only condition, cells were initially unable to activate the antibiotic-resistance gene, and an initial drop in fitness was observed. "Yet," the experimenters noted, "some cells must have had enough drug resistance to survive, because the growth rates of cultures started to recover after ~4 days" (González et al. 2015). Simultaneously, cells began to express *yEGFP::zeoR*. These strains showed no evidence of mutations in the gene for rtTA. Instead, mutations were found upstream of *yEGFP::zeoR* and elsewhere in the yeast genome as the cell lines developed alternate methods of regulating the antibiotic-resistance gene. In zeocin-doxycycline conditions, cells found a "sweet spot" to balance the conflicting pressures. They evolved both T mutations to weaken rtTA toxicity and G mutations to enhance antibiotic resistance.

Except in the 0.2 μg/ml doxycycline case, experimental results bore out the predictions on the type and timing of mutations in each condition. In the low-doxycycline condition, the effects were subtle, with mutations in both the gene coding for rtTA and in loci outside the engineered circuit. An increase in the variability of gene expression was also observed, indicative of "a unique example of noisy gene expression evolving under opposing selection pressures" (González et al. 2015). It is likely that this study will soon be followed by a host of other investigations of gene network evolution, with a focus on the delicate balance inherent in fitness tradeoffs.

*** *** ***

When one thinks of experimental evolution, one typically pictures a gigantic pile of petri dishes in Lenski's lab (Fig. 12.12). However, evolution can also be studied "experimentally" in computational models. These in silico studies are obviously vastly simplified compared to a project like the LTEE, but can still give valuable insights into evolutionary mechanisms. Early computational studies of evolutionary dynamics include the "typogenetics" model proposed by Douglas Hofstadter (1979) in *Gödel, Escher, Bach* and the simple gene network models used by Stuart Kauffman (1993)

Fig. 12.12 Zachary Blount (*left*) and Richard Lenski with just a few of the plates used for their citrate experiments. Reproduced from Fox and Lenski (2015) under a C.C. BY 4.0 Creative Commons Attribution License

in *The Origins of Order*. More recent works involve detailed agent-based computational simulations involving reproducing organisms, such as the study by de Aguiar et al. (2009). Other studies along these lines include work from my own research group, investigating phase transitions in evolutionary models on rugged (Dees and Bahar 2010) and neutral (Scott et al. 2013; Scott 2014; King et al. 2017) landscapes, as well as the structure of evolutionary lineages on a neutral landscape and mechanisms of recovery from mass extinction (King 2015).

Some of the most striking in silico evolution studies have been performed by Charles Ofria and colleagues at Michigan State University, using the Avida platform in which populations of computer programs compete and reproduce, with their fitness determined by their ability to perform certain tasks (Ofria and Wilke 2004). In Avida, populations of individual organisms (computer programs with a fixed-length "genome" of instructions) compose colonies (also referred to as multicells or groups). As organisms execute tasks contained in their genome (and are rewarded if they evolve to execute the Boolean logical functions NOT, NAND, AND, ORNOT, OR, ANDNOT, NOR, XOR and EQU), they acquire resources for their colony. Different logical functions may be associated with different amounts of resource. The total amount of resources in the ecosystem is limited, with a continual resource flow into and out of the system (a "digital chemostat"). When a colony has acquired a certain amount of resources, it reproduces clonally, generating offspring colonies seeded by organisms from the parent colony. New colonies can displace older ones, which are then removed from the simulation.

Avida has been used to investigate topics like adaptive radiation as a result of resource competition (Chow et al. 2004). Some of the most dramatic recent Avida studies have focused on the evolutionary origins the division of labor and the factors that maintain a balance between levels of selection. In 2012, Ofria, along with Heather Goldsby, Ben Kerr, and University of Arizona ant researcher Anna Dornhaus, used Avida to investigate the role of task-switching costs in promoting the evolution of the division of labor (Goldsby et al. 2012). From Adam Smith's *Wealth of Nations* to Dornhaus's own studies of ant behavior, the cost (in time, energy, and/or learning) of an individual switching between tasks has been found to favor the division of labor; it has even been suggested that the division of labor is impossible in the absence of task-switching costs. A classic example of this is the division of *Volvox* into somatic and germ-line cells, driven by requirement of decoupling the flagella from the basal bodies in order to perform mitosis (see Chap. 11).

Ofria's team set up replicate "worlds" containing 400 colonies of computer program organisms, each containing up to 25 identical (clonal) organisms. Mutations were allowed to occur only during colony reproduction. Since each function executed by an organism provides a bit of a limited resource, and colonies must amass a significant amount of resources (far greater than the total amount of any single resource) in order to reproduce, a colony will gain a reproductive advantage by collecting a variety of resources. This could be done in several ways. A colony could be made up of "generalist" organisms, which execute lines of code corresponding to a variety of Boolean functions. Or, a colony could be made up of "specialists", each of which performs a single (different) function. As Goldsby et al. (2012) explain, "[t]he specialist dynamic is analogous to honey bee colonies where bees specialize on collecting nectar from one type of flower but collectively gather nectar from all flowers in their habitat." As they evolved, Avida colonies could sense their location within the colony and send information to other organisms; these processes could be used in order to divide up the tasks within the colony.

Goldsby et al. (2012) simulated experiments in which various different task-switching costs were imposed. In a control set of experiments, no external costs were added. In this case, *intrinsic* task-switching costs still could be found: a non-zero number of CPU cycles was needed in order to switch between tasks, likely as a result of inefficiently evolved programs. In other experiments, moderate or high costs were imposed, in which organisms were made to wait 25 (or 50) CPU cycles before performing a new task. The degree of division of labor was quantified using a measure based on Shannon mutual information (Gorelick et al. 2004). The control condition favored the development of generalist colonies (low division of labor), while high task-switching costs favored specialists (high division of labor). With higher resource requirements, colonies evolved to perform more complex tasks. Some individual colonies developed strategies analogous to leafcutter ants, some removing leaves from trees, others of bringing cut leaves back to the nest, and still others tending the fungal gardens fed with the leaf material. Like the ants, a digital colony exhibited "problem decomposition and assembly line processing of task material" (Goldsby et al. 2012).

As they evolved division of labor under high task-switching costs, digital colonies assumed an increasing degree of individuality. For example, the insertion of an interloper organism into the colony had a detrimental effect on the performance of a complex set of tasks. "However," the authors found, "when the same perturbation was performed on different lineages evolved under low task-switching costs, fitness did not diminish… These data serve as preliminary evidence that making it costly for individuals to switch tasks not only favors division of labor but also favors a shift in individuality to a higher level."

In 2014, Ofria's team turned to the problem of the origin of somatic cells. They set out to test the *dirty work hypothesis*, proposed by Michod (1996) and then explored further by Bendich (2010). According to the dirty work hypothesis, since metabolic activities can produce highly reactive oxidative byproducts that can damage DNA,[16] an ensemble of cells could gain a selective advantage by keeping the germ line separate from highly metabolically active somatic cells. This idea can be contrasted with the suggestion of Leo Buss that germline sequestration is advanta-

---

[16] A similar process appears to be at work in higher-level collectivities. In eusocial insects, there can be several orders of magnitude difference in the lifetime of workers compared to queens, and this has been suggested to result from oxidative stress on the workers' metabolism (Corona et al. 2005; Remolina and Hughes 2008; Goldsby et al. 2014a). Young and Robinson (1983) suggested that the increased metabolic load imposed by foraging behavior in workers increases oxidative stress. The relative roles of RNA and DNA, which have significantly different levels of stability with respect to mutation, may represent a molecular division of labor "ensuring both high fidelity transmission of hereditary information and the execution of critical chemical work" (Goldsby et al. 2014a).

geous because it reduces the number of cell divisions undergone by germline cells (Buss 1987). Both ideas provide means by which germline DNA would be shielded from excess mutation; the two possibilities are certainly not mutually exclusive.

In order to simulate the dirty work hypothesis, Goldsby et al. (2014a) updated their previous simulation to include a line of code called *block_propagation*. If such an instruction were executed, the cell (and its clonal descendants) would be unable to propagate a new colony when the colony reproduced. Furthermore, in addition to mutations during colony replication, a "functional mutagen level", or FML, was now imposed on most functions. Thus, when a cell executed a computational task (analogous to metabolism), it incurred a possible mutational cost.

> When a cell performs a function, it acquires 5% of the associated resource, and also accu-
> mulates any mutagenic effects associated with the function. By performing more types of
> functions, a cell is able to collect resources more rapidly. However, each performance of
> one of these mutagenic functions may alter a cell's genetic material, potentially damaging
> its ability to self-replicate or to collect future resources. (Goldsby et al. 2014a)

Under these conditions, would germ-soma differentiation occur?

> The evolution of cells that are both metabolically active and propagule-ineligible (i.e.,
> soma) faces two seemingly insurmountable challenges: First, any cell that performs muta-
> genic functions will damage its genome… Second, if a cell becomes ineligible to be used
> as a propagule, it removes itself from the reproductive line of the multicell. For the multicell
> to thrive, it appears that a subset of cells must perform both damaging actions, while the
> genetic material that encodes these actions must persist, unexpressed, in other cells (the
> germ). (Goldsby et al. 2014a)

Mutations posed another problem still: by decreasing the similarity between cells within the colony, they increased intra-colony competition, shifting selection toward the individual, rather than the colony, level.

Simulations were performed over a range of FML values, with populations in which every cell was initially propagule-eligible. For intermediate values, colonies evolved with a significant proportion of propagule-ineligible cells which performed a "disproportionately large share of functions associated with mutagenic conse-quences" (Goldsby et al. 2014a). For one FML value, for example, propagule-ineligible cells performed an average of 64.17 mutagenic functions, while propagule-eligible cells performed an average of 0.59!

> Such division of labor was favored because somatic cells, which had removed themselves
> from consideration as propagule cells, performed the mutagenic functions necessary for the
> digital multicell to reproduce, while the germ cells maintained the multicell's genetic infor-
> mation in pristine condition. (Goldsby et al. 2014a)

How did the germ-soma division emerge? Did a population of propagule-ineligible cells arise first, and then take on a range of mutagenic functions? Or did the cells shoulder the mutagenic workload first? The second of these options "has some immediate benefits for the reproductive rate of the multicell but at the cost of putting the genetic information of the multicell in peril" (Goldsby et al. 2014a). On the other hand, the first option provides no initial selective benefit, and could pre-sumably arise only via genetic drift. In fact, the lineages at an intermediate FML value "followed an unanticipated variation" on the dirty-work-first option.

The multicell first split the workload heterogeneously among the propagule-eligible cells, where some propagule-eligible cells (which we call "pseudo-soma") performed many more mutagenic functions than others. This innovation protected the genetic material in some of the propagule-eligible cells while the pseudo-soma dramatically increased the multicell's replication rate. Next, the multicell evolved to make these pseudo-somatic cells into actual somatic cells by blocking their ability to be selected as a propagule and thus guaranteeing that only high-fidelity germ cells were used to produce the next generation of offspring multicells. (Goldsby et al. 2014a)

As the propagule-ineligible cells emerged in the population, the workload of the remaining cells dropped dramatically. In other simulations that blocked the system's ability to develop an initial heterogeneity, many fewer propagule-ineligible cells developed.

What about cheaters? If a propagule-ineligible cell undergoes a mutation that enables it to reproduce, its higher metabolism will still make its descendants more likely to suffer deleterious mutations. A colony founded by such a cell will be less likely to thrive than one founded by a "pristine" propagule-eligible cell that had not suffered the insults of a mutagenic workload. Thus, "[f]rom a multilevel selection perspective, selection within multicells favors the mutant, while selection between multicells disfavors the mutant" (Goldsby et al. 2014a).

The difficulty faced by cheaters within an Avida colony highlights an important problem in multilevel selection: potential antagonisms between levels. Such antagonisms can often be found in systems that exhibit division of labor, "where there is a within-group pressure to specialize on the role with the highest reward and a between-group pressure to perform a diverse suite of tasks" (Goldsby et al. 2014b). To investigate this, Ofria's team varied the level of antagonism between groups. Periodically, five colonies from within a population of 400 were randomly selected to compete against each other in a "tournament". The group performing the greatest variety of tasks was allowed to replicate. In this case, replication was performed by duplication of the original colony, with the addition of some mutation, rather than seeding with one organism from the parent colony. In one sense, this may seem artificial, since it involves deliberate selection for division of labor. Importantly, however, this scheme imposed a tunable degree of group selection on the system.

Four types of tournaments were held. In *between* tournaments, only between-group pressure was imposed: the winning colony was the one that executed the greatest range of tasks. In *within* tournaments, tournament winners were selected randomly, but within-group selection was imposed because different logic tasks performed by individual organisms had different reward values. In *both* tournaments, both the *between* and *within* conditions were applied. In the control case, *none* tournaments, all tasks were rewarded equally, and tournament-winning colonies were selected at random. As shown in Fig. 12.13, the investigators found that *between* and *both* tournaments evolved to perform a larger range of tasks (around 4.5 tasks[17] on average) than *none* or *within* tournaments (on average, less than one task).

---

[17] In these experiments, colonies could only perform up to a total of five tasks.

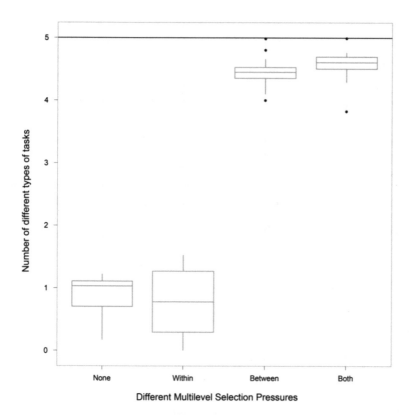

Fig. 12.13 Results of different types of multilevel selection pressure on the evolution of division of labor. Values shown are averages over 30 replicates in each case. Reproduced from Goldsby et al. (2014b) under a C.C. BY 4.0 Creative Commons Attribution License

In order to modulate the strength of group selection, the investigators varied the length of time between tournaments. Longer inter-tournament periods would allow the system more time to acclimate to changing conditions, weakening between-group selection. As expected, an increase in the inter-tournament period decreased the mean number of unique tasks performed by each colony, though this effect was qualitatively small. In another set of studies, group selection was weakened by allowing individuals from one group to migrate into another during the replication process. Changing the migration rate from 0 to 5% caused a drop from ~4.5 to 3 tasks; raising the migration rate to 20% decreased the tasks to two (Goldsby et al. 2014b). Variation in propagule size showed that, as the number of organisms in a propagule increased, so did the number of tasks performed by the evolved colony. In this case, division of labor was not favored by a bottlenecked life cycle.

Successful colonies exhibited a variety of lineage structures. Despite the propagule size result just mentioned, some colonies evolved from single lineages, engaging in a range of tasks by phenotypic plasticity (for example, having a range of possible functions in an organism's code that could be expressed depending on the conditions and interactions with other organisms). Other colonies contained a range of genetic lineages; some of these genetically diverse colonies also maintained phenotypic diversity. Knockout studies of phenotypically diverse colonies, in which individual lines of code were excised, showed that passage of epigenetic information from parent to offspring played a significant role in the "differentiation" of organisms within the colony. Within genetically diverse colonies, lineages were found to modulate their rate of replication in order to avoid "replicating over" other lineages. If between-group pressure was removed from these colonies, however, the lineages within them no longer modulated their replication rate.

The Avida experiments highlight the problem of balancing cooperation and competition, and how the tension between these complementary drives affects the levels at which collectives can emerge. We have encountered this pair of complementary factors repeatedly throughout the preceding pages. Now, we must we explore competition and cooperation in more depth and specificity.

References

Bendich AJ (2010) Mitochondrial DNA, chloroplast DNA and the origins of development in eukaryotic organisms. Biol Direct 5:42

Blount ZD, Borland CZ, Lenski RE (2008) Historical contingency and the evolution of a key innovation in an experimental population of *Escherichia coli*. Proc Natl Acad Sci U S A 105(23):7899–7906

Blount ZD, Barrick JE, Davidson CJ, Lenski RE (2012) Genomic analysis of a key innovation in an experimental *Escherichia coli* population. Nature 489:513–518

Boraas ME, Seale DB, Boxhorn JE (1998) Phagotrophy by a flagellate selects for colonial prey: a possible origin of multicellularity. Ecol Evol 12(2):153–164

Buss LW (1987) The Evolution of Individuality. Princeton University Press, Princeton

Celiker H, Gore J (2014) Clustering in community structure across replicate ecosystems following a long-term bacterial evolution experiment. Nat Commun 5:4643

Chow SS, Wilke CO, Ofria C, Lenski RE, Adami C (2004) Adaptive radiation from resource competition in digital organisms. Science 305:84–86

Corona M, Hughes KA, Weaver DB, Robinson GE (2005) Gene expression patterns associated with queen honey bee longevity. Mech Ageing Dev 126(11):1230–1238

de Aguiar MA, Baranger M, Baptestini EM, Kaufman L, Bar-Yam Y (2009) Global patterns of speciation and diversity. Nature 460(7253):384–387

Dees ND, Bahar S (2010) Mutation size optimizes speciation in an evolutionary model. PLoS ONE 5(8):e11952

Elena SF, Cooper VS, Lenski RE (1996) Punctuated evolution caused by selection of rare beneficial mutations. Science 272:1802–1804

Fiegna F, Yu YT, Kadam SV, Velicer GJ (2006) Evolution of an obligate social cheater to a superior cooperator. Nature 441(7091):310–314

Fox JW, Lenski RE (2015) From here to eternity—the theory and practice of a really long experiment. PLoS Biol 13(6):e1002185

Goldsby HJ, Dornhaus A, Kerr B, Ofria C (2012) Task-switching costs promote the evolution of division of labor and shifts in individuality. Proc Natl Acad Sci U S A 109(34):13686–13691

Goldsby HJ, Knoester DB, Ofria C, Kerr B (2014a) The evolutionary origin of somatic cells under the dirty work hypothesis. PLoS Biol 12(5):e1001858

Goldsby HJ, Knoester DB, Kerr B, Ofria C (2014b) The effect of conflicting pressures on the evolution of division of labor. PLoS ONE 9(8):e102713

González C, Ray JC, Manhart M, Adams RM, Nevozhay D, Morozov AV, Balázsi G (2015) Stress-response balance drives the evolution of a network module and its host genome. Mol Syst Biol 11(8):827

Gorelick R, Bertram SM, Killeen PR, Fewell JH (2004) Normalized mutual entropy in biology: quantifying division of labor. Am Nat 164(5):677–682

Gould SJ (1989) Wonderful Life: The Burgess Shale and the Nature of History. W. W. Norton and Company, New York/London

Gould SJ (2002) The Structure of Evolutionary Theory. The Belknap Press of Harvard University Press, Cambridge, MA/London

Hofstadter D (1979) Gödel Escher Bach: An Eternal Golden Braid. Basic Books, New York

Kauffman SA (1993) The Origins of Order: Self-Organization and Selection in Evolution. Oxford University Press, Oxford

Kawecki TJ, Lenski RE, Ebert D, Hollis B, Olivieri I, Whitlock MC (2012) Experimental evolution. Trends Ecol Evol 27(10):547–560

Kim W, Levy SB, Foster KR (2016) Rapid radiation in bacteria leads to a division of labour. Nat Commun 7:10508

King DM (2015) Evolutionary Dynamics of Speciation and Extinction. PhD Dissertation, University of Missouri at St. Louis

King DM, Scott AD, Bahar S (2017) Multiple phase transitions in an agent-based evolutionary model with neutral fitness. R Soc Open Sci 4(4):170005

Koschwanez JH, Foster KR, Murray AW (2011) Sucrose utilization in budding yeast as a model for the origin of undifferentiated multicellularity. PLoS Biol 9:e1001122

Leiby N, Marx CJ (2014) Metabolic erosion primarily through mutation accumulation, and not tradeoffs, drives limited evolution of substrate specificity in *Escherichia coli*. PLoS Biol 12(2):e1001789

Leiby N, Harcombe WR, Marx CJ (2012) Multiple long-term, experimentally-evolved populations of *Escherichia coli* acquire dependence upon citrate as an iron chelator for optimal growth on glucose. BMC Evol Biol 12:151

Lenski RE, Travisano M (1994) Dynamics of adaptation and diversification: a 10,000-generation experiment with bacterial populations. Proc Natl Acad Sci U S A 91:6808–6814

Lenski RE, Rose MR, Simpson SC, Tadler SC (1991) Long-term experimental evolution in *Escherichia coli*. I. Adaptation and divergence during 2,000 generations. Am Nat 138(6):1315–1341

Michod RE (1996) Cooperation and conflict in the evolution of individuality. II. Conflict mediation. Proc R Soc B 263(1372):813–822

Michod RE, Nedelcu AM, Roze D (2003) Cooperation and conflict in the evolution of individuality. IV. Conflict ediation and evolvability in *Volvox carteri*. BioSystems 69:95–114

Nevozhay D, Adams RM, Van Itallie E, Bennett MR, Balázsi G (2012) Mapping the environmental fitness landscape of a synthetic gene circuit. PLoS Comput Biol 8(4):e1002480

Ofria C, Wilke CO (2004) Avida: a software platform for research in computational evolutionary biology. Artif Life 10(2):191–229

Oud B, Guadalupe-Medina V, Nijkamp JF, de Ridder D, Pronk JT, van Maris AJ, Daran JM (2013) Genome duplication and mutations in *ACE2* cause multicellular, fast-sedimenting phenotypes in evolved *Saccharomyces cerevisiae*. Proc Natl Acad Sci U S A 110(45):E4223–E4231

Quandt EM, Gollihar J, Blount ZD, Ellington AD, Georgiou G, Barrick JE (2015) Fine-tuning citrate synthase flux potentiates and refines metabolic innovation in the Lenski evolution experiment. eLIFE 4:e09696

Rainey PB, Rainey K (2003) Evolution of cooperation and conflict in experimental bacterial populations. Nature 425:72–74

Ratcliff WC, Denison RF, Borello M, Travisano M (2012) Experimental evolution of multicellularity. Proc Natl Acad Sci U S A 109(5):1595–1600

Ratcliff WC, Pentz JT, Travisano M (2013a) Tempo and mode of multicellular adaptation in experimentally evolved *Saccharomyces cerevisiae*. Evolution 67(6):1573–1581

Ratcliff WC, Herron MD, Howell K, Pentz JT, Rosenzweig F, Travisano M (2013b) Experimental evolution of an alternating uni- and multicellular life cycle in *Chlamydomonas reinhardtii*. Nat Commun 4:2742

Ratcliff WC, Fankhauser JD, Rogers DW, Greig D, Travisano M (2015) Origins of multicellular evolvability in snowflake yeast. Nat Commun 6:6102

Remolina SC, Hughes KA (2008) Evolution and mechanisms of long life and high fertility in queen honey bees. Age (Dordr) 30(2–3):177–185

Scott AD (2014) Speciation Dynamics of an Agent-Based Evolution Model in Phenotype Space. PhD Dissertation, University of Missouri at St. Louis

Scott AD, King DM, Marić N, Bahar S (2013) Clustering and phase transitions on a neutral landscape. Europhys Lett 102(6):68003

Spiers AJ, Kahn SG, Bohannon J, Travisano M, Rainey PB (2002) Adaptive divergence in experimental populations of *Pseudomonas fluorescens*. I. Genetic and phenotypic bases of wrinkly spreader fitness. Genetics 161(1):33–46

Spiers AJ, Bohannon J, Gehrig SM, Rainey PB (2003) Biofilm formation at the air-liquid interface by the *Pseudomonas fluorescens* SBW25 wrinkly spreader requires an acetylated form of cellulose. Mol Microbiol 50(1):15–27

Turner CB, Blount ZD, Lenski RE (2015) Replaying evolution to test the cause of extinction of one ecotype in an experimentally evolved population. PLoS ONE 10(11):e0142050

Velicer GJ, Stredwick KL (2002) Experimental social evolution with *Myxococcus xanthus*. Antonie Van Leeuwenhoek 81(1–4):155–164

Velicer GJ, Yu YT (2003) Evolution of novel cooperative swarming in the bacterium *Myxococcus xanthus*. Nature 425(6953):75–78

Velicer GJ, Kroos L, Lenski RE (1998) Loss of social behaviors by *Myxococcus xanthus* during evolution in an unstructured habitat. Proc Natl Acad Sci U S A 95:12376–12380

Velicer GJ, Kroos L, Lenski RE (2000) Developmental cheating in the social bacterium *Myxococcus xanthus*. Nature 404(6778):598–601

Velicer GJ, Lenski R, Kroos L (2002) Rescue of social motility lost during evolution of *Myxococcus xanthus* in an asocial environment. J Bacteriol 184:2719–2727

Willensdorfer M (2009) On the evolution of differentiated multicellularity. Evolution 63:306–323

Wiser MJ, Ribeck N, Lenski RE (2013) Long-term dynamics of adaptation in asexual populations. Science 352:1364–1367

Young RG, Robinson G (1983) Age and oxygen toxicity related fluorescence in the honey bee thorax. Exp Gerontol 18(6):471–475

Part III
Beyond the Barricade

Chapter 13
Cooperation and Competition: One Level Sitting on Another

I will sit right down, waiting for the gift of sound and vision

David Bowie

THE EVOLUTION of multicellularity demonstrates the operation of selection at various levels; so does the evolution of eukaryotic cells themselves, suggested by Lynn Margulis to have resulted from the symbiosis of prokaryotes (Margulis 1981). In the chapters above, we have also considered situations where looser affiliations have functioned – or been classified – as a group. We have seen evidence suggesting that a delicate balance of competition and cooperation knits units at one level into a higher collective. In this chapter we will explore specific examples of this "essential tension", and investigate how a shift of dominance between cooperation and competition can solidify a level of selection. Central to this process is a transition between two different types of multi-level selection: multi-level selection 1 (MLS1) and multi-level selection 2 (MLS2) in the terminology used by Okasha (2006). MLS1 involves selection on a group trait that derives from an average of individual traits (for example, average height in a population). MLS2 traits are properties of the group itself, such as geographic range or overall phenotypic diversity. Michod and other have argued that *the transition from one level of selection to another involves a transition from selection predominantly at the MLS1 level to selection predominantly at the MSL2 level.* This can be viewed as a shift from *competition* to *conflict suppression.*

Before we explore the transition between MLS1 and MLS2, however, we must return to selection at the level of groups of organisms. While the action of natural selection on any sort of group could be called "group selection", the term is typically applied to selection on groups of individual organisms, either at the population or species level. It is here that the balance between competition and cooperation is most easy to spot, but it is also here that the idea of group selection has been most controversial.

Researchers have sparred for many decades over a central problem presented by the cooperation/competition balance: the evolution of altruism. Should altruism be explained by selection on *groups as entities in their own right,* or by the gene-level

© Springer Science+Business Media B.V. 2018
S. Bahar, *The Essential Tension*, The Frontiers Collection,
DOI 10.1007/978-94-024-1054-9_13

Fig. 13.1 Schematic
diagram of tradeoffs in the
Prisoner's Dilemma

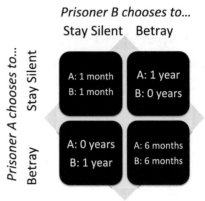

interpretation of kin selection? We have seen these arguments between Allee and Wynne-Edwards on the one hand, and Lack, Hamilton, Williams and Dawkins on the other. The controversy over group selection continues to rage to this day. In order to set the stage for the contentiousness surrounding current attempts to address multilevel selection, let us explore a few recent skirmishes in this ongoing intellectual conflict.

An early attempt to explain cooperation as a result of individual self-interest arose from the application of the "prisoner's dilemma" problem to interactions between organisms (Fig. 13.1). Here, individuals must decide whether to cooperate or defect (in the standard version of the problem, prisoners under interrogation must decide whether to protect another prisoner or to rat on them, while unaware of their imprisoned cohort's decision when subjected to the same choice). Such problems can be studied using game theory, developed by von Neumann and Nash. John Maynard Smith and George Price first applied it to the balance between cooperation and competition in an evolutionary context in a 1973 article entitled "The Logic of Animal Conflict". Maynard Smith later dubbed the stable outcome of such a game an "evolutionary stable strategy", or ESS (Maynard Smith 1974; Okasha 2005). Robert Axelrod and William Hamilton (1981) developed the idea further, and presented it as a model for the evolution of cooperation. Closely related to the concept of kin selection, the prisoner's dilemma approach is attractive to scientists who are drawn to the idea of selection at the individual rather than the group level, since this approach allows for the evolution of cooperation through a balance of individual selfish strategies.

Experimental studies have demonstrated the use of evolutionarily stable strategies in actual biological systems. Milinski (1987) found sticklebacks using a "tit for tat" strategy in the presence of a predator, and interpreted his results as "support [for] the hypothesis that cooperation can evolve among egoists". More recently, Brandl and Bellwood (2015) found evidence of direct reciprocity in pairs of rabbitfish living in coral reefs, in which one fish remained alert for predators while the other foraged.

Compared to solitary individuals, fishes in pairs exhibit longer vigilance bouts, suggesting that the help provided to the partner is costly. In turn, fishes in pairs take more consecutive bites and penetrate deeper into crevices than solitary individuals, suggesting that the safety provided by a vigilant partner may outweigh initial costs by increasing foraging efficiency. Thus, the described system appears to meet all of the requirements for direct reciprocity. We argue that the nature of rabbitfish pairs provides favourable conditions for the establishment of direct reciprocity. (Brandl and Bellwood 2015)

Likewise, Krama et al. (2012) demonstrated, in a paper delightfully entitled "You mob my owl, I'll mob yours: birds play tit-for-tat game", that breeding pairs of pied flycatchers exhibit reciprocity in mobbing a predator.

The prisoner's dilemma model and other aspects of game theory continue to be used in a wide range of studies of cooperation. For example, Zagorsky et al. (2013) considered situations in which a strategy of forgiveness can be advantageous in the long term. A recent study by Press and Dyson (2012), however, suggested the startling conclusion that strategy optimization can lead to extortion rather than cooperation. Stewart and Plotkin, after writing an initial commentary on the Press and Dyson work in 2012, dug more deeply into the model, and found that under some circumstances, cooperation remains an optimal strategy, but the outcome depends sensitively on the system's parameters (Stewart and Plotkin 2013, 2014). In the words of science journalist Emily Singer, "generosity and selfishness walk a precarious line. In some cases, cooperation triumphs. But shift just one variable, and extortion takes over once again"[1]

Szolnoki and Perc (2009) have suggested that multilevel selection can arise from a prisoner's dilemma game on coevolving random networks However, Boyd and Richerson (1992) argue that reciprocal strategies are insufficient to drive the evolution of eusocial behavior (as proposed by Trivers in 1971)[2] and that punishment is necessary in order for cooperation to evolve. Other critiques of the game theory approach point out that the "pair interaction" focus of the prisoner's dilemma is not always realistic in biological settings. Johnson et al. (2002) showed that individual differences, as when different individuals receive different payoffs for the same strategic choice, could destroy the expected outcome of the prisoner's dilemma game. The debate remains heated, especially in the field of sociobiology.

*** *** ***

Group selection remains such a controversial issue in large part because groups of organisms are looser affiliations than genes within a genome, or cells within an organism. As a result, it is harder for scientists to "see" them as a single unit, and thus it appears (and perhaps is) harder for natural selection to "see" them this way

---

[1] https://www.quantamagazine.org/20150212-game-theory-calls-cooperation-into-question/, retrieved 27 February 2016. This article, in the online Quanta Magazine, has an elegant summary of the studies by Press and Dyson and by Stewart and Plotkin, and places them in the context of the entire field.

[2] West et al. (2007) have suggested that the term "reciprocal altruism" is not really appropriate for this situation, since this sort of cooperation "provides a direct fitness benefit, [and thus] is mutually beneficial and not altruistic." They suggest using the term "reciprocity" or "reciprocal cooperation".

as well. This, however, does not mean that selection *never* sees them as such. Elliot Sober and David Sloan Wilson have cited a range of examples in which selection can be shown to act at the group or population level (Sober and Wilson 1998). But a prejudice remains against the very idea of group selection in the minds of many researchers, leading them to dismiss the idea out of hand, based on outdated arguments.

David Sloan Wilson and his then graduate student O. T. Eldakar surveyed a number of such arguments in a 2011 commentary published in *Evolution*, listing "eight criticisms not to make about group selection", in an attempt to spare future authors from referee reports asking them to "please remove all references to group selection", the twenty-first-century version of G. C. Williams's (1966) pronouncement that "group-related adaptations do not, in fact, exist".

Among the criticisms that are no longer supportable, Eldakar and Wilson argue, is that there is no empirical support for group selection. Sex ratio serves as a prime example of a group-level trait that, in Eldakar and Wilson's phrasing, "evolves purely by group selection". A study by Kerr et al. (2006) of phage virus infection of *E. coli* provides empirical support for Wynne-Edwards's hypothesis that populations can regulate their numbers in response to resource availability. Empirical studies such as those reviewed by Goodnight and Stevens (1997) regularly find that the between-group terms in the Price equation are far from negligible compared to the within-group terms. In fact, the Price equation approach of partitioning selection into between-group and within-group terms[3] strongly influenced Hamilton, who rethought his inclusive fitness theory along multi-level selection lines after reading Price's work.

Hamilton initially conceived of kin selection without any relation to group-level fitness. Hamilton later befriended Price, and helped him publish his work on the Price equation in *Nature*. Price's interest in altruism, movingly recounted in Oren Harman's 2010 book *The Price of Altruism: George Price and the Search for the Origins of Kindness*, was more than purely academic. A deeply principled but also troubled man, Price eventually gave away his possessions to the homeless, and died, by his own hand, in terrible (and partly self-imposed) poverty in 1975. Writing shortly after his death, Hamilton wrote that Price's "new presentation of natural selection effectively disposes of the problem dating back to Darwin of whether the individual or the group should be considered the unit of natural selection" (Harman 2010, p. 346). Using the Price equation, the problem of kin selection could literally be viewed as a mathematical balance between selection at the individual and group levels.

The publication of the "eight criticisms" paper did not have the desired effect of suppressing conflict in the field of evolutionary biology. In the same year the paper was published, David Sloan Wilson found himself unable to contain his frustration

---

[3] An alternative means of partitioning the effect of selection at different levels, called contextual analysis, was developed by Heisler and Damuth (1987) and Goodnight et al. (1992). Contextual analysis decreases the likelihood of misidentifying a correlation between a group-level trait and group-level fitness as being the result of selection in a case where the correlation is merely a byproduct of selection at the individual level.

with Jerry Coyne, a speciation expert who is far from keen on group selection. "When it comes to the topic of group selection," Wilson complained in a blog post,[4] Coyne

> hasn't written a single paper and there's little evidence that he's read the literature. Yet, that doesn't prevent him from holding forth on the topic and scolding others like a schoolteacher wagging his finger at truant students who haven't learned their lesson ... For me, this is like hearing Rip van Winkle mumbling in his sleep ... Jerry is the perfect example of a professional evolutionist who does not directly study group selection and perpetuates outdated views about it.

For better or worse, the blogosphere has become a frequent forum for arguments over group selection and altruism. When Martin Nowak, Corina Tarnita and E. O. Wilson published a paper on the evolution of altruism in 2010, sparks flew online almost immediately, ultimately prompting an article[5] in the *New York Times* about the controversy. In addition to a spate of impassioned Letters to the Editor of *Nature* (see the October 7, 2010 issue), Jerry Coyne, David Sloan Wilson, and Richard Dawkins all took to their blogs. In a piece titled "A Misguided Attack on Kin Selection", Coyne[6] pulled no punches.

> I don't know what's gotten into E. O. Wilson. He's certainly the world's most famous evolutionary biologist, and has gone from strength to strength over the years, winning two Pulitzer Prizes, writing great general books on not only ants but [also] conservation and social behavior. And he's kept his hands in the ant work, producing any number of technical papers and monographs. He's even written a novel! Frankly, I don't know how he does it. I haven't always agreed with what he says—I think he overreached with the sociobiology stuff, for instance—but you have to admire the guy's knowledge, breadth, dedication to conservation, and sheer workaholism. But now Wilson, along with some collaborators like David Sloan Wilson[7] and Martin Nowak, is definitely heading off on the wrong track. They're attacking kin selection, maintaining not only that it has nothing to do with the evolution of social insects, but that's it's also a bad way to look at evolution in general. And they're wrong—dead wrong. I'm baffled not only by Nowak et al.'s apparent and willful ignorance of the literature, but by statements that are just wrong. They flatly assert, for instance, that "inclusive fitness theory" is something different from "standard natural selection theory." But it's not: it's simply a natural extension of population genetics to the situation in which one's behavior affects related individuals. I could go on, but a little bird has told me that the big guns in the field will, soon and en masse, answer Nowak et al.'s arguments about both theory and data. I can't fathom any motive, either psychological or scientific, for Wilson and Company to repeatedly denigrate the importance of inclusive-fitness theory. It's just a shame that, this late in his career, Wilson has chosen to fight the wrong battle.

---

[4] http://scienceblogs.com/evolution/2011/09/11/jerry-coyne-on-group-selection. (retrieved December 23, 2015).

[5] Carl Zimmer, "Scientists Square Off on Evolutionary Value of Helping Relatives", August 30, 2010, http://www.nytimes.com/2010/08/31/science/31social.html?_r=0 (retrieved January 18, 2016).

[6] https://whyevolutionistrue.wordpress.com/2010/08/30/a-misguided-attack-on-kin-selection/. (retrieved January 18, 2016).

[7] The third author of the paper at issue is Corina Tarnita, not David Sloan Wilson.

Coyne concluded by awarding "a big raspberry" to "the folks at *Nature* who decided to publish such a strange paper in the interest of stirring up controversy. If they'd gotten decent reviewers, and followed their advice, it never would have seen print."

Richard Dawkins agreed,[8] writing that the Nowak, Tarnita and Wilson paper was "no surprise" since "Edward Wilson was misunderstanding kin selection as far back as [his seminal 1975 work] *Sociobiology*." David Sloan Wilson leapt into the mix, writing[9] an "open letter to Richard Dawkins", titled "Why Are You Still In Denial About Group Selection?" Mutual allegations of ignorance of the literature reappear as a common motif. "Your view is essentially pre-1975," wrote Wilson, "a date that is notable not only for the publication of *Sociobiology* but also a paper by W.D. Hamilton, one of your heroes, who correctly saw the relationship between kin selection and group selection thanks to the work of George Price."

The trigger for this wave of professional hostility was the suggestion by Nowak, Tarnita and Wilson that inclusive fitness theory (kin selection) was not sufficient to explain the evolution of altruism. This was particularly striking since E. O. Wilson was for many years a strong proponent of kin selection. Because of the role of relatedness in inclusive fitness theory, altruism was predicted to evolve more frequently in haploid organisms, such as many of the social insects (see Chap. 7). But in the past few decades, Nowak et al. pointed out, many diploid species (ambrosia beetles, several species of shrimp, and subterranean-dwelling mole rats) were found to also exhibit eusocial behavior. Another problem of inclusive fitness theory, they argued, was the rarity of eusocial species, as well as their

> odd distribution through the Animal Kingdom. Vast numbers of living species, spread across the major taxonomic groups, use either haplodiploid sex determination or clonal reproduction, with the latter yielding the highest possible degree of pedigree relatedness, yet with only one major group, the gall-making aphids, known to have achieved eusociality. For example, among the 70,000 or so known parasitoid and other apocritan[10] Hymenoptera, all of which are haplodiploid, no eusocial species has been found. Nor has a single example come to light from among the 4,000 known hymenopteran sawflies and horntails, even though their larvae often form dense, cooperative aggregations. (Nowak et al. 2010)

Nowak, Tarnita and Wilson pointed to evidence suggesting that a population of close kin could be detrimental for social insects. Cole and Wiernasz had published a 1999 *Science* article entitled "The Selective Advantage of Low Relatedness", showing that a polyandrous harvester ant, *Pogonomyrmex occidentalis*, exhibited low relatedness within colonies, and that relatedness was inversely proportional to colony growth rate. A 2004 study by Hughes and Boomsma predicted that genetic diversity, mediated by polyandry, in the leaf-cutting ant *Acromyrmex echinatior*, would increase resistance to a virulent fungal parasite. Other studies showed that

---

[8] https://whyevolutionistrue.wordpress.com/2011/03/24/dawkins-on-nowak-et-al-and-kin-selection/. (retrieved January 18, 2016).

[9] http://scienceblogs.com/evolution/2010/09/04/open-letter-to-richard-dawkins/. (retrieved January 18, 2016). David Sloan Wilson starts his letter as follows: "I do not agree with the cynical adage 'science progresses–funeral by funeral', but I fear that it might be true in your case for the subject of group selection". Not a bridge-building rhetorical device.

[10] The Apocrita are a sub-order of Hymenoptera, including bees and wasps.

nest temperature was more stable in more genetically diverse colonies of ants and honeybees, and suggested that phenotypic variability might facilitate the division of labor in ant colonies, thereby increasing fitness (Nowak et al. 2010). Inclusive fitness, Nowak, Tarnita and Wilson argued, was incapable of predicting, let alone explaining, such observations.

Even in its own domain, the prediction of "which of two strategies is more abundant at an average in the stationary distribution of an evolutionary process", inclusive fitness theory fell short, since it "requires stringent assumptions, which are unlikely to be fulfilled by any given empirical system" (Nowak et al. 2010). These include the assumption that interactions between organisms are additive and pairwise, thus excluding any situation with synergistic effects or where more than two organisms interact. Moreover, inclusive fitness is relevant only to a limited set of population structures. Without a specific model for the biological situation at hand,

> [i]t is possible to consider situations where all measures of relatedness are identical, yet cooperation is favoured in one case, but not in the other. Conversely, two populations can have relatedness measures on the opposite ends of the spectrum and yet both structures are equally unable to support evolution of cooperation. (Nowak et al. 2010)

In a massive Supplementary Online Information file much longer than the article itself, Nowak, Tarnita and Wilson presented a mathematical model using "standard natural selection theory to derive a condition for one behavioral strategy to be favoured over another". Their mathematical results, they wrote, reduced to the predictions of inclusive fitness under the extremely restrictive conditions where it could be applied. They argued that Hamilton's rule could be derived independently of inclusive fitness theory, and that it had never been empirically tested with precise measurements of relatedness, cost and benefit. Whether their mathematical results show what they assert, and whether inclusive fitness theory is truly as limited as they claim, is still being hotly debated.

Nowak, Tarnita and Wilson concluded that inclusive fitness theory was an essentially useless concept. "The exercise of calculating inclusive fitness does not provide any additional biological insight. Inclusive fitness is just another way of accounting, but one that is less general….there are no predictions that are specific to inclusive fitness theory." The authors proposed an alternative model for the evolution of eusociality. Such a process could be initiated by the formation of a group of organisms, whether as a result of offspring remaining near their parents, patchy food resources, or cooperation driven by "by simple reciprocity or by mutualistic synergism or manipulation". "What counts then," they wrote, "is the cohesion and persistence of the group" (Nowak et al. 2010). Such aggregation did not require relatedness, they emphasized, citing a striking experiment by Johns et al. (2009) in which multiple unrelated termite colonies were driven by ecological conditions to merge into a supercolony (Fig. 13.2). The causal arrow assumed in inclusive fitness theory, they wrote, should be reversed.

> Relatedness is better explained as the consequence rather than the cause of eusociality. Grouping by family can hasten the spread of eusocial alleles, but it is not a causative agent.

Fig. 13.2 The colony as superorganism. A mound constructed by the aptly named cathedral ter-mite, *Nasutitermes triodiae*, in the Northern Territory of Australia. Photograph by Yewenyi, repro-duced under a CC BY-SA 3.0 Creative Commons License, https://commons.wikimedia.org/w/index.php?curid=663058

> The causative agent is the advantage of a defensible nest, especially one both expensive to make and within reach of adequate food. (Nowak et al. 2010)

They second stage in Nowak, Tarnita and Wilson's proposed model involves the development of traits conducive to eusociality. These arise by selection at the indi-vidual level (though, rather obscurely, Nowak and colleagues also describe them as the result of adaptive radiation, "in which species split into different niches" where "some [species] are more likely than others" to develop such eusocial phenotypes).

Nowak et al. refer to these traits as "spring-loaded" pre-adaptations; as we will see in Chap. 15 below, Gould and Vrba suggest that this term should be discarded in favor of "preaptations". These traits include the tendency for solitary bees to behave eusocially when placed in a coerced partnership. Nowak et al. cite, among others, a study from Jennifer Fewell's lab at Arizona State University that followed forced pairings of bees from solitary and communal species (Jeanson et al. 2005). Strikingly, pairs of solitary bees were found to exhibit greater division of labor in tasks such as nest excavation than pairs of communal bees. Such studies suggest that

> division of labour appears to be the result of a pre-existing behavioural ground plan, in which solitary individuals tend to move from one task to another only after the first is completed. In eusocial species, the algorithm is readily transferred to the avoidance of a job already being filled by another colony member. (Nowak et al. 2010)

This is consistent with the "fixed threshold" model of the division of labor developed by Bonabeau et al. (1996), according to which individuals have different thresholds for undertaking certain behaviors; once an individual begins to perform a certain task, it inhibits other individuals from partaking in the behavior, leaving them available to perform other tasks, potentially even those for which they have a comparatively high threshold.

The third step in the process hypothesized by Nowak, Tarnita and Wilson is the development of "eusocial alleles". While these "have not yet been identified", the authors speculate that they might arise by even a single mutation; the description of *Volvox* above shows that this speculation is far from pure fantasy. Indeed, the wingless worker caste in social insects is simply a result of alterations in the genetic regulation of wing development "in such a way that some of the genes could be turned off under particular influence of the diet or some other environmental factor" (Nowak et al. 2010). In fire ants, alterations to the gene *Gp-9*[11] affect their ability to identify fertile queens and recognize intruders from other colonies. Now the stage is set for what Nowak, Tarnita and Wilson identify as the fourth step to eusociality: the shift of natural selection to act on emergent properties resulting from interactions between members of the colony. Although they do not explicitly state it in these terms, this represents a shift to MLS2 selection. Nowak et al. specify that, in this stage, natural selection is acting on a colony's fitness with regard to environmental factors. A fifth stage is identified as between-colony selection.

The stages identified by Nowak, Tarnita and Wilson for the evolution of eusociality could well be mapped onto the volvocine lineage hypothesis. This point was arguably more important than the wrangling over inclusive fitness. Yet hackles were raised and when, as Jerry Coyne had predicted, the "big guns" responded "en

---

[11] Gp-9 encodes a pheromone binding protein; a recent study by Lucas et al. (2015) shows that expression levels of *Gp-9* expression and another gene, *foraging*, correlate with the performance of various tasks. Another recent study in a different ant species shows that *foraging* expression correlates with age polyethism, the propensity of colony members of different ages to take on different tasks ("behavioral maturation"), rather than with divergent task performance among colony members of the same age (Oettler et al. 2015).

masse" to the Nowak et al. paper, the focus was on a defense of inclusive fitness theory. The big guns responded in more measured tones than the blog posts quoted above; several other smaller groups of researchers published separate replies in the same issue.

In the critique with the largest number of signatories, Abbot et al. (2011) argued that the criticisms regarding the prevalence of eusociality in diploid and haploid species were well known, and therefore Nowak et al. had contributed nothing new on that front. As for the contention that "standard natural selection theory" explains a broader range of phenomena than inclusive fitness theory, Abbot et al. argued simply, as Coyne had done in his blog post, that inclusive fitness theory is indeed *part* of "standard natural selection theory". They wrote that the "stringent limitations" under which Nowak et al. claimed inclusive fitness was constrained were, in fact, not essential. "Hamilton's original formulations did not make all these assumptions, and generalizations have shown that none of them is required," they wrote. "Inclusive fitness is as general as the genetical theory of natural selection itself. It simply partitions natural selection into its direct and indirect components."

Abbot et al. took issue with Nowak et al.'s statement that there was no empirical evidence supporting inclusive fitness theory, providing two detailed tables[12] listing behavioral phenomena explained or interpreted by inclusive fitness, as well as "[a]reas in which inclusive fitness theory has made successful predictions about behaviour in eusocial insects", such as altruistic helping, worker egg laying, sex allocation, and exclusion of non-kin. Many of these were the subject of experimental as well as correlational studies, and involved "interplay between theory and data" (Abbot et al. 2011). In response, Nowak, Tarnita and Wilson reiterated that

> [w]e do not know of a single study where an exact inclusive fitness calculation was performed for an animal population and where the results of this calculation were empirically evaluated. Fitting data to generalized versions of Hamilton's rule is not a test of inclusive fitness theory, which is not even needed to derive such rules. (Nowak et al. 2011)

In a separate response, Ferrière and Michod (2011) pointed out that, with the idea of inclusive fitness, Hamilton had introduced "[t]he idea that something other than the individual organism could be the fitness-maximizing unit"; this was "completely revolutionary at the time and opened new research areas that are still being developed, such as the study of transitions in units of evolution and individuality". Moreover, "[b]y opposing 'standard selection theory' and 'inclusive fitness theory'", they wrote,

> we believe that Nowak et al. give the incorrect (and potentially dangerous) impression that evolutionary thinking has branched out into conflicting and apparently incompatible directions. In fact, there is only one paradigm: natural selection driven by interactions, interactions of all kinds and at all levels. Inclusive fitness has been a powerful force in the development of this paradigm and is likely to have a continued role in the evolutionary theory of behaviour interactions. (Ferrière and Michod 2011)

---

[12] See Bourke (2011) for a similar table, with more detailed citations and references.

Herre and Wcislo (2011) accused Nowak, Tarnita and Wilson of setting up a false dichotomy, arguing that their proposed mathematical model contained assumptions quite as stringent as those they accused inclusive fitness of requiring. The Nowak model, Herre and Wcislo wrote, did not even allow for interaction of unrelated organisms as a control case.

A particularly salient critique came from Andrew Bourke (2011), who wrote that Nowak and colleagues mistakenly assumed that the interests of eusocial insect workers were subordinate to those of the queen. This revealed a fundamental problem with the "stringent" conditions they required as the basis of a supposedly general approach. In their Supplementary Information, Nowak et al. wrote that "there is no paradoxical altruism that needs to be explained" because "the eusocial gene is [at] the center of evolutionary analysis", which makes "the epicycles of kin selection and inclusive fitness disappear" (Nowak et al. 2010). Paradoxically, they claimed to be taking the gene's eye view *in contrast to* the kin selection approach, even though the gene's eye interpretation of kin selection was one of its major selling points for proponents of gene selectionism.

> We propose that kin selection among social insects is an apparent phenomenon which arises only when you put the worker into the center of evolutionary analysis. Kin selectionists have argued that a worker who behaves altruistically by raising the offspring of another individual, requires an explanation other than natural selection, and this other explanation is kin selection. (Nowak et al. 2010)

A first problem with this argument is the false opposition between natural selection and kin selection; the standard interpretation of kin selection is natural selection acting at the gene level.

The second problem, Bourke argued, lay in Nowak et al.'s "radical new suggestions for research on eusociality" that

> [t]he queen and her workers are not engaged in a standard cooperative dilemma. The reason is that the workers are not independent agents. Their properties are determined by the alleles that are present in the queen (both in her own genome and in that of the sperm she has stored). The workers can be seen as 'robots' that are built by the queen. They are part of the queen's strategy for reproduction. (Nowak et al. 2010)

Yet there are many ways in which the workers do not simply play a robotic role. First, the "inclusive fitness interests of workers and the mother queen do not coincide, because the two parties are differentially related to group offspring" (Bourke 2011). Moreover, a range of worker behaviors, such as eating the queen's eggs, manipulation of the colony's sex ratio by selectively destroying offspring, laying eggs "in response to perceived declines in queen fecundity" and occasional direct aggression toward the queen that all belie the "robot" assumption. "In the light of this proven lack of worker passivity," Bourke writes, "workers' reproductive self-sacrifice is paradoxical at first sight and this is the genuine problem of altruism that inclusive fitness theory has solved" (Bourke 2011).

Bourke's critique highlights another problem with Nowak et al.'s analysis. In laying out the conditions for their model in the Supplementary Information, they make confusing and contradictory statements about the level of selection at issue.

They claim to "put the gene at the center of the analysis" – contending that inclusive fitness does *not* do this, being instead concerned with "evolutionary games between the workers and the queen", despite the clearly gene-centered terminology of Hamilton's rule. This appears to be a misinterpretation of inclusive fitness theory, but also misrepresents Nowak et al.'s own approach. Later on the same page, they declare that "[o]ur model does not use standard multilevel selection. There is only one level of selection, the hymenopteran colony." Elsewhere in the same text they write that "[t]he target of selection is neither the phenotypic trait of the queen in particular, nor that of the colony, but the collectivity of traits that modify social behavior at both these levels" (Nowak et al. 2010).

There is a striking disconnect between the broad perspective taken in Nowak, Tarnita and Wilson's main article and the contradictory and restrictive requirements in the mathematical model in the Supplementary Information. The model purports to flesh out the sequence of steps proposed for the evolution of eusociality in the main article, but falls far short, hampered by internal contradictions over the level of selection at issue, incorrect assumptions about the behavior of workers, and restriction to the case of initial high relatedness among the organisms, with offspring failing to leave the nest.

Samir Okasha (who was a signatory to the Abbot et al. communication) attempted to cool things down with a commentary in *Nature* entitled "Altruism researchers must cooperate" (Okasha 2010). "Much of the current antagonism," he wrote,

> stems from the fact that different researchers are focusing on different aspects of the same phenomenon, and are using different methods. In allowing a plurality of approaches — a healthy thing in science — to descend into tribalism, biologists risk causing serious damage to the field of social evolution, and potentially to evolutionary biology in general. (Okasha 2010)

The choice between kin selection and a focus on ecological factors (and hence a group selection approach), Okasha argued, was a false dichotomy.

> Whether [scientists] stress the importance of one over the other will depend on the question they are asking. For example, relatedness has proved crucial to understanding conflicts between the queen and her workers over the production of male versus female offspring in ants, bees and wasps. For questions about how tasks are allocated to the workers in an ant colony or why the size of colonies differs across species, ecological factors are probably more relevant. (Okasha 2010)

In some cases, both processes may be at work, albeit possibly at different levels of selection. Further, Okasha argued, in many cases mathematical models of kin selection and multi-level selection provide identical results, and thus can serve as alternative *interpretations*, rather than alternative *explanations*, for the same phenomenon. One example of this is provided by Lehmann et al. (2007), who took a group selection result and showed that it could be reproduced exactly by a kin selection argument; see also Traulsen (2010), Marshall (2011), and Lehtonen (2016). Of course, such demonstrations do not lay bare the evolutionary history of the trait at issue, and therefore do not resolve underlying questions of causality.[13]

A broader demonstration of equivalence between multilevel selection and kin selection was provided by Wade (1980), who showed that Hamilton's rule can be derived as a special case of the Price equation. Wade expressed the change in gene frequency under kin selection as the sum of two terms, the change *within* groups and the change *between* groups, analogous to the two covariance terms in the Price equation. Hamilton's rule obtained when the change in gene frequency between groups was positive, and exceeded the absolute value of the change in gene frequency within groups. Thus a gene selection argument could be interpreted as a tension between selection at the individual and group level.

Okasha noted that alternative interpretations[14] in other fields of science, such as Lagrangian vs. Hamiltonian formulations of classical mechanics, or matrix vs. wave approaches to quantum mechanics, have not given rise to the level of infighting seen in evolutionary theory. (This may partly be a result of the role of causality in the alternative physics models: Lagrangian and Hamiltonian mechanics do not assume different underlying causes of the phenomena they describe, and in quantum mechanics causality is a total mystery to begin with, so hey). Political implications lie closer to the surface in evolutionary theory as well, as particularly evident in the critique of Gould's ideas regarding exaptations and spandrels (see Chap. 15). This is also likely to raise the temperature of the debate.

$$*** \quad *** \quad ***$$

It should come as no surprise that experimental studies purporting to show evidence for group selection, such as those of Pruitt and Goodnight (2014) on selection for group composition in a species of spider, have sparked heated exchanges in the pages of *Nature* (Grinsted et al. 2015; Gardner 2015; Pruitt and Goodnight 2015). Pruitt and Goodnight studied populations of the spider *Anelosimus studiosus*, which, though often solitary in the wild, is also found in multi-female colonies in certain regions. Working with populations from areas where multi-female colonies are prevalent, Pruitt and Goodnight showed that selection on colony-level traits "may drive adaptation to local conditions" (Linksvayer 2014).

---

[13] Inclusive fitness theory can be subject – justifiably or not – to the "mistaking bookkeeping for causality" criticism leveled at Dawkins's gene's eye view by Gould; Michael Doebeli makes precisely such a criticism in a Letter to the Editor of *Nature* in the wake of the Nowak, Tarnita and Wilson paper, writing that "[f]or eusocial insects, Nowak et al. convincingly argue that the basic mechanism of assortment is the formation of groups owing to ecological pressures, such as the need for nest defence. Despite the indignant response of the inclusive-fitness crowd, there can be no doubt about the fundamental tenet that, with or without the concept of inclusive fitness, in principle we have access to exactly the same amount of evolutionary knowledge. Personal modelling preferences may vary, but there is nothing magic about bookkeeping techniques" (Doebeli 2010). Note also the disparaging use of the term "accounting" by Nowak et al. (2010).

[14] It should be noted that "alternative interpretations" are distinct from the theory of "alternative facts", *sensu* Spicer and Conway (2017).

Different colonies were observed to have different proportions of docile and aggressive members. The proportion varied depending on the colony location. Pruitt and Goodnight hypothesized that the docility-to-aggressiveness ratio might be determined by selection at the group level. To test this, they constructed nests with different proportions of docile and aggressive females, and placed these nests at field sites where resources were either abundant or scarce. Further, some of the experimental colonies contained spiders local to the field site, and others were composed of "foreign" individuals native to a different area. Six sites were used, four in Tennessee and two in Georgia; three of these were considered low resource and three high resource. At each site, Pruitt and Goodnight established 53 colonies with random combinations of size (1–27 females) and composition (ranging from no aggressive spiders to a colony composed entirely of aggressives). Of the 53 colonies at each site, 37 were composed of spiders taken from the same site at which they were studied; the remaining 16 colonies were "taken from a paired site of opposing resource level."[15] Twenty local, naturally occurring colonies were monitored for comparison.

Pruitt and Goodnight then, as Timothy Linksvayer describes,

> tracked colony survival, composition and reproductive output over the next two generations, and found that the relationship between colony size and group composition strongly affects colony survival and reproductive success, and that sites with high or low resources consistently favor different relationships. Furthermore, they found that, after two generations, surviving colonies had shifted their size and composition to be more like their home site. These results suggest that the relationship between group size and colony composition is both heritable and locally adapted. Because whole colonies of these spiders survive or die depending on group traits, group selection is probably playing a central part in driving this local adaptation. (Linksvayer 2014)

Pruitt and Goodnight found that experimental colonies with a ratio of aggressive and docile members similar to the local, naturally occurring ratio tended to be nearly ten times more likely to survive than colonies that were "moderately dissimilar" to the natural colony proportions at the site. Colonies that were "extremely dissimilar" produced no offspring colonies at all. Moreover, as noted in Linksvayer's summary, colonies in their native area shifted their proportions to resemble the natural proportions in the site, suggesting that the environment had a selective influence on colony composition. Foreign colonies, however, shifted their composition to more closely resemble that of *their* home sites. This could mean that adaptation to local conditions, and thus selection at the group level, dominated for the local colonies, while a genetic predisposition toward a certain colony makeup, which could have arisen via a potentially vast number of selective events at multiple levels, led foreign colonies to shift back toward their home site proportions, regardless of the local selective pressure from the environment. Genetic predisposition could be acting on the

---

[15] Each high-resource site was "paired" with a low-resource site in the same state. Note that the experimental design results in all "foreign" colonies also being exposed to a different resource level than that under which they originally developed. This exposes "foreign" colonies to a variety of simultaneous environmental changes, providing a potentially confounding factor in the interpretation of the results.

local colonies as well, of course, and thus the behavior of the local colonies, while consistent with group adaptation to environmental conditions, cannot rule out other causal factors. Colony composition correlated with colony success; the foreign colonies tended to be dissimilar in composition to local colonies, and to fare worse under their native conditions. This suggests that a tendency to align with the composition of one's home site would be selected against by environmental factors when the home site composition was unfavorable to local conditions.

Pruitt and Goodnight's study left a number of unanswered questions. The way aggressiveness or docility affected an individual spider's ability to survive and/or reproduce, the behavior of the colonies over a longer time period than merely two generations, and the mechanisms by which colonies regulate their composition all remain unclear. For the last of these questions, Pruitt and Goodnight hypothesized that within-colony conflict and modulation of colony reproduction rate might drive the composition change. After only two generations, however, it is difficult to draw any robust conclusion. Also, as Pruitt and Goodnight point out, naturally occurring colonies of *A. studiosus* are composed of related individuals; the authors did not specify the relatedness of the experimental colonies. Grinsted et al. (2015) rasied the criticism that Pruitt and Goodnight did not measure individual fitness, and that *A. studiosus* is not always a social species, and indeed often takes a solitary lifestyle.[16] Grinsted et al. also expressed the concern that manipulation of colony composition might affect individual fitness, affecting the likelihood of colony extinction on that basis alone. To this, Pruitt and Goodnight (2015) countered that assuming individual fitness to be something that should be ruled out was essentially privileging it as a level of analysis (i.e., incorrectly applying Occam's razor). "Following their logic," Pruitt and Goodnight wrote, "all behavioural studies would be flawed because behaviour can be decomposed into physiology, genetics, applied physics, and so on. Thus, the arguments of Grinsted et al. aren't against group selection per se, but instead it seems they take issue with the word 'group'."

Another confounding issue with Pruitt and Goodnight's (2014) study is that measurements of composition and fitness in local colonies were, in many cases, taken over different years than the corresponding measurements in experimental colonies. Different environmental conditions in different years might render a direct comparison of the data sets invalid (Grinsted et al. 2015). Pruitt and Goodnight (2015) countered that the relation between group composition and group size nonetheless was typically positive at high resource sites and negative at low resource sites.

Other recent empirical investigations have provided evidence for group selection, such as the studies of water striders by Eldakar et al. (2009, 2010a, b), who revisited a problem first introduced by Garrett Hardin (1968) using the metaphor of "the tragedy of the commons". In a town with a common area for the grazing of sheep, for example, it is to each farmer's individual advantage to send more of his sheep to graze on the commons. Yet this could easily result in overexploitation and collapse of the community. Eldakar et al. considered a parallel situation in water striders, *Aquarius remigis*. In isolated communities, males use an aggressive mating

---

[16] Pruitt and Goodnight (2015) note that the species is typically colonial at the sites they studied.

strategy, which increases their individual fitness, but can decrease long-term female reproductive success, ultimately harming the community as a whole (Eldakar et al. 2010a). This can be viewed as the "overexploitation of females as a shared mating resource", and thus is a parallel situation to the tragedy of the commons (Eldakar et al. 2009).

To test how water strider communities deal with this potential problem, Eldakar et al. (2010a) set up a system of pools that could be either interconnected or isolated. Females moved away from the more aggressive males in interconnected pools, and the correlation between male aggression and mating success declined significantly compared to the isolated pools. Thus,

> selection strongly favored aggressive mating within isolated groups; however, in the multi-group population, the free movement of individuals amongst groups favored reduced aggression. Female dispersal created distribution patterns in which females clustered around less aggressive males, providing a favorable local sex ratio, while leaving more aggressive males with male-biased sex ratios and reduced mating opportunities. (Eldakar et al. 2010b)

Population structure was correlated with male aggressiveness: the "isolated pools... prevent aggressive male water striders from escaping the consequences of local exploitation" (Eldakar et al. 2010a). Decreased male aggressiveness in isolated pools can be viewed as the result of selection for a trait that is advantageous to the group, but not to the individual.

The results of Eldakar and colleagues are particularly interesting in that, in contrast to the Pruitt and Goodnight study, the species under study can choose its own group structure. The results are also notable because previous studies had suggested that group selection was only effective between isolated groups. Eldakar et al. (2010b) point out that dispersal between groups promotes inter-group differences as animals sort themselves out, resulting in an "increase [in] genetic and phenotypic variation among groups...in response to local conditions."

Host-parasite interactions can also lead to the tragedy of the commons. A parasite's success is enhanced by its ability to exploit the resources inside the host, and it can enhance its fitness relative to other parasites by competing with them for these resources. However, this only holds up to a certain point, beyond which the grazing grounds provided by the host will wear thin. Bashey and Lively (2009) investigated this problem from a group selection perspective, using *Steinernema carpocapsae*. This nematode infects insect larvae, mates inside them and produces offspring, which then kill and exit the host. Juvenile nematodes can survive in the soil in a "developmentally dormant" state between rounds of infection. *S. carpocapsae* has a symbiotic bacteria, *Xenorhabdus nematophila*, which lives in a specialized vesicle in the nematode's intestine, and which contributes to killing the host when the nematodes are ready to emerge.

Bashey and Lively infected the caterpillars of the wax moth *Galleria mellonella* with *S. carpocapsae*. Three replicate experiments were performed with three different laboratory stocks of the nematode in order to "determine whether the number of nematodes emerging from a host was a heritable trait". Bashey and Lively separated their nematode populations according to the number of juveniles that emerged from the host. "Although the number of nematodes emerging from a host is the sum of the individual fecundities," they noted, "there is also an emergent nature to this aggregate trait, as interactions among individual nematodes and their bacterial symbionts can influence nematode fecundity." In other words, this trait possessed aspects of both MLS1 and MLS2.

For each experiment, Bashey and Lively infected a number of hosts, and then counted the numbers of nematodes that emerged. After accounting for the mass of the host, they selected nematodes from the largest and smallest emerging populations, and used these to infect another set of hosts. Four such "high treatment" and "low treatment" passages were carried out in five replicates, for each of the three nematode stocks. For two of the stocks, the nematode population gradually declined over the course of the four passages. In the third stock, a statistically significant difference was observed between the high and low treatment nematode population sizes in the first passage, and this difference was maintained – though it did not increase – in the subsequent[17] passages. Bashey and Lively speculated that the differences between the three experiments may have been exacerbated by the fact that the third stock had only undergone eight passages through wax moth larvae before the start of the experiment, while the other two had undergone 21 passages. This might have left the third stock with a larger store of genetic variability, and thus better able to adapt to the high and low selection conditions.

In the third nematode population, other traits appeared to evolve "as correlated responses to group selection on population size" (Bashey and Lively 2009). Populations selected for large numbers emerged faster from their host, likely because they more quickly depleted their host's resources. Populations selected for small numbers emerged later, and were larger. In a natural environment, the "high treatment" nematodes would likely experience an individual fitness cost, since "[s]maller nematodes may have fewer energy reserves and thus may have lower survival when they are free-living in the soil … Moreover, once exposed to a host, smaller nematodes have been found to be less successful at colonizing and surviving to reproductive maturity." Bashey and Lively noted, however, that "in the current study, we saw no differences across treatments in number of nematodes successfully colonizing a host, and we saw a nonsignficant trend toward greater parasite success (i.e., probability of nematode emergence) in the High treatment."

The studies in nematodes, water striders, and spiders are only some of the most recent experimental studies of group selection. Many of the seminal studies in the field are reviewed in a 1997 paper by Goodnight and Stevens entitled "Experimental studies of group selection: what do they tell us about group selection in nature?" Michael Wade, for example, studied group selection in flour beetles, using *Tribolium castaneum* rather than Allee's *T. confusum* (Wade 1976, 1977). These were the first experimental studies conducted on group selection (Goodnight and Stevens 1997). Presaging the experimental design of Bashey and Lively, Wade took an initial common stock, and divided it into subpopulations of 16 individuals each. The populations that generated the most adults after a 37-day interval were then subdivided

---

[17] This population was put through an additional three passages, for a total of seven.

again, as were the populations that generated the fewest adults. Wade also established two control experiments, one (C) in which all populations contributed to the next generation, and one (D) in which random groups were chosen to found the next generation. After nine generations, the low selection condition showed population sizes significantly smaller than the controls, and the high selection condition showed population sizes significantly larger.

Wade noted that his results contained important information about the relative directions of selection at different levels. In the C control case, only individual selection was at issue. In the high selection treatment, individual selection led to a decrease in population (not only due to competition for resources, but also because *Tribolium* is cannibalistic), and thus operated in a direction opposite to that of the group selection imposed by the experimenter. In the low selection condition, however, individual selection and group selection were aligned in the same direction. Wade (1976) emphasized that "group selection in the opposite direction to individual selection can produce significant genetic change", and "group selection in the same direction as individual selection can produce results very different from individual selection acting alone".

*** *** ***

While some studies cast individual and group selection as antithetical, others isolate the points of balance between these complementary drives. Among the most interesting such works in recent years have come from Jeff Gore and colleagues at MIT. During postdoctoral work with Alexander van Oudenaarden, Gore investigated the dynamics of cooperation and cheating in a well-mixed[18] yeast population in the presence of sucrose. Like other yeast, *S. cerevisiae* preferentially metabolizes monosaccharides like glucose and fructose. It can, however, metabolize disaccharides such as sucrose, breaking them down into glucose and fructose using the enzyme invertase, coded for by the gene SUC2. Some of the glucose and fructose produced this way will diffuse away from the yeast cell, becoming available for consumption by neighboring cells. These monosaccharides thus become a "public good" for the entire yeast population. While a cell that breaks down sucrose can retain a greater portion of the output than a scavenging neighbor, the fact that hydrolyzed sugars become a public good opens to the door to cheating cells, which consume the output without making the metabolic effort to synthesize invertase themselves. Gore et al. (2009) investigated the interaction between a wild type ("cooperator") strain of *S. cerevisiae* and a "cheater" strain lacking the SUC2 gene, and therefore unable to produce invertase. Since the two strains each expressed a different fluorescent protein, their relative contributions to the total population could be easily visualized using flow cytometry.

Gore and colleagues first verified that there was a metabolic cost to producing invertase, by showing that inhibition of invertase synthesis in glucose-supplemented medium resulted in equal growth rates for cooperators and cheaters. They then used the fact that the cooperator strain also depended metabolically on histidine in order

---

[18] The spatial structure of a population is known to affect the outcome of competition between cooperators and cheaters (Greig and Travisano 2004; Hauert and Doebeli 2004).

to adjust the "cost" of cooperation, since a decrease in histidine concentration slowed the growth of cooperator cells relative to cheaters. Starting cultures with a range of different proportions of cooperators and cheaters, the investigators found that not only did a small fraction of cheaters spread within a cooperator population, but the reverse occured as well. The two populations were "mutually invasible", and reached a steady state ratio independent of their initial proportions, and depending only on the histidine concentration. Surprisingly, the equilibrium proportion of cooperators was quite small – typically around 15%. The interaction between cooperators and cheaters was indicative of a so-called *snowdrift game*.

> Our experimental observation of coexistence between the cooperator and cheater strains implies that the interaction is governed by what game theorists call the snowdrift game (also known as the hawk-dove game or the game of chicken). The snowdrift game derives its name from the potentially cooperative interaction present when two drivers are trapped behind a large pile of snow, and each driver must decide whether to clear a path. In this model of cooperation, the optimal strategy is the opposite of the opponent's (cooperate when your opponent defects and defect when your opponent cooperates). The snowdrift game is therefore qualitatively distinct from the prisoner's dilemma, in which all players have the incentive to cheat regardless of the strategies being followed by others. (Gore et al. 2009)

The snowdrift game has been suggested to be a better model for the evolution of human cooperation than the prisoner's dilemma (Kümmerli et al. 2007). Using game theory, Gore et al. (2009) identified the ranges of invertase expression cost and efficiency of monosaccharide capture over which coexistence of cooperators and cheaters were likely to occur. With a modulation of the efficiency term based on the observed nonlinear relation between glucose concentration and growth rate, the model fit the experimental data quite well. Further experimental studies showed that glucose concentration also affected the outcome. With more glucose present, cheaters had less need to rely on cooperators, and cooperators experienced a rising metabolic cost of invertase production. At a sufficiently high glucose concentration, their ability to produce invertase was inhibited completely. Their only option for survival was to adopt a cheater strategy. "We therefore see," the authors concluded,

> that the wild-type invertase production strategy is exactly what might be expected in a snowdrift game – wild-type cells pursue the strategy opposite to that of their opponents. It is possible that glucose repression of invertase is partly determined by these social considerations, helping to make a population of wild-type cells relatively immune to invasion by strains with alternate strategies. (Gore et al. 2009)

A few years later, Gore's own laboratory investigated the interaction between population dynamics and cooperation (Sanchez and Gore 2013; see also Allen and Nowak 2013). Evolutionary changes (such as allele frequency within a population) typically do not take place on the same timescale as changes in population dynamics, and therefore the study of "eco-evolutionary" feedback is comparatively rare. However, Sanchez and Gore noted, such feedback has been predicted to be very strong in species exhibiting cooperative growth and producing common goods. In the presence of sucrose, *S. cerevisiae* provided an ideal system in which to test this experimentally: it produced a public good, its population dynamics and evolutionary

Fig. 13.3 The *top* panel
shows an "eco-
evolutionary phase-space"
with the frequency of the
SUC2 gene plotted against
population density,
showing that different
populations follow
well-defined trajectories,
represented in different
colors. The *bottom* panel
shows a conceptual model
for these trajectories, with
cooperators represented by
gray circles, and cheaters
represented by *open
circles*. Reproduced from
Sanchez and Gore (2013)
under a Creative Commons
Attribution License

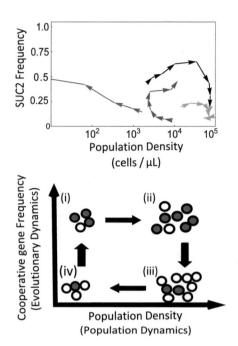

dynamics occurred on a similar timescale, and the breakdown of sucrose by invertase led

> to a cooperative transformation of the environment by the cells: at low population density,
> the cells are too dilute to effectively transform the sucrose environment into a glucose environment, so the cells grow slowly on what little glucose they retain following sucrose
> hydrolysis. At high population density, however, the cells are able to produce enough glucose for the population to grow rapidly. (Sanchez and Gore 2013)

Consistent with these expectations, populations of only invertase-producing cells either grew to a stable size, or collapsed, depending on the initial population density. A mixed population of cooperators and SUC2-deficient cheaters, however, showed much more complicated feedback dynamics. Plotting the system's behavior in a phase space with population density on one axis and the frequency of the SUC2 gene in the population on the other, the system was shown to exhibit complex trajectories, either approaching a point of collapse or exhibiting a spiraling trajectory toward a stable fixed point at which a large population of cheaters coexisted with a small proportion (~10%) of cooperators. The spiral path (Figs. 13.3 and 13.4) toward this fixed point indicated the presence of a transient feedback loop between the evolutionary dynamics (SUC2 axis) and the population dynamics (population density axis). Importantly, this also indicated that the system was bistable: it could approach two different outcomes (population collapse or a mixed equilibrium state) depending on the initial conditions (population density and proportion of cooperators). A Lotka-Volterra-type model of the dynamics was found to predict the observed experimental results quite well. The system's behavior was also critically deter-

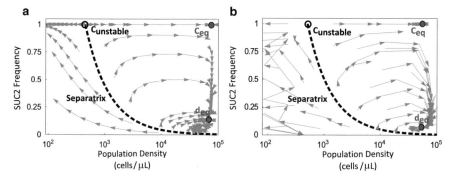

Fig. 13.4 "Eco-evolutionary trajectories" from simulation (**a**) and from experiment (**b**). At the d_{eq} point, both cheaters and cooperators coexist. Reproduced from Sanchez and Gore (2013) under a Creative Commons Attribution License

mined by the fact that the cooperators retained a slight advantage over the cheaters, being able to capture somewhat more of the monosaccharides they produced than the cheaters scavenging nearby. "An essential feature of the eco-evolutionary feedback in our system", Sanchez and Gore noted, "is the fact that cooperators have preferential access to the common good they produce… This preferential access creates the density-dependent selection that favors cooperators at low densities and cheaters at high densities, which is essential for the feedback loop."

As in the 2009 study, it was found that only a small percentage of cooperators are needed to sustain the population. However, this does not mean that the presence of cheaters has no detrimental effect. Sanchez and Gore found, both theoretically and experimentally, that the presence of cheaters made the population less able to recover from an environmental shock, such as a dilution (analogous to killing a large percentage of the population). Cooperators alone could survive the shock, while a mixed population under similar conditions would fall to extinction. Sanchez and Gore elegantly showed that this could be explained by the relative proximity of the equilibrium points in the cooperator-only vs. the mixed population case to the separatrix between the regimes of collapse and stability in the phase space.

In the same year, Gore's group also investigated the effect of range expansion on the relation between cooperators and defectors. Using the same *S. cerevisiae* model, they set-up a "stepping-stone" model for short-range population dispersal. A given well-mixed population was allowed to expand within a "habitat" of 12 wells on a 96-well plate. The population "spread" via the transfer ("migration") of a fraction to its nearest neighbor wells (Datta et al. 2013). With a population of pure cooperators, the population expanded as a travelling wave with "a characteristic profile in space consisting of a high-density bulk region and a low-density front". Mixed populations also advanced as travelling waves, but travelled more slowly. Most interestingly, they "observed significant spatial heterogeneity in allele frequencies within the mixed population wave." While the fraction of cooperators in the bulk population remained around 15%, they "observed that the frequency of cooperators was

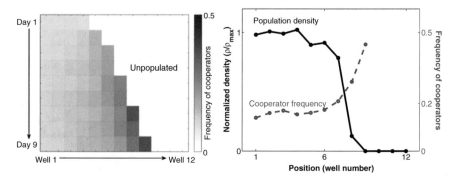

Fig. 13.5 *Left* panel shows cooperator frequency as a function of time (vertical axis) over the spatially expanding population of cooperators mixed with defectors. Note the increased frequency of cooperators at the leading edge of the wave front. *Right* panel shows the normalized population density (*black*) and cooperator frequency (*blue*) as a function of position, from day 9. Reprinted with permission from M.S. Datta et al., Range expansion promotes cooperation in an experimental microbial metapopulation, *Proc. Natl. Acad. Sci. U.S.A.* 110(18): 7354–7359, 2013

significantly larger on the low-density front of the expanding wave, reaching a frequency that was three times higher than that found in the bulk." This striking result it had a simple interpretation: *since cooperators could outcompete cheaters at low densities, they thrived in the sparsely populated wells at the front of the wave* (Fig. 13.5).

While the cooperators led the wave front, the cheaters rushed to catch up with them. Their "invasion velocity", however, did not equal the velocity of the cooperators. "We were intrigued", wrote Datta et al., "by this comparison, because it suggested that if a population of cooperators continued to migrate as it was invaded, the two populations might never completely mix." To explore this, they tuned the "severity" of the environment by changing the daily dilution factors of the population[19]; higher dilution factors would effectively subject the population to a greater stress. All velocities (the cooperator wave front velocity, the cheater invasion velocity, and the velocity of the bulk mixed population) decreased monotonically with the dilution factor, consistent with the density-dependent growth rate. However, the slopes of these monotonic decreases were not identical, and this had the effect of creating two regimes of relative cooperator/cheater spread, as the curves crossed one another. For lower dilution factors, the cooperators were able to "outrun" the cheaters, likely as a result of their growth advantage at low population densities (Fig. 13.6). For larger dilution factors, however, the cheaters were able to overtake them. This might at first seem counterintuitive, since cooperators have an advantage at low densities. Yet, as Datta et al. remarked, "defectors spread into wells that are already occupied by cooperators, in which population densities are high and growth

---

[19] In all these *S. cerevisiae* experiments, populations undergo daily dilution and regrowth; see Gore et al. (2009), Sanchez and Gore (2013), and Datta et al. (2013) for details of the experimental protocol.

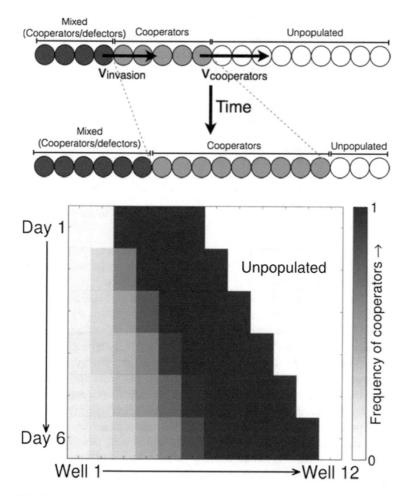

Fig. 13.6 Schematic (*top*) and experimental observation (*bottom*) "depicting the case in which cooperators can outrun defectors, in which the region occupied by the cooperators increases over time, even as the defectors invade". Reprinted with permission from M.S. Datta et al., Range expansion promotes cooperation in an experimental microbial metapopulation, *Proc. Natl. Acad. Sci. U.S.A.* 110(18): 7354–7359, 2013

conditions are favorable (glucose concentration is high)." Interestingly, no regime was observed in which the cooperators could move quickly enough to completely split off from the population of cheaters. The outrunning, even if under limited conditions and without splitting off, did offer "a plausible mechanism through which cooperation could be maintained in spatially extended populations", even in the absence of "spatial heterogeneity in environmental conditions generated by deteriorating conditions … or habitat destruction."

Celiker and Gore (2012) identified another ecological factor that promoted cooperation: interspecies competition. Cooperator and cheater strains of *S. cerevisiae* were co-cultured with an *E. coli* strain (DH5α) that could not metabolize sucrose. Instead, DH5α survived on arabinose, which was present in the culture medium but not utilized by the yeast. After 10 days in co-culture, the yeast cooperators had risen from 15 to 45% of the population. But were the *E. coli* simply out-cheating the yeast cheaters?

> A possible explanation for this increase in cooperator fraction within the yeast population is that bacteria behave as a 'superior' cheater strain by assimilating available free glucose, thus depriving cheater yeast cells of any sugar. In such a scenario, cooperator cells would do better than cheaters as they have at least some preferential access to the produced glucose. To test this, we competed yeast against a mutant strain of *E. coli* (JM1100) that has much reduced glucose and fructose uptake rates… We found a somewhat smaller albeit still significant increase in the cooperator fraction within the yeast population under the same conditions… Bacterial competition for the public good may therefore be a contributing factor toward increasing cooperator frequency in the yeast population, but there is another mechanism at work as well. (Celiker and Gore 2012)

The other mechanism, they soon identified, was cell density. The co-cultures exhibited a clear two-phase population pattern, in which the bacteria initially grew faster, before being overtaken by the yeast. This resulted in large part from the acidification of the medium caused by yeast sugar fermentation; *E. coli* do not thrive well in an acid environment. The investigators explored this further by using a PIPES (piperazine-N,N′-bis(2-ethanesulfonic acid)) buffer to adjust the acidity of the medium, essentially tuning the intensity of the competition. At higher PIPES concentration (and therefore higher pH), *E. coli* population density increased, thus simultaneously increasing the selection pressure on the yeast and decreasing the yeast density. As we have seen from previous experiments, cooperator cells do better in low density, since they can preferentially retain a portion of the monosaccharides they produce. As expected, as the PIPES concentration was increased, the cooperator fraction of the yeast population increased as well. In a few words, *interspecific competition drove selection for cooperator cells in one of the two species.* It is worth noting, of course, that while cooperator cells were being selected for, they were not really doing any actual cooperation per se, since the yeast densities were so low that each cooperator cell was essentially consuming the glucose it had produced, rather than benefitting substantially from sharing with its neighbors.

Interspecific competition does not always promote cooperation. Harrison et al. (2008) demonstrated that competition between two bacterial species (*S. aureus* and *P. aeruginosa*) drove selection for cheaters in one of the species. Celiker and Gore observed a similar phenomenon when they competed their yeast populations against *B. subtilis* rather than *E. coli*. Similar to the yeast cooperators, *B. subtilis* can secrete an enzyme that breaks down sucrose. In this case, however, Celiker and Gore observed selection for the cheater yeast strain.[20]

*** *** ***

---

[20] These studies can be compared with those of Popat et al. (2012) on cheater strains in biofilms, discussed in Chap. 9 above.

In this chapter, we have examined a recent controversy over the evolution of cooperative behavior (specifically, altruism), and surveyed a range of experimental studies of cooperation in the context of group selection in spiders (Pruitt and Goodnight), water striders (Eldakar and Wilson), nematodes (Bashey and Lively), and beetles (Wade), among other species. Lastly, we have explored some[21] studies conducted by Jeff Gore's laboratory on cooperation and cheating in yeast, and the role of interspecific competition in promoting cooperation. This last topic brings us to an important point we have skirted repeatedly in the preceding chapters but now must explore head-on. *What determines the level at which selection acts?* In Gore's yeast-bacteria experiments, for example, interactions with the bacteria clearly have some effect on individual yeast cells. Yet these interactions also affect the *proportion* of cooperators within the yeast population. This exemplifies the effect of interspecific competition on a group-level trait (related though it may be, of course, to action on all the individual yeast cells in the population). Can a formal distinction be made between these two types of selection – one acting on a number of individuals within a population, and another acting on a property of the population itself, as a whole? This is a problem that Samir Okasha explores in his book *Evolution and the Levels of Selection* (2006). Following the work of Heisler and Damuth (1987, 1988), as well as Richard Michod and others, Okasha concludes that these two types of selection are indeed fundamentally different. Though we will examine more subtle aspects of these definitions below, one can broadly equate *selection on a property of a group that can be described as an average property of all the group members* with **MLS1**, or multi-level selection of type 1. In contrast, *selection on a property of the group qua group* is **MLS2**, or multi-level selection of type 2. Michod has argued that transition from MLS1 to MLS2 involves an evolutionary transition *from cooperation to conflict suppression* that can be seen most vividly in aspects of metazoan evolution like germ-line sequestration. The transition from MLS1 to MLS2 corresponds a major evolutionary transition in the sense of Maynard Smith and Szathmáry (1995): the emergence of a new level of evolutionary individuality. And we have seen from examination of the volvocine algae how simple such a transition can be. Just a few genes can make all the difference.

Okasha approaches the problem of *Evolution and the Levels of Selection* from the assumption that evolutionary theory should be analyzed from a *diachronic* perspective, neither assuming that selection occurs only at one level (for in his view it clearly does not), nor assuming the existence of multiple levels a priori. "Ideally," he argues,

---

[21] Other studies include, for example, the role of the rate of host evolution in symbiotic interactions (Damore and Gore 2011); theoretical investigations of the application of Price's equation and Hamilton's rule to the problem of microbial cooperation (Damore and Gore 2012); the role of tipping points in pushing a "producer/freeloader" yeast ecosystem to the brink of collapse (Chen et al. 2014); the role of clustering in driving the structure of multispecies bacterial communities (Celiker and Gore 2014); computational modeling of the effect of slow switching between environments in small populuations (Tan and Gore 2012).

we would like an evolutionary theory which explains how the biological hierarchy came into existence, rather than treating it as a given. From this perspective, the levels of selection question is not simply about identifying the hierarchical level(s) at which selection *now* acts, which is how it was traditionally conceived, but about identifying the mechanisms which led the various hierarchical levels to evolve in the first place.... This new 'diachronic' perspective gives the levels-of-selection question a renewed sense of urgency. Some biologists were inclined to dismiss the traditional debate as a storm in a teacup – arguing that in practice, selection on individual organisms is the only important selective force in evolution, other theoretical possibilities notwithstanding. But as Michod (1999) stresses, multicelled organisms did not come from nowhere, and a complete evolutionary theory must surely try to explain how they evolved, rather than just taking their existence for granted. So levels of selection other than that of the individual organism must have existed in the past, whether or not they still operate today. From this expanded point of view, the argument that individual selection is 'all that matters in practice' is clearly unsustainable. (Okasha 2006, pp. 16–17)

To that end, Okasha devotes a large portion of his 2006 book, as well as other works (Okasha 2004a, b; Okasha and Paternotte 2012) to using the Price equation, which, as we have seen, can be applied to selection at any level, to characterize the simultaneous operation of selection at multiple levels. Such "causal decomposition" of the levels of selection can be used to analyze the shifting of selection from dominance at one level to dominance at another. While Okasha has been criticized for a less than transparent use of the concept of covariance (Waters 2011), he makes a resonantly successful analysis of Michod's model of the transition between levels of selection as a shift from MLS1 to MLS2.

The idea of parsing multi-level selection into MLS1 and MLS2 was originally proposed by Heisler and Damuth (1987) (see also Arnold and Fristrup 1982; Damuth and Heisler 1988). While we have briefly summarized the distinction between these two types of selection as relating to fitness of a group defined based on the average traits of the individual group members (MLS1) and selection relating to fitness defined with respect to a property or properties of the group as a whole, and thus irreducible to an average organismal fitness (MLS2), their more correct definitions are somewhat subtler. MLS1 properties can be thought of as aggregate characteristics, and MLS2 properties as emergent ones (Okasha 2006, p. 48). More precisely, consider the definition of fitness itself. This is typically taken to be the number of surviving offspring produced. Consider a hierarchy of biological units, where "each unit contains a number of smaller units" (Okasha 2006, p. 40). This could be a number of cells within a *Volvox*, a number of genes within a cell, or a number of insects within a colony. Indeed, as Okasha remarks, "it is not obvious which biological relation(s) are supposed to correspond to the abstract notion of containment". This allows the nesting to be considered in a very general sense. The groupings may be ecological or genealogical, and need not even involve interactions between the organisms.[22] However they are defined, units at any level in the hierarchy are assumed to be capable of living freely and capable of reproduction, or homologous

---

[22] As Okasha points out, McShea (1996, 2001a, b) argues that interaction among parts is necessary for a part-whole relation; Sober and Wilson (1998) specify that these interactions must affect fitness.

in some way to free-living, reproducing organisms (McShea 2001b; Okasha 2006, pp. 41–2); Weismann's battling internal organs do not qualify as units in this sense.

Various other definitions have been offered for a collective of organisms at some level of the evolutionary hierarchy. While the arguments of Okasha and Michod we are about to explore can be applied regardless of the definition of choice, it is worth briefly reviewing them, since the definition of a collectivity is a core theme of the present work. Michod and Nedelcu (2003) have argued that a collective of interacting units only becomes an evolutionary individual once it has evolved mechanisms for conflict suppression (see also Okasha 2006, p. 42). Elisabeth Vrba (1989) suggested that conflict between selection at the individual and species levels could serve as an acid test for higher-level selection. Szathmáry and Wolpert (2003) proposed that coordination of functions is essential within a group of organisms in order for it to exist as an evolutionary individual. As Okasha (2006, p. 42) points out, this would rule out most bacterial colonies. However, the sharing of public goods, quorum sensing, and formation of structures by cells in a biofilm could be seen as a sort of proto-coordination. David Sloan Wilson (1975) offered a more liberal set of minimal requirements to form an evolutionary individual. He argued that any organisms involved in an interaction that affected their fitness could be defined as belonging to a *trait group*; see also Sober and Wilson (1998) and Okasha (2006, p. 43 and throughout). Others have argued that spatial compartmentalization is an essential factor in promoting the positive association of cooperators, as in the putative enclosures that formed around the constituents of primitive enzymatic hypercycles (Szathmáry and Demeter 1987); Wilson also emphasized such positive association in the context of trait groups. As Okasha points out, the idea of a trait group might help avoid a logical fallacy pointed out by Griesemer in his critique of Dawkins. The requirement of functional organization among organisms is tantamount to requiring properties which could only arise *by* group selection, thus resulting in a chicken-and-egg problem (Okasha 2006, p. 43). (See Maliet et al. 2015 for a model developed by Michod's group that addresses this issue). Another consideration is whether the relation between organisms is ecological or genealogical (Eldredge 1985). In the case of genealogically related members of a group, such as the aspen tree ramets discussed in Chap. 2, interaction would seem to be unnecessary in order to have a part/whole relation: in this case, the ramets belong to the "whole" genet. Yet competitive interactions can exist within a group of ramets, which secrete auxins in order to suppress the growth of their neighbors (Wan et al. 2006). Cooperative interactions also exist within a clonal aspen grove. As we saw in Chap. 2, Tew et al. (1969) found that the roots can redirect nutrients toward parts of the system that have been exposed to stress.

Okasha considers a general two-level nested case, where the lower level units are called "particles" and the groups in which they are nested are called "collectives". Recalling Lewontin's tripartite analysis of the requirements for natural selection, if multi-level selection is to occur in this system, both the particles and the collectives must exhibit heritable variations that result in fitness differences.

This raises an overarching question: what is the relation between the characters, fitnesses, and heritabilities at each level? For example, how does the fitness of a collective relate to the fitness of the particles within it? Does variance at the particle level necessarily give rise to variance at the collective level? Does the heritability of a collective character depend somehow on the heritability of particle characters? The literature on multi-level selection has rarely tackled these questions explicitly, but they are crucial. (Okasha 2006, p. 47)

In precise terms, the problem is to quantify how character (or trait, if you prefer) z_i of particle i affects character Z of the collective to which particle i belongs. Certainly, the collective character Z can be an average of the individual characters z_i; Okasha gives the average frequency of a gene in a population as an example of such an aggregate collective character. Other characters of the collective can be better described as emergent: they may depend on, or relate to, the properties of the individual particle, but not in a simple manner. Okasha cites "the degree of morphological differentiation between castes, or the number of cell divisions before germline sequestration" as examples of such emergent characters. Another example would be the geographic range of a population. Okasha suggests that "aggregate and emergent are really opposite ends of a continuum,… rather than dichotomous alternatives" (Okasha 2006, pp. 48–49).

A further point to be considered is the relative reproductive timescales of the particle and the collective life cycles. Biologically, there are examples of particles that reproduce simultaneously with their collectives (chromosomes within the cell) and others that do not (mitochondria within the cell; cells within metazoan organisms). The timescales involved can affect the definition of fitness; a mitochondrion may reproduce multiple copies, but pass along a smaller number of copies to the progeny of the collective (the cell) within which it resides.

For the purposes of his argument, Okasha assumes that these two timescales will be equal, so that the definition of particle fitness will be straightforward: the number of offspring particles a parent particle leaves, unaffected by the relative reproductive timescale of the particle and the collective. The fitness of the collective is harder to define. It could be defined as an aggregate property, in *terms of the average (or total) fitness of the particles within the collective*. Alternatively, it could be defined as *the number of offspring collectives*. Okasha calls these possible definitions "collective fitness$_1$" and "collective fitness$_2$", respectively (Fig. 13.7). Collective fitness$_1$ is itself an aggregate trait of the type defined above, while collective fitness$_2$ is an emergent trait.[23]

Is the focal level of selection the particles (and the relative frequencies of their various characters)? In this case, the collectives can be viewed as "part of the environment", and we are dealing with what Damuth and Heisler (1988) and Okasha refer to as multi-level selection 1 (MLS1). "Alternatively," Okasha writes, "we may be interested in the collectives as evolving units in their own right, not just as part of

---

[23] Typically, collective fitness$_1$ is most naturally applicable to situations where reproduction of the levels occurs on a synchronous timescale, and collective fitness$_2$ is more applicable to the case the generation times of collectives and particles are asynchronous. However, these correspondences are not inevitable (Okasha 2006, p. 58).

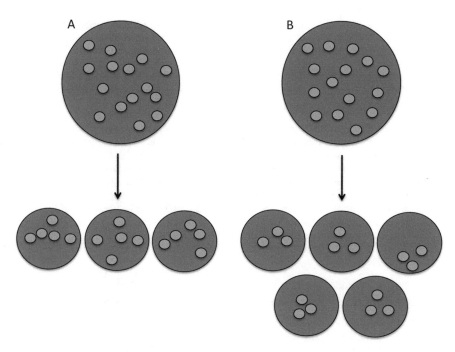

Fig. 13.7   Collectives A and B have the same collective fitness$_1$ (same number of offspring particles), but different collective fitness$_2$ (different numbers of offspring collectives). Drawing adapted
by the author from Figure 2.7 in Okasha (2006)

the particles' environment. If so, we will wish to track the changing frequency of
different particle-types *and* collective-types" (Okasha 2006, p. 56). Here, we have
multi-level selection 2 (MLS2). Importantly, in MLS1, *the collectives do not need
to be anything more than transient groups for sorting individual particles.* "The role
of the collectives in MLS1 is to generate population structure for the particles,
which affects their fitnesses. For MLS1 to produce sustained evolutionary consequences, collectives must 'reappear' regularly down the generations, but there is no
reason why the collectives themselves must stand in parent-offspring relations."
(Okasha 2006, p. 58)

In order to demonstrate how the relation between MLS1 and MLS2 maps onto
that between collective fitness$_1$ and collective fitness$_2$, Okasha turns to D. S. Wilson's
idea of trait groups in the context of the evolution of altruism.

Organisms are of two types in this model: selfish and altruist. They assort in groups for part
of their life cycle, during which fitness-affecting interactions take place, before blending
into the global population and reproducing. Within each group, altruists have lower fitness
than selfish types. But groups containing a high proportion of altruists have a higher group
fitness$_1$, that is, contribute more *individual* offspring to the global population, than groups
containing a lower proportion. So within-group selection favors selfishness, while between-
group selection favors altruism; the overall outcome depends on the balance between the
two selective forces. Wilson's model is thus designed to explain the changing frequency of
an *individual* trait – altruism – in the overall population. Although the explanation makes

essential appeal to group structure, and treats groups as fitness-bearing entities, it permits
no inference about the frequency of different types of group. Both levels of selection contribute to a change in a single evolutionary parameter. (Okasha 2006, pp. 56–57)

As an example of the "dovetailing" of collective fitness$_2$ and MLS2, Okasha cites
Jablonski's 1987 study of "heritability at the species level", which dealt with the
increase in geographic range of late-Cretaceous molluscs. Here,

species with large geographic ranges became more common, in a particular mollusc clade,
than those with smaller ranges. The suggested explanation is that species with larger geographic rages had greater fitness$_2$, that is, more offspring *species*, and that geographic range
was heritable. Note that this hypothesis permits no inference about the frequency of different types of *organism*, even though the species character in question – geographic range –
presumably depends on organismic characters, such as mobility and dispersal. Within each
species, these characters can evolve by selection at the level of the individual organism, so
there is potential interplay between the two levels of selection. But the key point is that fitnesses at each level are independently defined; so selection at each level leads to a different
type of evolutionary change, measured in different units. This is the hallmark of MLS2.
(Okasha 2006, p. 57)

Although "in Jablonski's example, the collective character subject to selection… is
emergent, [while] in Wilson's model it is aggregate", it does not necessarily follow

that the MLS1/MLS2 distinction always lines up with the aggregate distinction… *Variation
between collectives with respect to emergent characters could influence the number of offspring particles they leave – in which case MLS1 could operate on an emergent character.
Conversely, variation between collectives with respect to aggregate characters could influence the number of offspring collectives they leave – in which case MLS2 would operate on
an aggregate character.* This suggests that the MLS1/MLS2 distinction crosscuts the aggre-
gate/emergent distinction… (Okasha 2006, p. 57, my italics)

Indeed, even in the trait group analysis of altruism, the mechanism which favors
altruistic groups (and their individual members) is a black box. It could depend on
emergent group properties such as division of labor, or on individual properties, or
a combination of both.

Having parsed the relation between MLS1 and MLS2, and their intersection with
the two types of collective fitness, Okasha follows the ideas of Michod and colleagues (Michod 1997, 1999, 2005; Roze and Michod 2001; Michod and Nedelcu
2003) to analyze how major evolutionary transitions – the establishment of new
levels at which selection operates – can be viewed as shifts from MLS1 to MLS2.
Below the species level, such transitions are often characterized by the fact that, as
Maynard Smith and Szathmáry write, "entities that were capable of independent
replication before the transition can replicate *only* as a larger whole after it" (quoted
by Okasha 2006, p. 218, my italics). Like Michod, Okasha argues that "both types
of multi-level selection are relevant, but at different stages of a transition" (Okasha
2006, p. 219). Specifically, the dominant role shifts from MLS1 to MLS2 as the collective takes on an individuality of its own. This shift involves the development of
mechanisms for *conflict suppression* within the collective as "[c]ooperation exports
fitness from lower to higher levels" (Michod 2005). "Our theoretical work," Michod
writes,

proposes that the successful integration of previously independent evolutionary units into a new higher level individual involves a cycle of cooperation–conflict–conflict mediation. The point at which the cycle is entered depends on the nature of the initial ecological interaction associated with each ETI [evolutionary transition in individuality] (Michod and Nedelcu 2003). For example, the interaction may be conflictual to begin with, as with parasitic theories for the origin of the eukaryotic cell, or conflict may arise as a result of the evolution of selfish mutants, as can occur during the origin of multicellular groups. Furthermore, the nature of the subsequent interactions may differ among transitions; for example, kin selection may operate as a conflict mediator during the origin of multicellularity but not during the origin of the eukaryotic cell. In spite of these and other differences, a common framework involving cooperation, conflict and fitness reorganization can be used to understand ETIs. (Michod 2005)

From a formal perspective, Michod and colleagues used a population genetic model in which individual cells possess a single genetic locus, which can exist as one of two alleles, *cooperate* or *defect*. Cell growth rates may differ depending on whether the cell is a cooperator or a defector. Organisms are ensembles of individual cells, and reproduce by generating N-cell propagules. The fitness of an organism is the number of propagules it produces, while the fitness of an individual cell within the organism is determined by its individual cell replication rate. Thus, "an organism's fitness is not equal to the mean fitness of the cells within it, although these two quantities may be proportional to each other. This is the defining mark of an MLS2 theory" (Okasha 2006, p. 231).

Michod and Nedelcu (2003) argued that the essence of an evolutionary transition involves the *decoupling* of fitness at the level of the individual cell from that of the larger organism (or between the fitnesses at any two levels in a similar mutual relation).

Group fitness is, initially, taken to be the average of the lower-level individual fitnesses; but as the evolutionary transition proceeds, group fitness becomes decoupled from the fitness of its lower-level components. Indeed, the essence of an evolutionary transition in individuality is that the lower-level individuals must "relinquish" their "claim" to individual fitness in favor of the survival and reproduction of the new higher-level unit. The lower-level units still survive and may multiply, but in so doing they contribute to the fitness of the new higher-level unit. This transfer and reorganization of fitness components from lower to higher levels occurs through the evolution of cooperation and mediators of conflict that restrict the opportunity for within-group change and enhance the opportunity for between-group change. (Michod and Nedelcu 2003)

While Michod and Nedelcu do not explicitly use the MLS1-MLS2 terminology in their 2003 paper, Okasha notes that their argument "bears heavily on the MLS1 versus MLS2 issue. For in effect, Michod and Nedelcu are saying that in the early stages of a transition, *collective fitness is defined in the MLS1 way, but as the transition proceeds and fitness decoupling occurs, collective fitness in the MLS2 sense becomes relevant*" (Okasha 2006, p. 232, my italics).

To parse how such decoupling occurs, Okasha follows an argument proposed by Michod and Roze in 1999, who used both a population genetics approach (hence, a focus on the genic level), and the Price equation, which is applicable to multiple levels of selection. This melds the genic approach to the problem of major transitions in evolution taken by Maynard Smith and Szathmáry (1995) with the hierarchical approach advocated by Leo Buss (1987).

Consider organisms composed of cooperator and defector cells, which reproduce via N-cell propagules. They grow into adult organisms as the cells within them multiply. Each organism may have a different proportion of cooperating and defecting cells. The fitness of the ith organism is defined as the number of offspring organisms it produces. Michod and Roze (1999) and Okasha (2006) define this fitness as $W_i = 1 + \beta q_i'$, where q_i' is the frequency of cooperating cells in the adult organism, and β quantifies "the degree to which cooperation among cells benefits the organism" (Okasha 2006, p. 233). However, fitness will also be affected by the adult size of the organism, which will be proportional to the number of cells it contains.

> Since there is no separate germ-line, the greater the number of cells in the adult, the more offspring propagules it can send out. Also important is the size of the propagules themselves – the smaller they are, the more of them can be produced. (Okasha 2006, p. 233).

The definition of organismal fitness can thus be amended to $W_i = (1 + \beta q_i')k_i / N$, where N is the propagule size and k_i is the total number of cells in organism i. Since smaller propagules will lead to a larger number of offspring organisms, a *decrease* in N will *increase* W_i. By the same token, a larger organism will be able to produce comparatively more propagules, so an *increase* in k_i will *increase* W_i. Note, however, that k_i will typically be a function of N, since larger propagules will produce larger organisms.

Through k_i, the second equation for W_i *directly relates the fitness of cells to the fitness of the organism*, since fast-dividing cells will lead to a larger organism, which will lead to more propagules, and hence higher organismal fitness. "For fully fledged organisms such as ourselves," Okasha notes, "the number of cells we contain as adults does not directly affect our fitness – there is no particular advantage to being fatter" (Okasha 2006, p. 234). However, for simpler organisms "on the threshold of multicellular life", fitness *is* likely to depend on adult size. Recall, for example, the experiments of Travisano and colleagues selecting for larger clumps of yeast (Chap. 12), or the arguments for the correlation between fitness and size in the volvocine algae (Chap. 11). This suggests a

> lack of true individuality in these early cell groups, since there is a direct contribution of cell fitness to organism fitness … For individuality to emerge at the cell-group [organism] level, fitness at the new level must be decoupled from the fitness of the component cells. (Michod and Roze 1999)

This is a "cross-level by-product", a contribution of fitness at one level of selection to fitness at another. If such a situation were to be analyzed with the Price equation, it might lead to a spurious (i.e., non-causal) covariance between character and fitness at the organism level; as Okasha explains, Roze and Michod (2001) developed an alternate partitioning of the terms in the Price equation in order to avoid this problem. Their partitioning approach is temporal, breaking the fitness covariance into one component that occurs during development, corresponding to selection at the cell level, and another that occurs during reproduction, corresponding to adult size and/or adult functionality (such as germ line sequestration). "It is particularly striking," Okasha notes,

> that the problem cases for the Price approach, where cross-level by-products are in play, represent transitional stages en route to the evolution of new hierarchical levels. The original critiques of the Price approach made no mention of this point, since they operated with a synchronic rather than a diachronic formulation of the levels-of-selection question. (Okasha 2006, p. 236)

The transition thus occurs in three stages. First, organismal (or more generally, collective) fitness is defined as average cellular (or more generally, particle) fitness. This is a case of pure MLS1. Next, cooperation spreads between the particles until collective fitness is no longer defined by average particle fitness, but rather has some proportional relation to it. This is

> a transitional phase during which collective fitness in the MLS2 sense can apply, that is, collectives do produce offspring collectives, but where a collective's fitness is directly dependent on the average fitness of its constituent particles ... This represents a sort of grey area between MLS1 and MLS2: collective fitness is not defined as average particle fitness, as in MLS1, but it is proportional to average particle fitness; so the entirety[24] of the collective-level character-fitness covariance is due to a cross-level by-product. In Michod's terms, this means that the emerging collective lacks 'individuality', and has no collective-level functions of its own. As the transition proceeds, collective fitness is gradually decoupled from average particle fitness, and starts to depend on the functionality of the collective itself. MLS2 then occurs autonomously of MLS1, and the collectives can evolve adaptations of their own. Therefore, the relation between particle fitness and collective fitness itself undergoes a change, during a major evolutionary transition. (Okasha 2006, pp. 237–238)

Now, a stage has been reached where collective fitness is independent of average particle fitness, and the collective has achieved complete individuality. In fact, Michod (2005) argues that after the transition is complete, MLS1 fitness reduces to zero, since somatic individual cells cannot survive independently.

References

Abbot P et al (2011) Inclusive fitness theory and eusociality. Nature 471(7339):E1–E4
Allen B, Nowak MA (2013) Cooperation and the fate of microbial societies. PLoS Biol 11(4):e1001549
Arnold AJ, Fristrup K (1982) The theory of evolution by natural selection: a hierarchical expansion. Paleobiology 8(2):113–129
Axelrod R, Hamilton WD (1981) The evolution of cooperation. Science 211:1390–1396
Bashey F, Lively CM (2009) Group selection on population size affects life-history patterns in the entomopathogenic nematode *Steinernema carpocapsae*. Evolution 63(5):1301–1311
Bonabeau E, Theraulaz G, Deneubourg J-L (1996) Quantitative study of the fixed threshold model for the regulation of division of labour in insect societies. Proc R Soc Lond B 263:1565–1569

---

[24] I would argue that this grey area could be more broadly defined to include the case where only a portion of collective fitness was proportional to cellular fitness, and thus only a portion of the collective character-fitness covariance was driven by cross-level byproducts from cellular fitness.

Bourke AFG (2011) The validity and value of inclusive fitness theory. Proc R Soc B 278(1723):3313–3320

Boyd R, Richerson PJ (1992) Punishment allows the evolution of cooperation (or anything else) in a sizeable group. Ethol Sociobiol 13:171–195

Brandl SJ, Bellwood DR (2015) Coordinated vigilance provides evidence for direct reciprocity in coral reef fishes. Sci Rep 5:14556

Buss LW (1987) The Evolution of Individuality. Princeton University Press, Princeton

Celiker H, Gore J (2012) Competition between species can stabilize public-goods cooperation within a species. Mol Syst Biol 8:621

Celiker H, Gore J (2014) Clustering in community structure across replicate ecosystems following a long-term bacterial evolution experiment. Nat Commun 5:4643

Chen A, Sanchez A, Dai L, Gore J (2014) Dynamics of a producer-freeloader ecosystem on the brink of collapse. Nat Commun 5:3713

Cole BJ, Wiernacz DC (1999) The selective advantage of low relatedness. Science 285:891–893

Damore JA, Gore J (2011) A slowly evolving host moves first in symbiotic interactions. Evolution 65(8):2391–2398

Damore JA, Gore J (2012) Understanding microbial cooperation. J Theor Biol 299:31–41

Damuth J, Heisler IL (1988) Alternative formulations of multilevel selection. Biol Philos 3(4):407–430

Datta MS, Korolev KS, Cvijovic I, Dudley C, Gore J (2013) Range expansion promotes cooperation in an experimental microbial metapopulation. Proc Natl Acad Sci U S A 110(18):7354–7359

Doebeli M (2010) Inclusive fitness is just bookkeeping. Nature 467:661

Eldakar OT, Wilson DS (2011) Eight criticisms not to make about group selection. Evolution 65-6:1523–1526

Eldakar OT, Dlugos MJ, Pepper JW, Wilson DS (2009) Population structure mediates sexual conflict in water striders. Science 326(5954):816

Eldakar OT, Dlugos MJ, Holt GP, Wilson DS, Pepper J (2010a) Population structure influences sexual conflict in wild populations of water striders. Behaviour 147(12):1615–1631

Eldakar OT, Wilson DS, Dlugos MJ, Pepper JW (2010b) The role of multilevel selection in the evolution of sexual conflict in the water strider *Aquarius remigis*. Evolution 64(11):3183–3189

Eldredge NE (1985) Unfinished Synthesis: Biological Hierarchies and Modern Evolutionary Thought. Oxford University Press, New York

Ferrière R, Michod RE (2011) Inclusive fitness in evolution. Nature 471(7339):E6–E8

Gardner A (2015) Group selection versus group adaptation. Nature 524(7566):E3–E4

Goodnight CJ, Stevens L (1997) Experimental studies of group selection: what do they tell us about group selection in nature? Am Nat 150(Suppl. 1):S59–S79

Goodnight CJ, Schwartz JM, Stevens L (1992) Contextual analysis of models of group selection, soft selection, hard selection, and the evolution of altruism. Am Nat 140:743–761

Gore J, Youk H, van Oudenaarden A (2009) Snowdrift game dynamics and facultative cheating in yeast. Nature 459(7244):253–256

Greig D, Travisano M (2004) The Prisoner's Dilemma and polymorphism in yeast SUC genes. Proc R Soc B 271(Suppl 3):S25–S26

Grinsted L, Bilde T, Gilbert JD (2015) Questioning evidence of group selection in spiders. Nature 524(7566):E1–E3

Hardin G (1968) The tragedy of the commons. The population problem has no technical solution; it requires a fundamental extension in morality. Science 162(3859):1243–1248

Harman O (2010) The Price of Altruism: George Price and the Search for the Origins of Kindness. W. W. Norton & Company, New York

Harrison F, Paul J, Massey RC, Buckling A (2008) Interspecific competition and siderophore-mediated cooperation in *Pseudomonas aeruginosa*. ISME J 2(1):49–55

Hauert C, Doebeli M (2004) Spatial structure often inhibits the evolution of cooperation in the snowdrift game. Nature 428(6983):643–646

Heisler IL, Damuth J (1987) A method for analyzing selection in hierarchically structured populations. Am Nat 130:582–602

Herre E, Wcislo WT (2011) In defence of inclusive fitness theory. Nature 471(7339):E8–E9

Hughes WHO, Boomsma JJ (2004) Genetic diversity and disease resistance in leaf-cutting ant societies. Evolution 58:1251–1260

Jablonski D (1987) Heritability at the species level: analysis of geographic ranges of cretaceous mollusks. Science 238(4825):360–363

Jeanson R, Kukuk PF, Fewell JH (2005) Emergence of division of labour in halictine bees: contributions of social interactions and behavioural variance. Anim Behav 70:1183–1193

Johns PM, Howard KJ, Breisch NL, Rivera A, Thorne BL (2009) Nonrelatives inherit colony resources in a primitive termite. Proc Natl Acad Sci U S A 106(41):17452–17456

Johnson DD, Stopka P, Bell J (2002) Individual variation evades the prisoner's dilemma. BMC Evol Biol 2:15

Kerr B, Neuhauser C, Bohannan BJ, Dean AM (2006) Local migration promotes competitive restraint in a host-pathogen 'tragedy of the commons'. Nature 442(7098):75–78

Krama T, Vrublevska J, Freeberg TM, Kullberg C, Rantala MJ, Krams I (2012) You mob my owl, I'll mob yours: birds play tit-for-tat game. Sci Rep 2:800

Kümmerli R, Colliard C, Fiechter N, Petitpierre B, Russier F, Keller L (2007) Human cooperation in social dilemmas: comparing the Snowdrift game with the Prisoner's Dilemma. Proc R Soc B 274(1628):2965–2970

Lehmann L, Keller L, West S, Roze D (2007) Group selection and kin selection: two concepts but one process. Proc Natl Acad Sci U S A 104(16):6736–6739

Lehtonen J (2016) Multilevel selection in kin selection language. Trends Ecol Evol 31(10):752–762

Linksvayer T (2014) Survival of the fittest group. Nature 514(7522):308–309

Lucas C, Nicolas M, Keller L (2015) Expression of *foraging* and *Gp-9* are associated with social organization in the fire ant *Solenopsis invicta*. Insect Mol Biol 24(1):93–104

Maliet O, Shelton DE, Michod RE (2015) A model for the origin of group reproduction during the evolutionary transition to multicellularity. Biol Lett 11:20150157

Margulis L (1981) Symbiosis in Cell Evolution. W. H. Freeman and Company, San Francisco

Marshall JA (2011) Group selection and kin selection: formally equivalent approaches. Trends Ecol Evol 26(7):325–332

Maynard Smith J (1974) The theory of games and the evolution of animal conflicts. J Theor Biol 47(1):209–221

Maynard Smith J, Price GR (1973) The logic of animal conflict. Nature 246:15–18

Maynard Smith J, Szathmáry E (1995) The Major Transitions in Evolution. Oxford University Press, Oxford

McShea DW (1996) Metazoan complexity and evolution: is there a trend? Evolution 50:477–492

McShea DW (2001a) The minor transitions hierarchical evolution and the question of directional bias. J Evol Biol 14:502–518

McShea DW (2001b) The hierarchical structure of organisms: a scale and documentation of a trend in the maximum. Paleobiology 27:405–423

Michod RE (1997) Cooperation and conflict in the evolution of individuality I: multilevel selection of the organism. Am Nat 149(4):607–645

Michod RE (1999) Darwinian Dynamics: Evolutionary Transitions in Fitness and Individuality. Princeton University Press, Princeton

Michod RE (2005) On the transfer of fitness from the cell to the multicellular organism. Biol Philos 20(5):967–987

Michod RE, Roze D (1999) Cooperation and conflict in the evolution of individuality. III. Transitions in the unit of fitness. In: Nehaniv CL (ed) Mathematical and Computational Biology: Computational Morphogenesis, Hierarchical Complexity, and Digital Evolution. American Mathematical Society, Providence

Michod RE, Nedelcu AM (2003) On the reorganization of fitness during evolutionary transitions in individuality. Integr Comp Biol 43(1):64–73

Milinski M (1987) TIT FOR TAT in sticklebacks and the evolution of cooperation. Nature 325(6103):433–435

Nowak MA, Tarnita CE, Wilson EO (2010) The evolution of eusociality. Nature 466(7310):1057–1062

Nowak MA, Tarnita CE, Wilson EO (2011) Nowak et al. reply. Nature 471(7339):E9–E10

Oettler J, Nachtigal AL, Schrader L (2015) Expression of the foraging gene is associated with age polyethism, not task preference, in the ant *Cardiocondyla obscurior*. PLoS ONE 10(12):e0144699

Okasha S (2004a) Multilevel selection and the partitioning of covariance: a comparison of three approaches. Evolution 58(3):486–494

Okasha S (2004b) Multi-level selection, covariance, and contextual analysis. Br J Phil Sci 55:481–504

Okasha S (2005) Maynard Smith on the levels of selection question. Biol Philos 20(5):989–1010

Okasha S (2006) Evolution and the Levels of Selection. Oxford University Press (Clarendon Press), Oxford/New York

Okasha S (2010) Altruism researchers must cooperate. Nature 467(7316):653–655

Okasha S, Paternotte C (2012) Group adaptation, formal Darwinism and contextual analysis. J Evol Biol 25(6):1127–1139

Popat R, Crusz SA, Messina M, Williams P, West SA, Diggle SP (2012) Quorum-sensing and cheating in bacterial biofilms. Proc R Soc B 279(1748):4765–4761

Press WH, Dyson FJ (2012) Iterated Prisoner's Dilemma contains strategies that dominate any evolutionary opponent. Proc Natl Acad Sci U S A 109(26):10409–10413

Pruitt JN, Goodnight CJ (2014) Site-specific group selection drives locally adapted group compositions. Nature 514(7522):359–362

Pruitt JN, Goodnight CJ (2015) Pruitt & Goodnight reply. Nature 524(7566):E4–E5

Roze D, Michod RE (2001) Mutation, multilevel selection, and the evolution of propagule size during the origin of multicellularity. Am Nat 158(6):638–654

Sanchez A, Gore J (2013) Feedback between population and evolutionary dynamics determines the fate of social microbial populations. PLoS Biol 11(4):e1001547

Sober E, Wilson DS (1998) Unto Others: The Evolution and Psychology of Unselfish Behavior. Harvard University Press, Cambridge, MA

Stewart AJ, Plotkin JB (2012) Extortion and cooperation in the Prisoner's Dilemma. Proc Natl Acad Sci U S A 109(26):10134–10135

Stewart AJ, Plotkin JB (2013) From extortion to generosity, evolution in the Iterated Prisoner's Dilemma. Proc Natl Acad Sci U S A 110(38):15348–15353

Stewart AJ, Plotkin JB (2014) Collapse of cooperation in evolving games. Proc Natl Acad Sci U S A 111(49):17558–17563

Szathmáry E, Demeter L (1987) Group selection of early replicators and the origin of life. J Theor Biol 128(4):463–486

Szathmáry E, Wolpert L (2003) The transition from single cells to multicellularity. Chapter 13. In: Hammerstein P (ed) Genetic and Cultural Evolution of Cooperation. MIT Press, Cambridge, MA

Szolnoki A, Perc M (2009) Emergence of multilevel selection in the prisoner's dilemma game on coevolving random networks. New J Phys 11:093033

Tan L, Gore J (2012) Slowly switching between environments facilitates reverse evolution in small populations. Evolution 66(10):3144–3154

Tew RK, DeByle NV, Schultz JD (1969) Intraclonal root connections among quaking aspen trees. Ecology 50(5):920–921

Traulsen A (2010) Mathematics of kin- and group-selection: formally equivalent? Evolution 64(2):316–323

Trivers R (1971) The evolution of reciprocal altruism. Q Rev Biol 46:35–57

Vrba E (1989) Levels of selection and sorting with special reference to the species level. Oxf Surv Evol Biol 6:111–168

Wade MJ (1976) Group selection among laboratory populations of *Tribolium*. Proc Natl Acad Sci U S A 73(12):4604–4607

Wade MJ (1977) An experimental study of group selection. Evolution 31:131–153

Wade MJ (1980) Kin selection: its components. Science 210(4470):665–667

Wan X, Landhäusser SM, Lieffers VJ, Zwiazek JJ (2006) Signals controlling root suckering and adventitious shoot formation in aspen (*Populus tremuloides*). Tree Physiol 26(5):681–687

Waters CA (2011) Okasha's unintended argument for toolbox theorizing. Philos Phenomenol Res 82(1):232–240

West SA, Griffin AS, Gardner A (2007) Evolutionary explanations for cooperation. Curr Biol 17:R661–R672

Williams GC (1966) Adaptation and Natural Selection: A Critique of Some Current Evolutionary Thought. Princeton University Press, Princeton

Wilson DS (1975) A theory of group selection. Proc Natl Acad Sci U S A 72(1):143–146

Zagorsky BM, Reiter JG, Chatterjee K, Nowak MA (2013) Forgiver triumphs in alternating Prisoner's Dilemma. PLoS ONE 8(12):e80814

Chapter 14
Evol = f(Evol)

...with a wild surmise...

Keats

DESPITE THE claims of Dawkins, evolution is far more complex than the "survival of the stable". In order for any evolutionary change to occur at all, in order for life to explore the vast landscape of possible forms, variation is essential. Perhaps we are lucky this is so, since in our imperfect world mistakes are inevitable, however accurate a proofreading enzyme might be. The inevitability of variation, coupled with its "value" as a source of new evolutionary possibilities, provided Arthur H. Sturtevant with the answer to a question he posed in an influential 1937 article in the *Quarterly Review of Biology*. Sturtevant, who had assembled the first genetic linkage map of a chromosome (the *Drosophila* X chromosome) in 1913, asked *why mutation rates do not evolve to zero*. His paper laid the groundwork for the study of what is now called the *evolution of evolvability* or, somewhat less poetically, *second-order selection* (Tenaillon et al. 2001).

Evolvability can be defined as *the potential to respond to selective pressure*, or *the capacity to generate adaptive solutions when faced with a varying environment*. The topic, as we will see, remains controversial. It is important, however, not only for an understanding of evolution itself, but also for more practical concerns like dealing with invasive species (Gilchrist and Lee 2007; Wagner and Draghi 2010). Massimo Pigliucci (2008) has suggested that evolvability (and the evolution thereof) "may constitute one of several pillars on which an extended evolutionary synthesis will take shape during the next few years". Wagner and Draghi (2010) state flatly that arguments against the evolution of evolvability "have not been rigorously examined by their proponents and are thus a self inflicted blind spot in evolutionary biology."

One of the primary reasons for resistance to the idea of the evolution of evolvability is that evolvability is a trait at the population-level or above, and thus an unattractive concept for those opposed to the very notion of group selection. Moreover, the evolution of evolvability[1] seems at first glance to require the paradox

---

[1] Though it is beyond the scope of this chapter, it is important to note that some parameters descriptive of evolvability can be quantified. One of these measures is the additive genetic variance-

© Springer Science+Business Media B.V. 2018
S. Bahar, *The Essential Tension*, The Frontiers Collection,
DOI 10.1007/978-94-024-1054-9_14

of evolutionary forethought: something selected for its future benefits. The paradox vanishes when evolvability is seen to result from the unintended consequences of traits evolved in response to other selection pressures, before becoming subject to selection at a level higher than the individual.[2] In a 2007 critique, Lynch suggested that evolvability and other emergent features of biological systems "may be nothing more than indirect by-products of processes operating at lower levels of organization". This is one of those cases when nothing more turns out to be quite a lot.

*** *** ***

Useful though they may be for exploring an evolutionary landscape, Sturtevant recognized that most mutations are deleterious. Using genetic studies of the fruit fly *Drosophila melanogaster* as a case study, he noted that "[i]n general... one may conclude that each normal... individual of *D. melanogaster* is subject to a lowering of its potential reproductive value through the occurrence of new mutations in the germ tract" (Sturtevant 1937). However,

> a body of data has gradually been accumulating that shows that different strains may deviate significantly from the usual values for mutation rate... It appears that, for the X chromosome at least, some strains give lethal mutation rates as much as five or six times as high as those shown by other strains... There are, then, genes that affect general mutation rate, and stocks differ in their constitution with respect to such genes. It follows that, in wild populations, such genes must be subject to selection. (Sturtevant 1937)

Certainly, there would be selection against high mutation rates, and this would be affected by environmental conditions.

> Any external agents that increase the general mutation rate will evidently increase the average loss in potential reproductive value, and will therefore increase the intensity of the selection for the genes that lower the rate. Such selection will always be present, but it must be more intense in regions with high natural ionization, and also in regions with high temperatures. (Sturtevant 1937)

Sturtevant observed that artificial laboratory conditions seemed to raise the mutation rate, and wondered whether "all mutation rates so far determined

---

covariance matrix G, which gives a measure of the short-term response to selective pressure (Lande 1979; Wagner and Altenberg 1996; Pigliucci 2008; Wagner and Draghi 2010). Another is the mutational covariance matrix M, introduced by Lande in 1980, which is "more relevant for exploring the medium-term evolutionary responses after the standing genetic variation has been exhausted" (Wagner and Draghi 2010). The ability of a lineage to generate novel phenotypes, however, does not yet have a quantitative measure (Wagner and Draghi 2010).

[2] A population can also adapt to recurring environmental changes; the ability to perform such adaptation is a type of evolvability. As Wagner and Draghi (2010) note, "genotypes that adapted quicker to the last environmental change of the same kind can also be evolvable in response to a similar change in the future. The opportunity for evolvability to evolve therefore depends on the correlation between past and future environmental changes, and is an empirical question, not a logical one. Similarly, biotic interactions are also predictable in many cases, for instance, a parasite can 'know' that the host immune system will most likely attack antigenic residues on its surface rather than, say, the GC content of its genomic DNA."

experimentally [are] too high to be applicable for natural populations". It was clear that mutation rate genetically tunable, though the mechanism remained obscure.

> The method of action of genes affecting the mutation rate is still unknown. Probably they act in various ways, but it remains possible that they have something in common, such for example as an influence on the relative duration in time of specific stages of mitosis or some other general property. In this case it might well be that such effects would have greater selective value through other results than their effects on mutation rate. (Sturtevant 1937)

Despite being less prevalent that deleterious mutations, favorable mutations clearly do occur, and Sturtevant wondered how selection might act in their favor.

> It seems at first glance that there should be a counter-selection, due to the occurrence of favorable mutations. It is true that favorable mutations furnish the only basis for the improvement of the race, and must be credited with being the only raw material for evolution. It would evidently be fatal for a species, in the long run, if its mutation rate fell to zero, for adjustments to changing conditions would then not long remain possible. While this [counter-selection] effect may occur, it is difficult to imagine its operation. It is clear that the vast majority of mutations are unfavorable, and Fisher (1930) has shown that the rare favorable ones must, in general, be [in] genes with slight effects. In other words, for every favorable mutation, the preservation of which will tend to increase the number of genes in the population that raise the mutation rate, there are hundreds of unfavorable mutations that will tend to lower it. Further, the unfavorable mutations are mostly highly unfavorable, and will be more effective in influencing the rate than will the relatively slight improvements that can be attributed to the rare favorable mutations. (Sturtevant 1937)

Sturtevant was left, then, with a conundrum:

> This raises the question – why does the mutation rate not become reduced to zero? No answer seems possible at present, other than the surmise that the nature of genes does not permit such a reduction. In short, mutations are accidents, and accidents happen. (Sturtevant 1937)

As Sturtevant concluded from the evidence available to him at the time, deleterious mutations are known today to be far more prevalent than beneficial mutations. Denamur and Matic (2006) cite values of up to 8×10^{-4} deleterious mutations per genome per replication in *E. coli*, in contrast to 2×10^{-9} beneficial mutations. Recent arguments from biochemistry and bioenergetics are consistent with Sturtevant's suggestion of a lower bound on mutation rates (Sniegowski et al. 2000). Another factor that may set such a lower limit is the *cost of fidelity*. Optimization of transcription fidelity is likely to incur not only a metabolic cost, but a temporal cost as well, since excessive time spent correcting transcriptional errors will delay the rate of replication (Kimura 1967; Sniegowski et al. 2000). In other words, the best is the enemy of the good.

The limits on mutation rates from above raise another set of problems. As Sturtevant noted, too high a mutation rate has obvious negative consequences. But how high is too high? Might there be some optimal mutation rate, providing "just the right" amount of variability? Studies of microbial genetics initially found similar mutation rates across a wide range of genome sizes (Drake 1991; Drake et al. 1998). However, as with most biological problems, exceptions to the rule were quick to follow, and evidence for a universal mutation rate in eukaryotes has not been forthcoming, as shown in Fig. 14.1 (Sniegowski et al. 2000). Indeed, Sturtevant himself mentioned observations of higher intrinsic mutation rates in *D. melanogaster* than in *D. funebris* (Sturtevant 1937).

Other studies suggest that observed mutation rates, whether optimal or not, are certainly not minimal. Perhaps more importantly, mutation rates are *variable*. Evidence now suggests that mutation rate not only evolved, but is continually adjustabe. Experiments in organisms as diverse as bacteriophage T4 (Drake et al. 1969; Schapper 1998), *E. coli* (Schapper 1998) and *Drosophila melanogaster* (Nöthel 1987) have succeeded in driving mutation rates below the wild type under selection pressure. Other experimental evolution studies, from Lenski's group at Michigan State, have shown that selection pressure can drive mutation rates in *E. coli* above wild type. Sniegowski et al. (1997) subjected 12 populations of *E. coli* derived from the same ancestral clone to a low-glucose environment. Three of these 12 populations developed a mutator phenotype, with mutation rates between one and two orders of magnitude higher than that of their ancestor.

Antibiotics provide an obvious source of selection pressure on bacteria. A number of studies show that "mutator" bacterial strains, with high mutation rates compared to wild type, have greater antibiotic resistance than wild type, and thus have a clear selective edge in the presence of an antibiotic (Denamur and Matic 2006). Experimental evolution studies demonstrated the induction of mutator strains as a response to antibiotics. Mao et al. (1997) applied a sequence of antibiotic treatments to an experimental *E. coli* population with an initial population of one mutator cell (deficient in mismatch-repair mechanisms) in 100,000. After only four selective cycles, *all* surviving cells exhibited at mutator phenotype. An analogous process may be at work in metazoans such as *Daphnia*, which undergo sexual reproduction during times of stress, resulting in increased genetic diversity.

Despite the bacterial studies just cited, there is no simple correlation between high mutation rate and antibiotic resistance. For example, Denamur et al. (2005) found that strains of naturally occurring *E. coli* with intermediate mutation rates survived exposure to multiple antibiotics better than those with high or low mutation rates. This is consistent with a computational study suggesting that, even though they spread more slowly through the population, weak mutator strains are more stable than strong mutators (Taddei et al. 1997). Weak mutator strains might be more robust with respect to environmental change, occupying a broad fitness plateau rather than a narrow peak. The value of such broad fitness peaks has been dubbed "survival of the flattest" (Wilke et al. 2001).

Mutator strains occur even in the absence of an identifiable selective drive. Strains having higher mutation rates than the norm are also common within a group of isolates of any given species of bacteria. Naturally occurring mutator strains have been identified in *Salmonella enterica*, *Neisseria meningitides*, *Staphyllococcus aureus*, and *E. coli*, among others (Denamur and Matic 2006).

In some bacterial mutator strains, the source of the increased mutation rate has been identified. Often, mutator strains arise from inactivation of *mutS* or *mutL* genes in the DNA mismatch repair system. This can result from a range of mutations, including insertions, deletions, premature stop codons and shifts in the reading

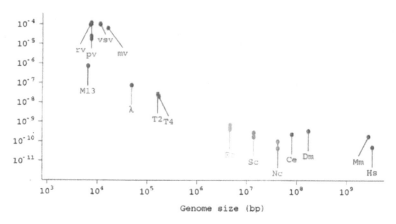

Fig. 14.1 Mutation rate per genome (vertical axis) as a function of genome size. Points in pink correspond to RNA viruses (rhinovirus (rv), poliovirus (pv), vesicular stomatitis virus (vsv), measles (mv)). The point in yellow represents *E. coli* (Ec). Points in red are DNA-based bacteriophages. *S. cerevisiae* (Sc) and *Neurospora crassa* (Nc) are shown in green. Blue points indicate *C. elegans* (Ce), *Drosophila melanogaster* (Dm), *Mus musculus* (Mm) and *Homo sapiens* (Hs). Multiple symbols for the same species indicate different measurements at different loci. Reprinted with permission from Sniegowski et al. (2000). Figure © John Wiley & Sons, Inc

frame (Denamur and Matic 2006). The protein coded by *mutS* recognizes all but one of the eight possible base pair mismatches; it can also bind small groups of unpaired bases that may occur as a result of misalignment of the reading frame. The protein coded by *mutL* helps to link MutS with other repair proteins (Denamur and Matic 2006); recent experimental studies suggest that MutL traps MutS at the mismatched site to serve as a marker while the repair process is carried out (Qiu et al. 2015).

Not all changes in mutation rate are modulated by the mismatch repair machinery. Genetic "hot spots" or "contingency loci" are highly mutable regions that include genes coding for antigens on the surface of pathogenic bacteria like *Neisseria meningitides* and *Helicobacter pylori*. Mutations in these proteins enable the bacteria to escape the immune response, at least for a while. Tenaillon et al. (2001) point out that "[i]n the long run, the localised mutator strategy would be more advantageous than the generalised one because it avoids a high mutation rate in housekeeping genes and would therefore bear almost no cost". Radman et al. (2000) compare this situation to the hypermutability in regions of the lymphocyte genome involved in antibody production, where mutations occur up to a million times more often than at other loci.

Even though they already provide a constitutively high level of mutations, contingency loci can be enhanced by the mutator phenotype. Contingency loci often contain *microsatellite repeats* (Moxon et al. 2006). As Radman et al. (2000) explain,

all known DNA polymerases skip, or add, one or a few motifs when copying such sequences. This polymerase slippage error is up to 10^4 times more likely than a base substitution error but it is very effectively corrected by the mismatch repair system... Thus, in wild-type cells, but much more so in mismatch-repair-deficient mutants, such microsatellite sequences will make genes highly susceptible to frameshift mutagenesis. (Radman et al. 2000)

Mutation rate can also be increased under stress via the SOS response, first identified by Radman in 1975 in *E. coli*, in which the expression of mutator genes is triggered. Silent under normal conditions, these genes include *umuC* and *umuD*, which code for proteins that combine to form the DNA polymerase V complex, and *dinB*, which codes for DNA polymerase IV. A similar set of SOS response genes are found in yeast. The expression of such genes under stress conditions can trigger a high rate of mutations through interrelated mechanisms. For example, DNA polymerase IV appears to induce mutations in undamaged DNA. The polymerase V complex is capable of copying damaged DNA, and thus reproduces new mutations as well as fixing the DNA damage that initiated the SOS response (Radman 1999). The SOS response is now know to be triggered by a cell's detection of single-stranded DNA, and the molecular mechanisms by which the response is induced have been well studied (Baharoglu and Mazel 2014). While the occurrence of single-stranded DNA might sound like a highly specialized trigger for such a complex event, many forms of DNA damage – such as those caused by antibiotics, naturally occurring toxins, UV irradiation, or exposure to reactive oxygen species – can result in unpaired portions of DNA strands. A wide range of stressors can thus trigger a cascade of events that, while repairing damaged DNA, also increase the frequency of mutations, and transcribe mutated sequences with wild abandon. Metzgar and Wills (2000) have suggested that such hypermutability might not have evolved under direct selection, but instead might be a spandrel, resulting from selection for repair mechanisms at the individual level. Population-level selection could have preserved and further refined these processes due to their role in modulating evolvability.

Mutator genes can spread through a bacterial population via a process called hitchhiking. Since bacteria reproduce asexually, they lack the extensive molecular machinery for genetic recombination present in sexually reproducing species. As a result, linkage between nearby genes is less likely to be disrupted during reproduction. If a mutator gene lies near some other gene on the chromosome, these two genes are likely to remain associated through multiple cell divisions. If that other gene is beneficial to the organism, it is likely to increase in frequency in the population, since organisms carrying it will have a selective advantage. The mutator gene will travel right along, whether in the chromosome during replication or in smaller genetic parcels during horizontal gene transfer (Tenaillon et al. 2001). This has been demonstrated experimentally for horizontal gene transfer in *E. coli*, by Funchain et al. (2001). Sniegowski et al. (1997) suggested that the development of the three mutator *E. coli* strains they obtained as a result of glucose deprivation may have been a result of hitchhiking.

Despite their propensity to hitchhike, mutator mismatch repair genes have another, seemingly contradictory property: they have a "hyper-recombination" phenotype, promoting genetic recombination. As suggested by Denamur and Matic (2006), "when selective pressure for increased genetic variability is no longer present, the hyper-recombination phenotype of mismatch repair-deficient strains might facilitate the reacquisition of the functional mismatch repair genes via horizontal gene transfer. This can 'save' the adapted mutator genome from being overburdened with deleterious mutations", and perhaps also save it from losing beneficial mutations gained during the period of hypermutability.

Prokaryotic cells have one more reason to follow mutator strains across the fitness landscape toward new heights: the robustness conferred by their large population sizes. While various definitions of the term robustness are still being debated, and the term continues, unfortunately, to be used too often as a buzzword, *mutational robustness* can be defined as "the ability to preserve a constant phenotype despite the genomic mutational load", where this load is the "burden" on the genome resulting from deleterious mutations (Elena and Sanjuán 2005).

In multicellular organisms, mutational robustness is most likely to derive from redundancy in the genome (Krakauer and Plotkin 2002) which can provide structural modularity and serve as a template for evolutionary experimentation. In addition to genetic modularity, sources of redundancy include

alternative metabolic pathways, or chaperone proteins that buffer against mutation-induced proteins in other enzymes. [In the presence of mutational load]… these mechanisms would produce phenotypes similar or identical to that of the unmutated wild type, such that individuals would have similar chances for survival and reproductive success. (Elena and Sanjuán 2005)

With their large population sizes, bacteria and viruses possess robustness based on numbers alone. Minimal redundancy renders bacterial cells highly susceptible to deleterious mutations. But if they die, there is a swarm of other bacterial cells waiting in the wings, and at least some of these are likely to have beneficial mutations. Importantly, this is an example of robustness – and, as a result, the ability to take advantage of mutator phenotypes – at the population, rather than at the individual, level (Elena and Sanjuán 2005).

The differential role of population size and the ability of mutator genes to hitchhike in asexually reproducing cells are only a few of the differences in evolvability between eukaryotes and prokaryotes. Prokaryotic and eukaryotic genomes are subject to different selection pressures as a result of their very architecture (Poole et al. 2003). The circular prokaryotic genome is constrained by the fact that it only has one replication origin per chromosome; this limits the replication rate. Since organisms that produce offspring faster have a selective advantage, this creates a pressure *against replication fidelity*, since a more sloppily replicated genome is a more quickly replicated genome. Replication rate can also be favored by reduced genome size, which imposes a selection pressure *for increased fidelity*, since smaller genomes allow for less redundancy, and therefore fewer backup options if a gene should suffer a mutation that renders its protein nonfunctional. Functions necessary to respond to diverse conditions may be too costly to carry in the genome, due to the increased replication time they would entail; this can be obviated by periodically-selected functions (PSFs), which may be lost in individuals but maintained within a population and regained when needed through lateral gene transfer;

genome size varies by ~20% in some bacterial species, indicating the degree to which genomes gain and lose genes (Poole et al. 2003). The limited nature of the prokaryotic genome also speaks to the related evolutionary histories of prokaryotes and eukaryotes. It is often assumed that there is a trend toward increased complexity, implying that eukaryotes evolved from prokaryotes. However, the reductions in the prokaryote genome and the diversity of eukaryote genomes, including elements that pre-date the last universal common ancestor (LUCA), suggest that evolutionary history is far more tangled and complex than the simple hypothesis of a prokaryote-to-eukaryote trajectory would allow.

Mutability in prokaryotes is also driven by increased competence (ability to take up DNA from the external environment) induced by stress. Like hypermutability induced by the SOS-response, increased competence allows prokaryotes to quickly explore the evolutionary landscape during times of starvation. DNA uptake has been suggested to be favored between related strains of bacteria, approximating a form of sex. However, the need for genome reduction,

> due to constraints on replication rate during exponential growth, suggests that any sequences taken up will only be fixed if they confer a selective advantage on the organism. Greater promiscuity permits greater sampling of environmental DNA, potentially bestowing a greater propensity to adapt to environmental change (greater evolvability). (Poole et al. 2003)

In contrast to prokaryotes, eukaryotic genomes have multiple sites of replication initiation per chromosome. This decouples genome size from replication rate, and facilitates increased redundancy.[3] Indeed, genome size in eukaryotes varies 80,000-fold! This reduced constraint allows for the insertion of "selfish" genetic elements that could be co-opted for a functional role in the host cell. Adding to eukaryotic genetic variability is the fact that mRNA can undergo post-transcription splicing and editing. DNA methylation also adds another source of (potentially heritable) variability. This leads to what Poole et al. describe as a "more complex relation between genotype and phenotype" in eukaryotes than in prokaryotes. They emphasize that while all these factors may contribute to eukaryote evolvability, none of them were selected *for* that purpose, any more than were prokaryotic stress responses. Kirschner and Gerhart emphasized this point in an influential 1998 paper on evolvability.

The ecological niches occupied by metazoan species provide further constraints on evolvability. This can limit the degree of experimentation that is possible without causing potentially deleterious effects, especially in reptiles and amphibians, which may occupy a range of ecological niches during their lifetimes.

> Many amphibian and reptile taxa experience dramatic shifts in their environment during development, essentially having to function in different niches. For instance, the komodo dragon (*Varanus komodoensis*) begins life as an arboreal predator of small insects, progressively moves onto larger insects, small vertebrates and eggs, then larger vertebrates and eventually fills a terrestrial large predator/scavenger niche... Mutations providing a

---

[3] Prokaryotes can gain some degree of redundancy by containing multiple copies of their genomes (Maldonado et al. 1994; Åkerlund et al. 1995; Bendich and Drlica 2000; Poole et al. 2003).

potential fitness advantage at any point along this continuum may be deleterious some-where else during growth. This effect is less in mammals and birds because they typically feed their young until they can occupy the adult niche. Compared with other vertebrates, mammals and birds are also notable for an increased emphasis on homeostasis, particularly endothermy... Both effects, reducing the range of niches during development and stabiliz-ing internal conditions, should enhance morphological evolvability. Indeed, while mam-mals and birds have diversified into widely different niches and morphologies from their ancestors that shared the planet with dinosaurs 65 million years ago, amphibians, turtles, lepidosaurs (snakes and lizards) and crocodilians typically have not... (Poole et al. 2003)

Kirschner and Gerhardt (1998) argue that metazoan evolvability derives from cellular and developmental processes conserved at least since the Cambrian explo-sion. In other words, some of the most rigidly conserved aspects of metazoan biol-ogy are precisely those that enable the unique characteristics of metazoan evolvability. "How," they asked, "can all this commonality serve as the platform for the immense diversification of metazoa?" Redundancy in the eukaryotic genome is crucial, since "[f]unction is frequently conserved even when sequences differ sub-stantially... Genotypic variation is often not matched by an equal functional varia-tion." This provides proteins that can easily be co-opted into other functions, while retaining their original role; the crystallins[4] provide a dramatic example of this. Likewise, the opposite also holds, and small changes in gene sequence can produce large changes in functionality; this can enable a wide range of phenotypes to result from a few genetic changes.

Kirschner and Gerhart point out that, while many core cellular processes are highly conserved, the regulatory mechanisms that control them are quite diverse. Much of the regulation in biochemical pathways takes place by inhibition (and by activation resulting from the inhibition of an inhibitor). Inhibitory molecules are quite versatile and "sticky", able to bind to a range of substrates. Calmodulin is a prime example of such versatility, being "effective without having to be highly spe-cific" (Kirschner and Gerhart 1998). Small mutational changes can trigger unique – and possibly advantageous – new interactions.

Like gene regulatory networks, biochemical pathways possess what Kirschner and Gerhart call "weak linkage", in contrast to the strong linkage present in meta-bolic networks, where "[s]teric requirements are high, and the complementary fit of surfaces of interacting components is precise." In systems with weak linkage, the components often exist simply in an on or off state. "Such regulatory organization," Kirschner and Gerhart explain, "facilitates...accommodation to novelty (to new activating or inhibiting signals) and reduces the cost of generating variation." This can be seen in the modularity of neural systems and in the regulation of gene tran-scription in eukaryotes. In the latter case, in contrast to the single prokaryote initia-tion site, "the eukaryotic system admits to many transcriptional inputs from proteins bound at short enhancer sequences of different kinds, which can be at almost any distance within 50,000 bases from the initiation site and at either orientation" (Kirschner and Gerhart 1998). Likewise, enhancer proteins may have low sequence

---

[4] See Chap. 15.

specificity and thus be able to bind rather promiscuously. As a result, it should be "easy to add and subtract regulatory elements to eukaryotic genes...increas[ing] the evolvability of the system" (Kirschner and Gerhart 1998).

Evolvability in metazoans is driven by what Kirschner and Gerhart call "exploratory mechanisms": a proliferation of variation, followed by selective pruning. Such mechanisms involve adaptive immunity, the establishment of contacts between microtubules and the kinetochore during mitosis, the development of neural connections, cell migration in embryonic development, the development of local angiogenesis, and the adaptation of neural crest cells to their local environment within the embryo. "The exploratory nature of these systems is very deconstraining," write Kirschner and Gerhart.

> Evolutionary modification of a vertebrate limb shape and size is reduced mostly to the mutational modification of the cartilaginous condensations [which are directed by *Hox* genes and lay down the basic pattern for limb structure]. It need not be simultaneously accompanied by mutationally derived changes in the muscle, nerve, and vascular systems, which can accommodate to any of a wide range of limb sizes and shapes. (Kirschner and Gerhart 1998)

Compartmentation of gene expression also enhances evolvability by providing both flexibility and robustness of the entire system against total disruption by a mutation in one component. Compartmentation can be spatial, as in cell differentiation within a metazoan, or temporal, as a result of sequential gene expression. The same gene can play different roles in different compartments, and thus "different conditional responses of a single gene's expression can be selected independently... A system with such great flexibility of use in one individual would seem exquisitely suited to generate, by modest mutation, different patterns in different individuals in evolution" (Kirschner and Gerhart 1998).

Temporal compartmentation occurs in the development of *Drosophila* bristles; the positions of the sensory organ precursor (SOP) cells that eventually form the bristles are specified before bristle cytodifferentiation. As a result, "a modification of the SOP by mutation or developmental imprecision will never compromise the function of the bristle" (Kirschner and Gerhart 1998). This is an example of an individual-level property that may have had the exaptive effect of increasing evolvability at the lineage level.

> Although genomic compartmentation and redundancy may have been selected for the physiologic robustness they confer to the development and physiology of complex metazoa, they also facilitate evolutionary change by making various cell populations independent, reducing the chance of lethal mutation and increasing the independence of variation and selection within a compartment. Because of compartmentation, changes in extracellular or intracellular signals are more likely to result in local elaboration of new morphologies than in a catastrophic failure of global organization. (Kirschner and Gerhart 1998)

Compartmentation is also seen in the segmentation of body plans within any given phylum, and the comparative stability of the phylotypic[5] stage across a wide range

---

[5] As Michael Richardson explains, "[t]he phylotypic stage is a key concept in evolution and development. It can be defined as the time point in the development of an animal when it most closely resembles other species. In vertebrates, this corresponds to the organogenetic period, when numerous, undifferentiated organ primordia are appearing" (Richardson 2012).

of families, genera and species. A similar hierarchy was identified in regulatory networks by He and Deem (2010), who suggested "that the slow evolution of the top components and fast evolution of the bottom components of the hierarchy is a universal phenomenon in evolution". One reason for this, Kirschner and Gerhart would argue, is that the "top" components are extremely robust, as a result of compartmentation and patterns that can be easily triggered by self-activating signalling cascades. In *Drosophila* development, for example, initial egg polarity is determined only by four gene products, but quickly "elaborates into the 50 or 60 different spatial compartments of the body plan" (Kirschner and Gerhart 1998). The same compartmentation that allows experimentation at the branches imposes rigidity at the root.[6] Kirschner and Gerhart sketch out a hypothetical process of selection for evolvability at the clade[7] level.

> In each round of variation, selection, and fixation, it appears that during the fixation period there was a selection for properties of robustness and versatility of the new-found processes. These properties then facilitated the generation of phenotypic variation used in the next round, and, we think, ensured the conservation of the flexible robust process itself, which was further selected with those new variations. Organisms with processes lacking robustness and versatility presumably lost out in the next round because of their inability to retain as much genetic variation and to generate selectable phenotypic novelty with as few mutational changes. Added to this is the preemptive argument that a clade of organisms still perfecting its body plan at a time when others (with perhaps less perfect plans) had already started originating appendages and mouthparts was now at a selective disadvantage... (Kirschner and Gerhart 1998)

∗∗∗ ∗∗∗ ∗∗∗

It has been argued that mutator genes can play only a limited role in metazoan evolution due to the difficulty of hitchhiking in genomes that undergo extensive recombination during meiosis. Nevertheless, experimental evidence for evolvability in eukaryotes has been obtained. Not surprisingly, these claims for evolvability in the eukaryotic genome have been highly controversial. Suzannah Rutherford and

---

[6] Significant evolutionary malleability does occur at developmental stages preceding the phylotypic stage. Kirschner and Gerhart note that "there have been extensive diversifications of the egg and early stages of development before the phylotypic stage is formed. Thus, evolution since the mid-Cambrian has involved modifications before and after the phylotypic stage but not of the stage itself." This is likely driven by strong selective pressure on the egg, and is possible because "the network of reactions that characterizes the body plan of the phylotypic stage makes few demands on the reactions at earlier developmental stages for its initiation. The early embryo must generate some initial polarities, but they can be simple, *ad hoc*, and diverse. Once orientated, placed, scaled, and activated by these reactions, the compartments of the phylotypic stage become self-perpetuating... We suggest that a conserved property of the body plan, attributable to its selected constrained circuitry, is ease of formation." (Kirschner and Gerhart 1998).

[7] It might be more correct to pose such an argument as selection at the group, population, or other level, since there are a number of arguments against the possibility of selection at the clade level, not least of which being the fact that a clade contains its own offspring; see Okasha (2006), pp. 212ff.

Susan Lindquist made the first experimental demonstration of eukaryotic evolvability in 1998. They showed that when the gene (*Hsp83*) encoding the *Drosophila* heat shock protein Hsp90 (Fig. 14.2) was mutated by even a single base substitution, such as a cysteine for an arginine, the flies exhibited a wide array of phenotypic disturbances, as shown in Fig. 14.3 (Rutherford and Lindquist 1998).

Hsp90 stabilizes key proteins in various signal transduction pathways. As Wagner et al. (1999) explain,

> Hsp90 is a highly conserved molecular chaperone that, in addition to participating in the cell stress-response system by helping to refold denatured proteins, also has a more specific role in signal transduction. Through repeated low-affinity interactions, Hsp90 keeps inherently unstable proteins involved in signal transduction pathways poised for action. (Wagner et al. 1999)

Clearly, disruptions to this protein could have major effects. Pennisi (2013) elegantly explains how this could provide a "capacitor for cryptic variation". In its role as a molecular chaperone, she writes, Hsp90

> ensures [that] proteins take on and maintain their correct shape, even if their amino acid sequence varies slightly because of mutations. By doing this for proteins important for development,… [it] could theoretically hide the existence of accumulated mutations. Reduce its presence in an embryo, and the effects of those mutations would appear – providing grist for natural selection. (Pennisi 2013)

This provides a mechanism for what Waddington (1942, 1953) referred to as "decanalization". He had proposed that the robustness of developmental pathways provided a sort of channeling, or canalization, in which genetic variation could occur without disrupting the developmental process. With a mechanism of canalization removed, the underlying variation would be cryptic no longer, and would have potentially selectable effects on the organism.

In their *Drosophila* experiments, Rutherford and Lindquist observed a shockingly large array of developmental defects, ranging from deformation of eyes and legs to abnormal wing vein patterns in *Hsp83* mutants. They obtained similar deformations when Hsp90 was inhibited by feeding flies with the drug geldanamycin. Rutherford and Lindquist speculated that normal development in these flies might be caused by

> …three possibilities of increasing interest. First, the mutants might be more sensitive to the environment: as a stress-response factor, wild-type Hsp90 might simply buffer against noise caused by random micro-environmental effects with little or no genetic basis. Second, Hsp90 mutants might exhibit an increased mutation rate: Hsp90 might be directly or indirectly involved in the fidelity of DNA replication. Third, cryptic genetic variation might be expressed to a greater extent [in Hsp90 mutants]: because it is a chaperone for signal-transduction elements, Hsp90 might normally suppress the expression of genetic variation affecting many developmental pathways. (Rutherford and Lindquist 1998)

The third conjecture is consistent with the normal role of Hsp90. Indeed, the fact that the observed effect – if the third interpretation is correct – is mediated by a heat shock protein suggests that, as in prokaryotes, hypermutability is triggered by stress.

Crossing *Hsp83* mutants with normal flies often produced offspring with the same defects as their parents, suggesting these defects were heritable. The persis-

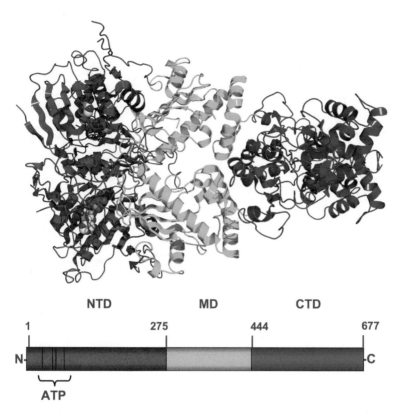

Fig. 14.2 Structure of the Hsp90 protein in its dimeric form. NTD: amino terminal domain; MD: middle domain; CTD: carboxyl terminal domain; ATP: ATP-binding region. Public domain (https://en.wikipedia.org/wiki/File:Hsp90_schematic_2cg9.png)

tence of the defect in subsequent generations allowed Rutherford and Lindquist to conclude that the third and "most interesting" hypothesis was correct. Studies of specific eye and wing abnormalities confirmed this conclusion. Most strikingly, lines with deformed eyes showed decoupling of the Hsp90 mutation from the deformed eye trait. By the sixth and seventh generations of crosses, more than 80% of the flies exhibited the deformed eye phenotype. But none of the flies' genomes showed evidence of the original Arg → Cys mutation in *Hsp83*. This suggested that the original mutation had reverted to its wild type, but that the mutations it had "catalyzed" remained in the genome. Hsp90 thus played the role of a buffer or "capacitor" for "silent polymorphisms".

> We have provided what is, to our knowledge, the first evidence for an explicit molecular mechanism that assists the process of evolutionary change in response to the environment. We suggest that in nature, transient decreases in Hsp90 levels resulting from its titration by stress-damaged proteins could uncover morphological variants for selection to act upon… Once the frequency of [some] trait is increased in this manner, given a moderate fitness advantage, selection could increase the frequency of genetic polymorphisms affecting the

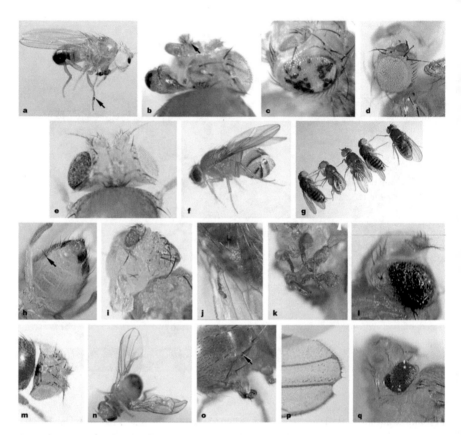

Fig. 14.3 An astounding range of developmental abnormalities are triggered by altered Hsp90 function. Reprinted by permission from Macmillan Publishers Ltd., S.L. Rutherford and S. Lindquist, Hsp90 as a capacitor for morphological evolution, *Nature* 396: 336–342, 1998, copyright 1998

trait to a point at which it no longer depends on reduced Hsp90 function to be expressed in the population. (Rutherford and Lindquist 1998)

Unsurprisingly, this study elicited a major reaction within the evolutionary community. Writing to *Nature*, Dickinson and Seger (1999) railed against the perceived teleological fallacy of selection acting for the future benefit of evolvability. Such a result could only come about through group selection, they wrote, and, since this was virtually impossible, explanations based on individual fitness differences were required – and presumably unattainable. They based their critique on the assumption that Rutherford and Lindquist had posited that selection acted on Hsp90 *as* an evolutionary "capacitor"; Lindquist (2000) responded that she and Rutherford "did not…claim that these properties evolved to this end". As an example of the tone of these arguments, here is an excerpt from Dickinson and Seger:

The need to see 'purpose' in evolution, or at least some internal drive to help the blind processes of random variation and natural selection, is remarkably resilient. Recent manifestations in the scientific literature imagine evolved mechanisms that actively promote further evolution or that facilitate rapid response to changed conditions... Such interpretations seem to call for the evolution of properties that anticipate future needs. But selection lacks foresight, and no one has described a plausible way to provide it. In principle, group selection might produce results that seem to escape this limitation. For example, increased mutation rates may indeed allow populations to adapt more quickly to changed conditions, even though they harm most individuals. The evolutionary problem is that such group benefits are usually weaker than individual costs, in a well-defined sense that makes group selection effective only under very restrictive conditions [here they cite G. C. Williams]. So, in general, we need explanations that are based on individual fitness differences... This is not a semantic quibble. Cosmic rays affect evolution by causing mutations, but we would not claim that they exist for that purpose. Similarly, developmental buffering and variable mutation rates may influence the course of evolution, but this does not mean that they evolved to that end. (Dickinson and Seger 1999)

In unpacking Dickinson and Seger's many objections, it is important to note, in the context of their reference to group selection, that evolvability is (sometimes tacitly) assumed that to be a population-level trait. From this viewpoint,

[a] population with a large amount of heritable variation for fitness can certainly be considered more evolvable than one with very little heritable variation for fitness. Similarly, a population with a larger amount of heritable variation for a phenotypic character will respond more quickly to natural or artificial selection on that character than one with a smaller amount of such variation. (Sniegowski and Murphy 2006)

It is also well worth noting that evolvability *can* be interpreted through its effects at the level of individual fitness. As Wagner and Draghi (2010) point out, if one genotype has a higher evolvability than another, then the fitness of individuals of that genotype will increase faster. As a result, at some given time point, an organism of a genotype with higher evolvability will have greater individual fitness than an organism of a genotype with lower evolvability. Thus, they argue, "simple Darwinian selection, acting on phenotypic differences among individuals in a population, is sufficient to explain selection for evolvability" (Wagner and Draghi 2010).

A more immediate difficulty faced by Lindquist and Rutherford was that all the traits identified in their *Drosophila* studies were manifestly deleterious. Wagner et al. (1999) pointed out that if the cryptic variation uncovered by *Hsp83* mutations "was purely deleterious one might expect that the regulatory proteins would have evolved to become independent of Hsp90, which seems plausible given that proteins vary in their dependence on Hsp90, or that Hsp90 through gene duplication and divergence [would have] evolved to decouple its two functions". This argument aside, the fact remained that the observed traits *were* deleterious. Absent evidence for positive traits generated by the same mechanism, it was difficult to see how this role of Hsp90 could have arisen through organismal selection, especially in an organism lacking the prokaryotic hitchhiking mechanism. Even if a plausible mechanism for the evolution of this role of Hsp90 were proposed, there would still be the "foresight" hurdle to get over. As Sniegowski and Murphy (2006) acidly remarked, quoting Sydney Brenner, one cannot retain a gene simply because it might "come in handy in the Cretaceous".

It is possible that the evolvability-curating role of a protein such as Hsp90 might have arisen as what Stephen Jay Gould would call a spandrel – an unintended side-consequence of, say, its protein chaperoning effects – which later served another, and quite useful, purpose. The same may be true of the DNA polymerases involved in prokaryote evolvability: selected for their ability to repair DNA damage, the production of increased variability might be a "nonselected byproduct" (Chicurel 2001).

One possibility raised by Sniegowski and Murphy, though it barely avoids the teleological portcullis slamming down on its tail, is that environmental uncertainty has favored *variability* – defined as the ability to generate variation – as a survival strategy. One could avoid the teleological fallacy more carefully by limiting this to a selective advantage for populations maintaining a standing amount of *variation*,[8] though this would still fail to please Dickinson and Seger, since it is a group-level trait.

One might conclude that, however attractive the concept of evolvability, one should distrust everything, above all what Lovejoy (1964) would call one's "intellectual pathos". "How do we know," write Sniegowski and Murphy,

> when it is necessary – rather than just appealing – to invoke evolvability differences in order to explain evolutionary histories? ... Invoking variability as a retrospective explanation for why one clade has diversified or changed more than another does not rule out the possibility that the clades evolved differently for reasons unrelated to variability. And finding isolated examples of evolutionary novelties related to distinctive evolutionary mechanisms – for example, mutations of major phenotypic effect caused by transposable elements – provides only anecdotal evidence for the importance of such variability mechanisms in evolution. (Sniegowski and Murphy 2006)

Are we not perilously close to the "just-so" stories beloved by proponents of the adaptationist programme and derided by Gould and Lewontin in their 1979 "Spandrels of San Marco" paper?

*** *** ***

Less than two years after the *Drosophila* Hsp90 paper, Lindquist, in collaboration with Heather True, published an even more striking study on the capacity for eukaryotic evolvability. This time, the focus was on single-celled eukaryotes, and a hidden reservoir of phenotypic diversity was uncovered not by a mutation but by a prion.

In normal *S. cerevisiae* cells, the Sup35 protein mediates the fidelity with which ribosomes terminate protein synthesis upon encountering a stop codon (Fig. 14.4). The activity of Sup35 is controlled by its carboxyl (C) terminal region. Its middle (M) and amino-terminal (N) regions, are not involved in the termination of protein

---

[8] Sniegowski and Murphy remark that "there are grounds for doubting that the selective conditions necessary to favor increased *variability* are generally important in nature" (2006, my italics).

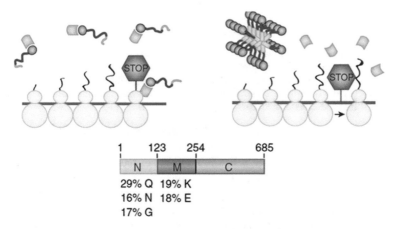

Fig. 14.4 The role of the [psi⁻] (top left) and [PSI⁺] (top right) forms of the protein Sup35 in transcription termination. The *blue, purple* and *grey* regions are the N, M and C domains, respectively, corresponding to the domain architecture shown at the bottom. In [psi⁻] state, the proteins bind to Sup45 (*orange*), and correctly recognize stop codons. In the [PSI⁺] form, Sup35 is insoluble, and does not bind to Sup45 or recognize stop codons. Reprinted by permission from Macmillan Publishers Ltd., P.M. Tessier and S. Lindquist, Unraveling infectious structures, strain variants and species barriers for the yeast prion [PSI⁺]. *Nat. Struct. Mol. Biol.* 16(6): 598-605, 2009, copyright 2009

synthesis and can be deleted without damaging the cell (True and Lindquist 2000; Partridge and Barton 2000). However, the N and M regions can misfold, allowing the protein to take on a prion conformation. Once misfolded, the protein can no longer reliably terminate translation from mRNA to protein, and produces irregular protein products – a source of potential damage to the cell, but also a potential source of useful variation.

The switch between prion ([PSI⁺]) and non-prion ([psi⁻]) states can occur spontaneously, at a rate of about one protein in a million (True and Lindquist 2000). In the manner of all prions, [PSI⁺] can convert properly folded Sup35 from the [psi⁻] to the [PSI⁺] form, effectively "reproducing" itself. The fact that the N and M regions have been retained in evolution,[9] and are present in yeast species even distantly related to *S. cerevisiae*, suggested to True and Lindquist that "the unusual ability of Sup35 to produce a heritable[10] conformation and phenotypic switch may provide a selective advantage" (True and Lindquist 2000).

To examine the effect of the prion on yeast phenotype, True and Lindquist created isogenic pairs of cell populations, which differed only in whether they had the

---

[9] Citing work by Bailleul et al. (1999), True and Lindquist note that the N and M regions do bind to an actin-binding protein, Sla1, suggesting a possible non-pathological role for these regions.

[10] It is now known that the [PSI⁺] prion exists in a variety of different strains – a virtual "cloud of variants", as described by Bateman and Wickner (2013). Each strain appears to "breed true", converting normal Sup35 into its own form.

[PSI⁺] or [psi⁻] prion phenotype. They examined the behavior of these pairs, at comparable growth phases, in more than 150 different sets of conditions,

> including growth on fermentable and non-fermentable carbon sources..., on simple and complex nitrogen sources in the presence of salts and metals..., in the presence of inhibitors of diverse cellular processes, such as DNA replication, signalling, protein glycosylation, and microtubule dynamics..., under general stress conditions, and at different temperatures... (True and Lindquist 2000)

Differences in growth patterns and in colony morphology between [PSI⁺] and [psi⁻] phenotypes were observed in almost half of the conditions tested (Fig. 14.5). In marked contrast to the entirely deleterious effects of mutant Hsp90 in *Drosophila*, improved growth was observed in the [PSI⁺] phenotype in 25% of the cases. Growth changes were almost as widely varied as the experimental conditions themselves. Here is True and Lindquist's description of just a fraction of their results (note that the term "background" refers to the particular yeast strain used):

> Each strain exhibited a unique constellation of phenotypes. In some cases very modest changes in conditions produced large growth differences between isogenic [PSI⁺] and [psi⁻] cells. In others, conditions that inhibited growth affected [PSI⁺] and [psi⁻] cells similarly. Simple patterns were not readily discernible. For example, in both the 10B-H49 and 5V-H19 backgrounds, [PSI⁺] inhibited growth on several nitrogen sources. However, under many other test conditions the phenotypic changes induced by [PSI⁺] in 10B-H49 diverged from those in 5V-H19. Moreover, in the SL1010-1A and 74D-694 backgrounds, [PSI⁺] enhanced growth on some nitrogen sources. In the presence of the alkali metal caesium [at 25 mM]..., [PSI⁺] had little effect on most strains, but strongly inhibited growth in the D1142-1A background. In the BSC783/4C background, [PSI⁺] strongly enhanced growth, but only when the concentration of caesium was raised to 100 mM. In the presence of the alkali metal lithium, [PSI⁺] inhibited growth in 74D-694, but enhanced growth in a concentration-dependent manner in both the 5V-H19 and 10B-H49 backgrounds. (True and Lindquist 2000)

[PSI⁺] strains were more tolerant to stressors such as heat and ethanol, consistent with the hypothesis that [PSI⁺] prion may activate the heat shock response, which is triggered by misfolded proteins. Phenotypic changes were also observed in strains where the N and M segments had been deleted, consistent with the idea that the N and M regions play a role beyond just inducing prion formation.

True and Lindquist concluded that [PSI⁺] provides "a means to unveil silent genetic information to produce new heritable phenotypes." They suggested, further, that "the epigenetic and metastable nature of [PSI⁺]... potentiates survival in a fluctuating environment and provides a conduit for the evolution of new traits." Spontaneous conversion to the [PSI⁺] phenotype kept the genetic richness constantly available, but if, after [PSI⁺] phenotypes had flourished in one set of environmental conditions, the environment shifted back to its [psi⁻]-favoring state, the cell was buffered against that as well, since the remaining [psi⁻] proteins would still be available. This would "allow cells to occupy a new niche without foregoing their capacity to occupy the old." In short, the best of both worlds. True and Lindquist suggested that traits uncovered by [PSI⁺], advantageous in multiple environmental contexts, could become fixed, and thus uncoupled from the [PSI⁺] phenotype in

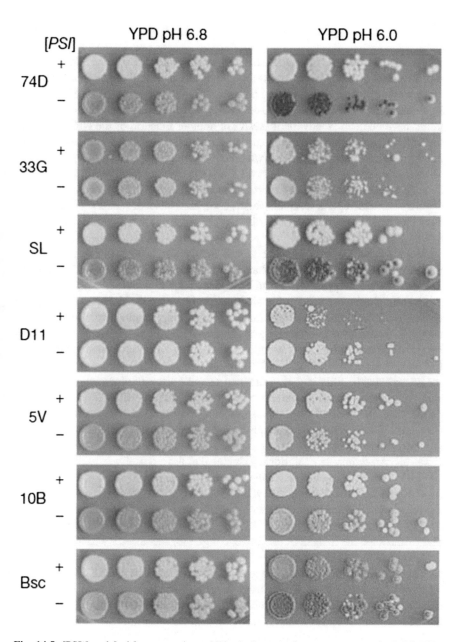

Fig. 14.5 [PSI⁺] and [psi⁻] yeast strains exhibit similar growth patterns on standard rich YPD media at pH 6.8 (*left column*), but show significant differences at pH 6.0 (*right column*). Reprinted by permission from Macmillan Publishers Ltd, H.L. True and S.L. Lindquist, A yeast prion provides a mechanism for genetic variation and phenotypic diversity. *Nature* 407: 477–483, 2000, copyright 2000

which they had arisen. Thus, like Hsp90 in *Drosophila*, [PSI⁺] could facilitate not just transient increases in variability, but also permanent evolutionary change.

How did the role of [PSI⁺] in enhancing protein variability evolve? True and Lindquist suggested that the failure of [PSI⁺] to to effectively recognize stop codons might have been an "inadvertent" result of another function of the NM portion of the protein, which was then favored under selection. In other words, evolvability may have arisen as an exaptation. Two other evolutionary scenarios – albeit at vastly different timescales – could have favored this inadvertent result on the basis of its benefits at the population level. First, for yeast occupying fluctuating environments ("warm, nutrient-replete summers versus cool, nutrient-poor winters") [PSI⁺] could have facilitated alteration between phenotypic states optimal for each set of conditions. Second, more broadly, "the natural environment occupied by *S. cerevisiae* during its evolution may have been sufficiently erratic to provide the pressure required to maintain a global mechanism for exploiting genome-wide variation to produce new phenotypes" (True and Lindquist 2000).

Unsurprisingly, the yeast prion study was met with a backlash, again based on the incorrect assumption that True and Lindquist were assuming evolutionary "foresight". Even the *News and Views* piece accompanying the yeast prion study, written by Partridge and Barton, was written in a tone that "hover[ed] between dismissal and marginalization" (Dover 2000). "We feel," they wrote, "that it is simpler to see the increased variability in these examples as a side effect of disrupted gene expression, rather than as an adaptation to facilitate evolution" (Partridge and Barton 2000). Lindquist replied that "this is obvious, and we never suggested otherwise" (Lindquist 2000). True and Lindquist never presented the variability caused by [PSI⁺] as a direct adaptation. Rather, although they did not use Gould's terminology, it was an *exaptation*, an evolutionary change that proved useful in a role – and at a level – for which it was not originally selected.

The view of traits favoring evolvability as exaptations preserved by selection at the population level is consistent with the arguments made by Poole et al. (2003) and by Kirschner and Gerhart (1998). Philosopher Todd Grantham (2004) argues that Gould used the inverse of this logic to argue for the necessity of a hierarchical theory of selection. As Grantham explains, Gould argued that differences in evolvability, which is by definition a property of *lineages*, are "crucial to explaining patterns of macroevolution." Gould then suggested that differences in evolvability could only be explained using a hierarchical view, and incorporating spandrels – initially nonadaptive byproducts of drift, developmental constraint, or selection acting on something else – arising at a lower level of selection as exaptations at the level of the lineage.

*** *** ***

Rutherford and Lindquist's Hsp90 studies were conducted in experimental populations. Could Hsp90 serve as a capacitor of cryptic evolutionary variation in a wild population? A recent study in which Lindquist collaborated with Nicolas Rohner and colleagues at Harvard Medical School demonstrated the role of Hsp90 in the

Fig. 14.6 Cryptic genetic variation in typical *Astyanax mexicanus* can quickly lead to eye loss in cave-dwelling populations. Photograph by H. Zell, reproduced under a CC BY-SA 3.0 license

evolved blindness of cavefish (Rohner et al. 2013; Pennisi 2013). Cavefish (*Astyanax mexicanus*) are closely related to surface-dwelling species, but lose their eyes (though maintaining the orbital structure) and their pigmentation in the cave environment (Fig. 14.6). Rohner et al. speculated that the cave environment might cause stress that would affect the Hsp90 system, revealing cryptic variation including a great variation in eye size. Small, and then vanishingly small, eyes would have a selective advantage because of the metabolic savings in not producing an unnecessary organ.

To test this hypothesis, Rohner et al. inhibited Hsp90 using the specific inhibitor Radicicol in laboratory and surface-dwelling strains fish strains. Larval *A. mexicanus* treated with Radicicol exhibited a greater than normal variation in eye size (normalized to body length), as did adult fish treated with Radicicol and raised in the dark. This variation was genetically based, since breeding of fish with small eyes produced small-eyed offspring. In contrast, application of Radicicol to fish already dwelling in caves produced no significant change in orbit size, suggesting that the related alleles had already been selected for in this population.

The effect of Radicicol on eye and orbit size is consistent with an Hsp90 effect in the natural development of blindness in cavefish. But does the cave environment affect Hsp90 directly? Rohner and colleagues found that the conductivity of water in the cave habitat was significantly lower than in the surface river habitat (230 µS compared to 1300 µS). Low conductivity had been previously shown to elicit a heat shock response in fish. Rohner et al. raised surface fish in a low conductivity environment, and found up-regulation of Hsp90 and other genetic markers of stress response.

Thus, the environment encountered by these fish during their evolutionary transition from surface to cave stresses the protein homeostasis mechanisms of the organism in a manner similar to a specific stress on HSP90 chaperone activities. Adult river fish placed in low conductivity during larval development displayed statistically significant increases in eye and orbit size variation of 50%... This demonstrates that a cave-specific environmental stress can elicit similar changes in morphological eye development as biochemical inhibition of HSP90. (Rohner et al. 2013)

There were other morphological variations in the cave-dwelling fish that were not explained by these studies, however. No Hsp90-related cryptic variation was observed in body size or number of neuromast cells (Rohner et al. 2013) or in pigmentation (Pennisi 2013). Nonetheless, the Rohmer et al. results were hailed as "a superb example of a full circle, starting from lab results, making a controversial hypothesis, and testing it in the wild" (Pennisi 2013).

Even more recently, Peuß et al. (2015) found that downregulation of Hsp90 is mediated by social cues in red flour beetles, raising "the exciting question of whether evolvability might be regulated through the use of information derived from the social environment". When in the neighborhood of wounded conspecifics, uninjured beetles begin to downregulate Hsp90. This study caught the attention of the popular press, and in November 2015 Ed Yong wrote an article in *The Atlantic* about the Peuß et al. study that opened with a flour beetle version of a battle from Game of Thrones:

> Imagine that you wake up in a pit, surrounded by people who are all wounded and bleeding. Something [has] clearly gone horribly wrong. Maybe you panic. Maybe you tend to the wounded. Maybe you team up to plan an escape. But if you're a red flour beetle, you do none of these things. Instead, you quietly become more evolvable.

References

Åkerlund T, Nordström K, Bernander R (1995) Analysis of cell size and DNA content in exponentially growing and stationary-phase batch cultures of *Escherichia coli*. J Bacteriol 177(23):6791–6797

Baharoglu Z, Mazel D (2014) SOS, the formidable strategy of bacteria against aggressions. FEMS Microbiol Rev 38(6):1126–1145

Bailleul PA, Newnam GP, Steenbergen JN, Chernoff YO (1999) Genetic study of interactions between the cytoskeletal assembly protein Sla1 and prion-forming domain of the release factor Sup35 (eRF3) in *Saccharomyces cerevisiae*. Genetics 153(1):81–94

Bateman DA, Wickner RB (2013) The [*PSI+*] prion exists as a dynamic cloud of variants. PLoS Genet 9(1):e1003257

Bendich AJ, Drlica K (2000) Prokaryotic and eukaryotic chromosomes: what's the difference? BioEssays 22(5):481–486

Chicurel M (2001) Can organisms speed their own evolution? Science 292(5523):1824–1827

Denamur E, Matic I (2006) Evolution of mutation rates in bacteria. Mol Microbiol 60(4):820–827

Denamur E, Tenaillon O, Deschamps C, Skurnik D, Ronco E, Gaillard JL, Picard B, Branger C, Matic I (2005) Intermediate mutation frequencies favor evolution of multidrug resistance in *Escherichia coli*. Genetics 171(2):825–827

Dickinson WJ, Seger J (1999) Cause and effect in evolution. Nature 399:30

Dover G (2000) Results may not fit well with current theories. Nature 408:17

Drake JW (1991) A constant rate of spontaneous mutation in DNA-based microbes. Proc Natl Acad Sci U S A 88:7160–7164

Drake JW, Allen EF, Forsberg SA, Preparata R-M, Greening EO (1969) Genetic control of muta-tion rates in bacteriophage T4. Nature 221:1128–1132

Drake JW, Charlesworth B, Charlesworth D, Crow JF (1998) Rates of spontaneous mutation. Genetics 148:1667–1686

Elena SF, Sanjuán R (2005) Adaptive value of high mutation rates of RNA viruses: separating causes from consequences. J Virol 79(18):11555–11558

Fisher RA (1930) The Genetical Theory of Natural Selection. Oxford University Press, Oxford

Funchain P, Yeung A, Stewart J, Clendenin WM, Miller JH (2001) Amplification of mutator cells in a population as a result of horizontal transfer. J Bacteriol 183(12):3737–3741

Gilchrist GW, Lee CE (2007) All stressed out and nowhere to go: does evolvability limit adaptation in invasive species? Genetica 129(2):127–132

Gould SJ, Lewontin RC (1979) The spandrels of San Marco and the Panglossian paradigm: a cri-tique of the adaptationist programme. Proc R Soc B 205:581–598

Grantham TA (2004) Constraints and spandrels in Gould's *Structure of Evolutionary Theory*. Biol Philos 19:29–43

He J, Deem MW (2010) Hierarchical evolution of animal body plans. Dev Biol 337(1):157–161

Kimura M (1967) On the evolutionary adjustment of spontaneous mutation rates. Genet Res 9:23–34

Kirschner M, Gerhart J (1998) Evolvability. Proc Natl Acad Sci U S A 95:8420–8427

Krakauer DC, Plotkin JB (2002) Redundancy, antiredundancy, and the robustness of genomes. Proc Natl Acad Sci U S A 99(3):1405–1409

Lande R (1979) Quantitative genetic analysis of multivariate evolution, applied to brain:body size allometry. Evolution 33(1):402–416

Lindquist S (2000) …but yeast prion offers clues about evolution. Nature 408:17–18

Lovejoy AO (1964) The Great Chain of Being: A Study of the History of an Idea. Harvard University Press, Cambridge, MA/London

Lynch M (2007) The frailty of adaptive hypotheses for the origins of organismal complexity. Proc Natl Acad Sci U S A 104(Suppl 1):8597–8604

Maldonado R, Jiménez J, Casadesús J (1994) Changes of ploidy during the *Azotobacter vinelandii* growth cycle. J Bacteriol 176(13):3911–3919

Mao EF, Lane L, Lee J, Miller JH (1997) Proliferation of mutators in a cell population. J Bacteriol 179(2):417–422

Metzgar D, Wills C (2000) Evidence for the adaptive evolution of mutation rates. Cell 101(6):581–584

Moxon R, Bayliss C, Hood D (2006) Bacterial contingency loci: the role of simple sequence DNA repeats in bacterial adaptation. Annu Rev Genet 40:307–333

Nöthel H (1987) Adaptation of *Drosophila melanogaster* populations to high mutation pressure: evolutionary adjustment of mutation rates. Proc Natl Acad Sci U S A 84:1045–1049

Okasha S (2006) Evolution and the Levels of Selection. Oxford University Press (Clarendon Press), Oxford/New York

Partridge L, Barton NH (2000) Evolving evolvability. Nature 407:457–458

Pennisi E (2013) Cavefish study supports controversial evolutionary mechanism. Science 342:1304

Peuß R, Eggert H, Armitage SA, Kurtz J (2015) Downregulation of the evolutionary capacitor Hsp90 is mediated by social cues. Proc Biol Sci 282(1819):20152041

Pigliucci M (2008) Is evolvability evolvable? Nat Rev Genet 9(1):75–82

Poole AM, Phillips MJ, Penny D (2003) Prokaryote and eukaryote evolvability. BioSystems 69(2–3):163–185

Qiu R, Sakato M, Sacho EJ, Wilkins H, Zhang X, Modrich P, Hingorani MM, Erie DA, Weninger KR (2015) MutL traps MutS at a DNA mismatch. Proc Natl Acad Sci U S A 112(35):10914–10919

Radman M (1975) SOS repair hypothesis: phenomenology of an inducible mutagenic DNA repair pathway in *Escherichia coli*. Basic Life Sci 5A:355–367

Radman M (1999) Enzymes of evolutionary change. Nature 401(6756):866–869

Radman M, Taddei F, Matic I (2000) Evolution-driving genes. Res Microbiol 151:91–95

Richardson MK (2012) A phylotypic stage for all animals? Dev Cell 22(5):903–904

Rohner N, Jarosz DF, Kowalko JE, Yoshizawa M, Jeffery WR, Borowsky RL, Lindquist S, Tabin CJ (2013) Cryptic variation in morphological evolution: HSP90 as a capacitor for loss of eyes in cavefish. Science 342:1372–1375

Rutherford SL, Lindquist S (1998) Hsp90 as a capacitor for morphological evolution. Nature 396:336–342

Schapper RM (1998) Antimutator mutants in bacteriophage T4 and *Escherichia coli*. Genetics 148:1579–1585

Sniegowski PD, Murphy HA (2006) Evolvability. Curr Biol 16(19):R831–R834

Sniegowski PD, Gerrish PJ, Lenski RE (1997) Evolution of high mutation rates in experimental populations of *E. coli*. Nature 387:703–705

Sniegowski PD, Gerrish PJ, Johnson T, Shaver A (2000) The evolution of mutation rates: separating causes from consequences. BioEssays 22:1057–1066

Sturtevant AH (1937) Essays on evolution I. On the effects of selection on mutation rate. Q Rev Biol 12:464–467

Taddei F, Radman M, Maynard-Smith J, Toupance B, Gouyon PH, Godelle B (1997) Role of mutator alleles in adaptive evolution. Nature 387(6634):700–702

Tenaillon O, Taddei F, Radman M, Matic I (2001) Second-order selection in bacterial evolution: selection acting on mutation and recombination rates in the course of adaptation. Res Microbiol 152:11–16

Tessier PM, Lindquist S (2009) Unraveling infectious structures, strain variants and species barriers for the yeast prion [*PSI+*]. Nat Struct Mol Biol 16(6):598–605

True HL, Lindquist SL (2000) A yeast prion provides a mechanism for genetic variation and phenotypic diversity. Nature 407:477–483

Waddington CH (1942) Canalization of development and the inheritance of acquired characters. Nature 3811:563–565

Waddington CH (1953) Genetic assimilation of an acquired character. Evolution 7(2):118–126

Wagner GP, Altenberg L (1996) Complex adaptations and the evolution of evolvability. Evolution 50(3):967–976

Wagner GP, Draghi J (2010) Evolution of Evolvability. In: Pigliucci M, Müller GB (eds) The extended synthesis. MIT Press, Cambridge, MA

Wagner GP, Chiu C-H, Hansen TF (1999) Is Hsp90 a regulator of evolvability? J Expt Biol (Mol Dev Evol) 285:116–118

Wilke CO, Wang JL, Ofria C, Lenski RE, Adami C (2001) Evolution of digital organisms at high mutation rates leads to survival of the flattest. Nature 412(6844):331–333

Chapter 15
Spandrels, Exaptations, and Raw Material

The river to the ocean goes
A fortune for the undertow

R.E.M.

STEPHEN JAY Gould and Richard Lewontin introduced the spandrel as a biological metaphor in a 1979 paper entitled "The spandrels of San Marco and the Panglossian Paradigm: a Critique of the Adaptationist Programme". Architecturally, the curved triangular spaces, called pendentives,[1] formed above and between two neighboring arches in a vaulted roof supported by four arches, are a type of spandrel (Fig. 15.1). Technically, a spandrel is a "space left over" between portions of an architectural structure, such as the vertical spaces between steps on a staircase or the "spandrel courses" on the sides of high-rise buildings, between the windows of one floor and the windows of the next (Fig. 15.2). As Gould explained after he had delved further into the architectural details, many architects use the term only to refer to two-dimensional spaces, while a continental European school does consider three-dimensional forms, like the San Marco pendentives to be spandrels (Gould 1997, 2002, p. 1250).

The spandrels of San Marco are, as Gould and Lewontin described them, "necessary architectural byproducts of mounting a dome on rounded arches". They are so beautifully decorated in some medieval and Renaissance religious spaces that one might think of them as deliberately designed structures. But that, Gould and Lewontin argue, "would invert the proper path of analysis".

> Anyone who tried to argue that the structure exists because the alternation of rose and portcullis makes so much sense in a Tudor chapel would be inviting the same ridicule that

---

[1] When they wrote the "Spandrels" paper, Gould and Lewontin did not use this terminology, and simply applied the general term "spandrel"; this was clarified by civil engineer and architect Robert Mark (1996), in an article entitled "Architecture and Evolution" in *The American Scientist*. Gould and Lewontin's terminological vagueness was noted gleefully by Daniel Dennett in his 1995 book *Darwin's Dangerous Idea*, which took such a stridently pan-adaptationist view that H. Allen Orr remarked, "Dennett does not so much champion adaptationism as excoriate those biologists who dare question it." (Orr 1996). See also Ahouse (1998) for a cogent critique of Dennett's "a priori selectionism".

© Springer Science+Business Media B.V. 2018
S. Bahar, *The Essential Tension*, The Frontiers Collection,
DOI 10.1007/978-94-024-1054-9_15

Fig. 15.1 A view of the mosaics in the Basilica San Marco, looking east. Photograph by Ricardo André Frantz. Reproduced under a Creative Commons Attribution-Share Alike 2.5 Generic license

Voltaire heaped on Dr. Pangloss[2] [in *Candide*]: 'Things cannot be other than they are… Everything is made for the best purpose. Our noses were made to carry spectacles, so we have spectacles. Legs were clearly intended for breeches, and we wear them.' Yet evolutionary biologists, in their tendency to focus exclusively on immediate adaptation to local conditions, do tend to ignore architectural constraints and perform just such an inversion of explanation. (Gould and Lewontin 1979)

Lewontin and Gould equated the "Panglossian paradigm" with the extreme version of adaptationism they found prevalent in the evolutionary biology

---

[2] Voltaire's Dr. Pangloss, and his idea that this is "the best of all possible worlds" (despite a deluge of events in Candide's life that prove the opposite) was at least in part a caricature of Leibniz and his philosophy. Voltaire had been deeply moved by the horrific tragedy of the 1755 Lisbon earthquake – after which the terrified residents rushed from their burning and collapsed homes toward the beach, only to be drowned by a tsunami triggered by the earthquake. This event was seminal in turning Voltaire into a bitter opponent of "philosophical optimism".

Fig. 15.2 These, too, are spandrels. Art deco terra cotta spandrels on the Powhatan building in Chicago. Photograph by Mary Nitsch

community. In the adaptationist view (or at least the Gould-Lewontin interpretation of it, which picks the more extreme examples in order to make a point), every evolutionary feature was "designed" by natural selection as an adaptation. If no adaptative value of was immediately obvious, one could make up a "just so" story to explain it. To this end, Lewontin and Gould mocked the hypotheses offered by Michael Harner and E. O. Wilson to explain Aztec human sacrifice – a result of a genetic predisposition for carnivory? Or perhaps a solution to a chronic meat shortage?

Lewontin and Gould laid the blame for the exaggerated adaptationist view squarely at the feet of Alfred Russel Wallace and August Weismann (in particular, Weismann's insistence on the omnipotence, or *allmacht*, of natural selection). The Adaptationist Programme, they wrote, could be summarized by the a set of restrictive rules. First, an organism should be "atomized" into traits. Then attempts should be made to find an explanation for why each trait is optimal for the conditions under which the organism lives. "After the failure of part-by-part optimization," they wrote, "interaction is acknowledged via the dictum that an organism cannot optimize each part without imposing expenses on others. The notion of 'trade-off' is introduced, and organisms are interpreted as best compromises among competing demands." However, while allowing for sub-optimality,[3] strict adaptationists (in Gould and Lewontin's view) assume that sub-optimal structures are the way they are solely as a result of a compromise driven by natural selection. The notions of structural constraints and historical contingency are not even considered.

Lewontin and Gould particularly objected to this extreme form of the adaptationist programme because it was *unfalsifiable*. "Often evolutionists use *consistency* with natural selection as the sole criterion and consider their work done when they concoct a plausible story. But plausible stories can always be told." Lewontin and Gould conceded that some adaptationists did consider possible alternative explanations such as drift and structural constraint. Yet "[t]he admission of alternatives in principle does not imply their serious consideration in daily practice". In practice, they found that many scientists limited themselves to various "styles of argument", which include "if one adaptive argument fails, try another"; "in the absence of a good adaptive argument in the first place, attribute failure to imperfect understanding of where an organism lives and what it does"; "emphasize immediate utility and exclude other attributes of form". As an example of this last approach, they cited a card next to a *Tyrannosaurus rex* skeleton at the Boston Museum of Science that suggests its tiny front legs were used "to help the animal rise from a lying position". Perhaps we should not be so hard on ourselves for being drawn to adaptationist explanations; the child's eternal question "why" points to a fundamental predilection for the adaptationist view that we may never be able to shake.

Darwin himself emphasized that, while natural selection was the most important evolutionary mechanism, it was by no means the only one. Gould and Lewontin cited a letter Darwin wrote to the editor of *Nature* in 1880 in response to an argument made by Sir Wyville Thomson, who had been studying deep marine life as chief scientist aboard the HMS Challenger. Thomson argued that "[t]he character of the abyssal fauna refuses to give the least support to the theory which refers the evolution of species to extreme variation guided only by natural selection." "Can Sir Wyville Thomson name any one," Darwin retorted, "who has said that the evolution of species depends only on natural selection?" (Darwin 1880).

Likewise, Gould and Lewontin cited a passage from the last edition of the *Origin of Species* where Darwin made this point in more detail. "As my conclusions," Darwin wrote,

---

[3] Voltaire mercilessly mocked the idea of such trade-offs. As Gould and Lewontin remind their readers, Dr. Pangloss told poor Candide not to worry about suffering from a venereal disease. It was "indispensable in this best of worlds. For if Columbus, when visiting the West Indies, had not caught this disease, which poisons the source of generation, which frequently even hinders generation, and is clearly opposed to the great end of Nature, we would have neither chocolate nor cochineal."

have lately been much misrepresented and it has been stated that I attribute the modification of species exclusively to natural selection, I may be permitted to remark that in the first edition of this work, and subsequently, I placed in a most conspicuous position – namely at the close of the Introduction – the following words: 'I am convinced that natural selection has been the main, but not the exclusive, means of modification.' This has been to no avail. Great is the power of steady misinterpretation.[4] (quoted by Gould and Lewontin 1979)

Darwin emphasized the co-option of previous functions for different uses in his argument with St. George Mivart over the origin of new evolutionary structures. In an 1871 book provocatively titled *On the Genesis of Species*, Mivart had challenged Darwin to explain the "incipient stages" of new evolutionary functions. Darwin had already dealt with this problem in the first edition of the *Origin*, in his chapter entitled "Difficulties of the Theory", writing that "[i]n considering transitions of organs, it is so important to bear in mind the probability of conversion from one function to another" (Darwin, *Origin of Speices*, p. 235; see also Gould 2002, p. 1219). Darwin expanded on this theme in his response to Mivart, which he included in a new chapter appended to the final edition of the *Origin*.

Even in the early editions of the *Origin*, in a passage that echoes a quote from Owen (see Chap. 3), Darwin cited the unfused sutures in the skulls of infant mammals as a prime example of such conversion from one function to another, though skull sutures might be easily mistaken for an adaptation if one failed to examine the situation closely (Fig. 15.3).

> The naked skin on the head of a vulture is generally considered as a direct adaptation for wallowing in putridity; and so it may be or it may possibly be due to the direct action of putrid matter; but we should be very cautious in drawing any such inference, when we see the skin on the head of the clean-feeding male Turkey is likewise naked. The sutures in the skulls of young mammals have been advanced as a beautiful adaptation for aiding parturition, and no doubt they facilitate, or may be indispensible for this act; but as sutures occur in the skulls of young birds and reptiles, which have only to escape from a broken egg, we may infer that this structure has arisen from the laws of growth, and has been taken advantage of in the parturition of the higher animals. (Darwin, *Origin of Species*, p. 250)

Several pages later, Darwin addressed the role of history in preserving the

> many structures that are now of no direct use to their possessors, and may never have been of any use to their progenitors; but this does not prove that they were formed solely for beauty or variety. No doubt the definite action of changed conditions, and the various causes of modifications, lately specified [i.e., discussed earlier in the *Origin*], have all produced an effect, probably a great effect, independently of any advantage thus gained. But a still more

---

[4]The power of steady misinterpretation has not yet relinquished its hold on power: creationists still use Darwin's pluralistic comments as "evidence" that he repudiated his own theory! One recent example of this is a book by Randall Hedtke entitled "Secrets of the Sixth Edition", with a photograph of an elderly, white-bearded Darwin gracing the cover... with the photoshopped addition of a finger raised to his lips in a "ssshhhh" gesture. The book is published by a house that specializes in "homeschool resources".

Fig. 15.3 Skull sutures
from a human adult. In
infants, the sutures lack the
winding, fractal quality
shown here. Figure from
Di Ieva et al. (2013)

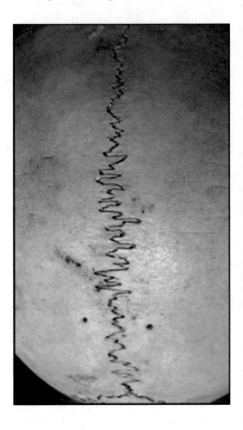

important consideration is that the chief part of the organisation of any living creature is due
to inheritance; and consequently, though each being assuredly is well fitted for its place in
nature, many structures have now no very close and direct relation to present habits of life.
Thus, we can hardly believe that the webbed feet of the upland goose or of the frigate-bird
are of special use to these birds; we cannot believe that the similar bones in the arm of the
monkey, in the fore-leg of a horse, in the wing of the bat, and in the flipper of the seal, are
of special use to these animals. We may safely attribute these structures to inheritance. But
webbed feet no doubt were as useful to the progenitor of the upland goose and of the
frigate-bird, as they now are to the most aquatic of living birds. So we may believe that the
progenitor of the seal did not possess a flipper, but a foot with five toes fitted for walking or
grasping; but we may further venture to believe that the several bones in the limbs of the
monkey, horse, and bat, were originally developed, on the principle of utility, probably
through the reduction of more numerous bones in the fin of some ancient fish-like progenitor of the whole class. It is scarcely possible to decide how much allowance ought to be
made for such causes of change, as the definite action of external conditions, so-called
spontaneous variations, and the complex laws of growth; but with these important exceptions, we may conclude that the structure of every living creature either now is, or was
formerly, of some direct or indirect use to its possessor. (Darwin, *Origin of Species*,
pp. 252–253)

It is important to unpack several aspects of this extended quote. While Darwin's statements echo some of the key elements discussed by Gould and Lewontin, the focus of his analysis is quite different. Darwin described a number of structures that are, essentially, vestigial: a result of history, perhaps useful at one time, but no apparent use now or in the future. His focus is primarily on the traces left by history on – and in – organisms, and rightly so; he was arguing *for a theory of evolution*, and nothing was more crucial to his argument than the traces of history. Darwin was also arguing *for a theory of adaptation*, in a cultural context where the "belief that organic beings have been created beautiful for the delight of man" (Darwin, *Origin of Speices*, p. 253) was still prevalent.[5] He allowed for the action of external conditions, for spontaneous variations, for complex laws of growth, and for accidents of history, calling them "important exceptions". But Darwin's intellectual agenda was fundamentally different from that of Gould and Lewontin. While fully aware of features co-opted for different uses, and of the role of historical contingency, Darwin's primary concern was to construct an argument for adaptation by natural selection. In contrast, Gould and Lewontin wanted to emphasize the importance of "quirky functional shift". Only in a rare passage like the one about skull sutures does Darwin describe a structure co-opted for a *new* use. Nonetheless, he *does* describe it. Here is the extended context of his statement about "conversion from one function to another". Darwin noted the role of redundancy in facilitating functional transitions, writing that

> two distinct organs, or the same organ under two very different forms, may simultaneously perform in the same individual the same function, and this is an extremely important means of transition: to give one instance, – there are fish with gills or branchiae that breathe air dissolved in the water, at the same time that they breathe free air in their swim bladders… In all such cases one of the two organs might readily be modified and perfected so as to perform all the work, being aided during the process of modification by the other organ; and then this other organ might be modified for some other and quite distinct purpose, or be wholly obliterated. (Darwin, *Origin of Species*, pp. 233–234)

This same theme is widely explored today in the context of gene duplication (see Andersson et al. 2015 for review). After citing the hypothesis that insect wings developed from the trachea, Darwin wrote:

> In considering transitions of organs, it is so important to bear in mind the probability of conversion from one function to another, that I will give another instance. Pedunculated cirripedes [a type of barnacle] have two minute folds of skin, called by me the ovigerous frena, which serve, through the means of a sticky secretion, to retain the eggs until they are hatched within the sack. These cirripedes have no branchiae, the whole surface of the

---

[5] Darwin responds to such arguments with the following – utterly charming – argument: "If beautiful objects had been created solely for man's gratification, it ought to be shown that before man appeared, there was less beauty on the face of the earth than since he came on the stage. Were the beautiful volute and cone shells of the Eocene epoch, and the gracefully sculptured ammonites of the Secondary period, created that man might ages afterwards admire them in his cabinet?" (Darwin, *Origin of Species*, p. 253)

body and the sack, together with the small frena, serving for respiration. The *Balanidae* or sessile cirripedes, on the other hand, have no ovigerous frena, the eggs lying loose at the bottom of the sack, within the well-enclosed shell; but they have, in the same relative position with the frena, large, much-folded membranes, which freely communicate with the circulatory lacunae of the sack and body, and which have been considered by all naturalists to act as branchiae. Now I think no one will dispute that the ovigerous frena in the one family are strictly homologous with the branchiae of the other family; indeed, they graduate into each other. Therefore it need not be doubted that the two little folds of skin, which originally served as ovigerous frena, but which, likewise, very slightly aided in the act of respiration, have been gradually converted by natural selection into branchiae, simply through an increase in their size and the obliteration of their adhesive glands. If all pedunculated cirripedes had become extinct, and they have suffered far more extinction than have sessile cirripedes, who would ever have imagined that the branchiae in this latter family had originally existed as organs for preventing the ova from being washed out of the sack? (Darwin, *Origin of Species*, pp. 235–236)

Having emphasized Darwin's awareness of non-adaptive explanations, and listed a litany of adapatationist excesses, Gould and Lewontin concluded their spandrels paper with an impassioned plea to give non-adaptive explanations a fair hearing. Among these possible explanations, they included the possibility of *no adaptation or selection at all*, i.e., drift. Another possibility was *no adaptation or selection on the particular feature under study*; examples of this included what Darwin called "correlations of growth", such as the allometric relation between body size and the size of various organs. This last example might well have been recast as *no selection at the particular level under study*; though they did not explicitly discuss it in the context of levels of selection, Gould and Lewontin cited the case of selection on the timing of maturation in beetles. Is this selection for the larval stage, they asked, or for speeding up the rapid cycling of generations?

Yet another alternative explanation, Gould and Lewontin suggested, was the *decoupling of selection and adaptation*. For example, selection can occur in a non-adaptive context, as in the selection for a higher number of offspring when conditions are resource-limited; adaptation can occur without selection, as in the shapes of marine organisms that are molded by the ocean currents that surround them. A fourth alternative explanation did not eliminate the possibility of adaptation, but emphasized that there might be *different evolutionary solutions to the same problem* (multiple peaks in the fitness landscape). Here, there would be "no selective basis for differences among adaptations"; several adaptations might have virtually identical fitness, but have been reached by different evolutionary pathways as a result of the unique history of each lineage under investigation. Lastly, Gould and Lewontin urged their colleagues to consider cases where adaptation and selection have occurred, but "*the adaptation is a secondary utilization of parts present for reasons of architecture, development, or history*". In other words, spandrels. In later works,

Gould emphasized the same point in slightly different terms, exhorting[6] his colleagues not to conflate *historical origin* with *current utility*.[7]

The term "adaptationism" does not mean the same thing to every scientist, nor does it mean the same thing in every context. As Grantham (2004) notes, various scholars such as Godfrey-Smith, Reeve and Sherman, and Sober, have suggested various definitions of adaptationism. For example, Godfrey-Smith defines empirical

---

[6] The tone of the spandrels article rubbed many people the wrong way. Its rhetorical flourishes inspired a virtual cottage industry of criticism, including a collection of essays about the article, entitled *Understanding Scientific Prose* (University of Wisconsin Press 1993), which dissected the paper from a variety of angles. The tone of some of the discussion of *The Spandrels of San Marco*, however, has become strangely personal in a way rarely seen in professional scientific writing. In a review of *Understanding Scientific Prose* (titled "The Scandals of San Marco", and published in the *Quarterly Review of Biology* in 1994), Gerald Borgia concluded that "[p]olitical bias remains as the only plausible explanation [sic!!] for Gould's attacks on adaptation and sociobiology". Borgia ended his review by declaring "selection the clear winner in the 'Spandrels' debate". As for Gould and Lewontin, Borgia applauded the fact that "fortunately, their elitist attempt to misdirect science failed in the end". I barely need to point out that these are not <u>arguments</u>.

Spandrels continues to irritate many in the evolutionary biology community nearly four decades after its publication. A 2011 blog post by Jeremy Fox is entitled "Why the Spandrels of San Marco Isn't a Good Paper". In his deliberately provocative blog post, Fox argued that the tone of *Spandrels* is more suited to that of a "deliberately-provocative blog post" (https://oikosjournal. wordpress.com/2011/08/26/why-the-spandrels-of-san-marco-isnt-a-good-paper/, retrieved on January 14, 2016). Fox accused Gould and Lewontin of "cherry-picking" quotes from Darwin in order to bolster their argument; the very clear and deliberate statements by Darwin in this matter, such as those quoted above, show the cherry-picking claim to be utterly spurious. Among other complaints, Fox cited the work of the architectural engineer Robert Marks, who noted that there were various ways known to construct a dome on four arches at the time San Marco was constructed, and the particular method used was the only one stable enough to support such a large dome; the spandrels are, therefore, adaptive. This, of course, *does not make for an argument that current utility is aligned with historical origin*. Neither does the parody title of David Queller's article entitled "The Spaniels of St. Marx and the Panglossian Paradox: A Critique of a Rhetorical Programme", published in the *Quarterly Review of Biology* in 1995, or the quasi-Gilbert-and-Sullivan song ("I am the very model of a science intellectual" [sic]) included therein. One can dislike someone for being glib, widely read, and at times bombastic. But that doesn't make him wrong.

[7] In *The Structure of Evolutionary Theory*, Gould pointed out that none other than Friedrich Nietzsche himself emphasized this point in *The Genealogy of Morals*. Nietzsche argued that it was critically important to separate current use and ethical interpretation of moral constraints in contemporary society from their historical origin (which was, in his view, the will to power). He wrote "that the origin or the emergence of a thing and its ultimate usefulness, its practical application and incorporation into a system of ends, are *toto coelo* separate; that anything in existence, having somehow come about, is continually interpreted anew, requisitioned anew, transformed and directed to a new purpose" (quoted in Gould 2002, p. 1216). Nietzsche's argument included biological as well as social structures. "No matter," he wrote, "how perfectly you have understood the usefulness of any physiological organ (or legal institution, social custom, political usage, art form or religious rite) you have not thereby grasped how it emerged…for people down the ages have believed that the obvious purpose of a thing, its utility, form and shape are its reason for existence: the eye is made to see, the hand to grasp. So people think punishment has evolved for the purpose of punishing" (quoted in Gould 2002, p. 1217).

adaptationism as the assertion that natural selection is, empirically, the primary driver of evolutionary change. In contrast, explanatory adaptationism takes a rather more circular argument, holding "that natural selection is the most important force because explaining adaptation is the central problem of evolutionary biology and natural selection is the only explanation for adaptation" (Grantham 2004). Gould and Lewontin took aim squarely at this latter argument, which is certainly easier to dismantle. In later works such as *The Structure of Evolutionary Theory*, Gould emphasized that even empirical adaptationism should be recognized as limited in its capacity to explain evolutionary change. Grantham argued that even Gould's arguments regarding spandrels can be interpreted as "compatible with at least some versions of adaptationism", though this may require the presence of adaptation at multiple levels of selection in order to explain observed patterns of evolutionary change. We will return to Grantham's critique below.

Three years after the publication of the Spandrels paper, Gould and Elisabeth Vrba took up the theme again, with a more measured approach, introducing the concept of *exaptations* in an article published in *Paleobiology* in 1982. They suggested a clarification "in the taxonomy of evolutionary morphology" in order to resolve a conceptual blurring of lines that had, they argued, become increasingly problematic. The term adaptation, they wrote, had come to hold two quite distinct meanings. Etymologically, adaptation implies something moving toward being fit (*ad*, toward; *aptus*, fit). However, the term was often used to describe a character used for something for which it was not originally intended. They noted that none other than G. C. Williams had addressed this problem in his 1966 book *Adaptation and Natural Selection*, noting that the term adaptation should be reserved for the case where one can "attribute the origin and perfection of this design to a long period of selection for effectiveness in this particular role." In such a case, the role of the adaptive character could correctly be referred to as its *function*. Other characters have fortuitous *effects* for which they were not shaped by natural selection. These characters are often referred to as adaptations. This terminology is incorrect, Gould and Vrba argued, because it leads, from observation of current usage, to an incorrect conclusion about historical origin. This misinterpretation was exacerbated by the fact that characters with fortuitous but unselected-for effects *had no name*. This led people to incorrectly describe them as "adaptations", and thus to tacitly assume that the historical origin of real adaptations was paralleled in these orphan characters as well. "What is to be done," Gould and Vrba asked, "with useful structures not built by natural selection for their current role?"

To resolve this problem, they proposed a rigorous nomenclature and the introduction of new terms. Following Williams, they suggested that we continue to "designate as an *adaptation* any feature that promotes fitness and was built by selection for its current role (criterion of *historical genesis*). The operation of an adaptation is its *function*" (Gould and Vrba 1982, their italics). Likewise, they proposed to follow Williams in designating "the operation of a useful character not built by selection for its current role as an *effect*". Then they proposed two new terms, *exaptation* and *aptation*.

But what is the unselected, but useful character itself to be called?... Its space on the logical chart is currently blank. We suggest that such characters, evolved for other usages (or for no function at all), and later "coopted" for their current role, be called *exaptations*... They are fit for their current role, hence *aptus*, but they were not designed for it, and are therefore not *ad aptus*, or pushed towards fitness. They owe their fitness to features present for other reasons, and are therefore *fit* (*aptus*) by reason of (*ex*) their form, or *ex aptus*. Mammalian [skull] sutures are an exaptation for parturition. Adaptations have functions; exaptations have effects. The general, static phenomenon of being fit should be called aptation, not adaptation. (Gould and Vrba 1982, their italics)

Gould and Vrba then described five examples of exaptations, showing how dignifying the concept with a name could help to prevent scientists from automatically assuming an adaptive role for any observed feature. Feathers in birds, for example, are hypothesized to have arisen via natural selection as a means to help control body temperature; it has been suggested that the large feathers along the arms of *Archaeopteryx* were enlarged by selection in order to increase effectiveness in catching insects. Further selection, including "changes in skeletal features and feathers, and for specific neuromotor patterns, resulted in the evolution of flight". Wings used for flight, however, can be used for other purposes as well. Gould and Vrba cite observations of the black heron (*Egretta ardesiaca*) using its wings as a canopy both to prevent glare on the water as it searches for fish and to attract fish to the shaded water (Fig. 15.4). In this complex example, feathers were initially an *adaptation* for insulation, but in this capacity served also as an *exaptation* for "the simplest feats of flight" and for prey capture. Further selection led to the full development of wings, which then serve a further exaptive role for the black heron as it hunts for food. In the development of wings, we see that an exaptive effect

Fig. 15.4 A black heron, *Egretta ardesiaca*, canopy fishing near Abuko, The Gambia. Photograph by Steve Garvie from Dunfermline, Fife, Scotland. Creative Commons Attribution-Share Alike 2.0 Generic License., https://commons.wikimedia.org/w/index.php?curid=11461458

can be acted upon by natural selection and refined as an adaptation (see Darwin's argument with Mivart above). This process of exaptations followed by secondary adaptations, Gould and Vrba wrote, was likely to occur often during the evolution of complex features.

> The evolutionary history of any complex feature will probably include a sequential mixture of adaptations, primary exaptations and secondary adaptations... [C]omplex features are a mixture of exapations and adaptations. Any coopted structure (an exaptation) will probably not arise perfected for its new effect. It will therefore develop secondary adaptations for the new role. The primary exaptations and secondary adaptations can, in principle, be distinguished. (Gould and Vrba 1982)

In 1961, Frederick G. E. Pautard suggested that bone had originally served the primary purpose of phosphate storage. The later structural use of bone would then be, using Gould and Vrba's terminology, an exaptation.

> Calcium phosphates, laid down in the skin of the earliest vertebrates, evolved initially as an adaptation for storing phosphates needed for metabolic activity. Only considerably later in evolution did bone replace the cartilaginous endoskeleton and adopt the function of support for which it is now most noted... The metabolic mechanism for producing bone per se can thus be interpreted as an *exaptation* for support. The metabolic mechanisms for depositing an increased quantity of phosphates and for mineralization, as well as the arrangement of bony elements in an internal skeleton, are then *adaptations* for support. (Gould and Vrba 1982, my italics)

The evolution of mammalian lactation provides another example of exaptation. Lysozyme, which hydrolyzes bacterial membranes, has significant homology with α-lactalbumin, essential for synthesis of lactose in mammalian milk. Despite similarities in sequence and in structure, lysozyme and α-lactalbumin clearly have radical differences in function. It was suggested as early as the 1960s that α-lactalbumin evolved from a duplicated lysozyme gene, which was then subject to selection for its function in producing lactose.[8] As with the hypothetical picture of the evolution of bone, this presents a scenario where an exaptation was subsequently subject to natural selection, ultimately becoming fully *ad*apted to a new function.

When sexual mimicry in hyenas was first investigated by Hans Kruuk, evolutionary biology still lacked a well-defined concept of exaptation. Gould and Vrba suggest that this, in part, resulted in scientists developing an incorrect adaptive explanation for their observations. As is well known, female spotted hyenas are larger than males, and behave dominantly toward them; they also have external genitalia that are, at first glance, indistinguishable from males'. Their fused labia look like a scrotal sac, and their enlarged, cylindrical clitoris looks like a penis. Sniffing genitalia is a crucial part of a greeting ritual used by hyenas when they return from hunting. It has been argued that the large genitalia in the female evolved "for" this ritual. Gould and Vrba quote Kruuk's statement that "[i]t is impossible to think of any other purpose for this special female feature than for use in the meeting ceremony". However, a perfectly reasonable alternate explanation is that the female

---

[8] See Irwin et al. (2011) for a recent survey of the evolution of the mammalian lysozyme gene family, including lactalbumin.

genitalia are an exaptation resulting from the unusually high concentration of andro-
gens in females that accompanies their larger stature and dominant social role in this
species. This feature would then serve as an exaptation in the greeting ceremony.
Gould and Vrba note with surprise that blood androgen levels in spotted hyenas
were not actually measured until 1979, when Ryan and Skinner found identical
concentrations in females and males. "In the other two species of the family
Hyaenidae, however," Gould and Vrba explain, "androgen levels in blood plasma
are much lower for females than for males. Females of these species are not dominant over males and do not develop peniform clitorises or false scrotal sacs."

The last example Gould and Vrba cite is that of repetitive DNA, such as the satellite DNA that appears as tandem noncoding repeats in centrosomes. An adaptive
hypothesis holds that this DNA was selected for a regulatory or structural role.
Another "tradition…holds that repetitive DNA must exist because evolution needs
it so badly for a flexible future – as in the favored argument that 'unemployed,'
redundant copies are free to alter because their necessary product is still being generated by the original copy" (Gould and Vrba 1982). Interestingly, Gould and Vrba
dismiss this possibility using the "future benefits" argument that we encountered in
the previous chapter. "While we do not doubt," they write, "that such future uses are
vitally important consequences of repeated DNA, they simply cannot be the causes
of its existence, unless we return to certain theistic views that permit the control of
present events by future needs." This argument is notable for two reasons. First,
recall that a higher-level selection argument can be made in favor of characteristics
that promote evolvability; Gould and Vrba fail to raise such a possibility here.
Second, satellite DNA in centrosomes is not easily accessible to such tinkering,
since it is tightly packed as a constituent of heterochromatin and not transcribed.
Other repetitive DNA sequences, of course, would indeed be accessible for transcription.[9] Gould and Vrba suggested that excess satellite DNA was a result of the
accumulation of selfish genetic elements, exapted for structural purposes. "Such
'selfish DNA' may be playing its own Darwinian game at a genic level, but it represents a true nonadaptation at the level of the phenotype… When used to great
advantage in [the] future, these repeated copies are exaptations."

The discussion of excess DNA raises, albeit obliquely, a critically important concept we will return to shortly below: cross-level spandrels. According to this concept, which echoes the more general idea of cross-level byproducts discussed by
Okasha (2006), exaptive characters at one level of selection provide a function at
another. Assuming the excess DNA under discussion was actually accessible for
transcription, Gould and Vrba present a scenario in which a feature at the gene level
acts exaptively at the individual level, by providing an additional genetic reservoir
of quirky – and potentially useful – functional shift. Although it is not considered by
Gould and Vrba, the scenario they describe also conceals a second cross-level interaction: an adaptation at the group or species level (additional genetic material available as a source of evolvability) acts exaptively at the individual level, again by

---

[9] See, for example, S. Ohno, *Evolution by Gene Duplication* (1970) for an important early discussion of this topic.

providing a reservoir of quirky functions, such as providing structural stability. A further important point is that there is a distinction between the exaptive/adaptive history of the excess DNA depending on its historical origin. If the selfish genetic element origin proposed by Gould and Vrba is correct, then increased copy number was not adaptive, but rather a result of genic selection. If the evolvability argument holds, its increased copy number is indeed adaptive, but at the group or species level. See Plohl et al. (2012) for a more recent review of evolutionary mechanisms acting on satellite DNA.

In summing up the importance of the exaptation concept, Gould and Vrba noted that the new terminology would remove the incorrect assumption of teleology in the use of the term "preadaptation". "The recognition of exaptation solves the dilemma neatly, for what we now incorrectly call 'preadaptation' is merely a category of exaptation considered before the fact," they wrote. They suggested that the term "preadaptation" be replaced by *preaptation*, defining a category of "potential, but unrealized exapations". Gould and Vrba also distinguished between exaptations of different origin. While all exapations originate randomly "with respect to their effects", some originate as adaptations for other functions, and some as "non-aptive structures". They suggest that "the enormous pool of nonaptations must be the reservoir of most evolutionary flexibility… this nonaptive pool is an analog of mutation – a source of raw material for further selection." While both types of exaptation provide "an enormous pool of variability at a level higher than mutations, for cooption as exaptations", it is the nonaptive pool that had not yet been explored in the literature.[10] "If all exaptations began as adaptations for another function in ancestors," wrote Gould and Vrba, "we would not have written this paper." The great unexplored mine of variability arising from nonaptations is "the missing concept". These members of the larger class of exaptations "are not covered by the principle of preaptation, for they were not adaptations in ancestors. They truly have no name, and concepts without names cannot be properly incorporated in thought." Ironically, Gould and Vrba initially left these nonaptive precursors defined by what they were not, a "curious negative definition" that was remedied later when Gould formally identified a crucial subset of them with the informal term he and Lewontin had already proposed: spandrels.

In the years following Gould and Vrba's paper, the term exaptation gradually became adopted as a scientific term, as they had hoped. Additional examples of exaptations accrued as well. For example, as described by E. N. Arnold in 1994, and reviewed by Gould (2002), certain tropical lizard species are able to crawl into extremely narrow crevices by literally pulling their eyes back into their heads. The aerodynamic features of this flattened shape were later co-opted for use in aiding the lizards in gliding from one tree to another; phylogenetic analysis suggests that "flat-

---

[10] Gould and Vrba suggest that the literature of sociobiology would undergo a "constructive collapse", which would "vastly broaden our range of hypotheses" if it incorporated the idea of exaptations. Certainly it would eliminate "unprovable reveries about primal fratricide on the African savanna or dispatching mammoths at the edge of great ice sheets", among other just-so stories giving causal evolutionary explanations for aspects of modern human behavior.

tening first developed in the context of crevice use and was only later co-opted for gliding" (Arnold et al. 1994; see also Gould 2002, p. 1236).

Exaptations play another role in the evolutionary history of flattened lizard heads; in order to achieve the temporarily flattened structure, the eyes must be retracted into the head, giving the poor lizards what Gould calls "a mouthful of eye". More technically, in Arnold's words,

> when a lizard flees into a narrow crevice the eyes must be accommodated within the depth of the flattened head. They are most usually pushed downwards by the ceiling of the crevice as the lizard moves deeper into it, so their upper margins are flush with the skull roof and their lower sections bulge through the palate into the buccal cavity. (Arnold et al. 1994; see also Gould 2002, p. 1236)

In some lizard species (lacertids and scincids), the eyes are pushed down into the suborbital foramen, a cavity below the eye. Phylogenetic analysis shows that this opening existed in ancestral species before the lacertid and scincid lizards began using narrow crevices as hideaways. The suborbital foramen thus appears have been exapted for accommodation of the eyes, though it may have been subsequently enlarged by an adaptive process in order to fit them better. Further supporting the idea that the suborbital foramen is exaptive in this role is the observation that related lizard species, the cordylids, use a different cavity, the interpterygoid vacuity (an open space between the plates of the palate), as a temporary home for their eyes during crevice-crawling. This space, which is much smaller in the lacertids and scincids, was more easily exapted in the cordylids for this purpose.

Another dramatic example of exaptation occurs in lizard sand-diving. Arnold et al. (1994) and Arnold (1995) reviewed six different cases in which lizards developed the ability to "dive" quickly into the sand in order to escape predators. In each case, the lizard lineage faced a similar technical problem, but each lineage had a different evolutionary toolkit to work with. In two cases, the ability to sand-dive appeared to have arisen by a direct adaptive process; in the other four, it appeared to have been exapted from other traits in the ancestral lineages, such as a mechanism for drilling into hard substrates or from the mechanisms used by some lizards to bury themselves in the ground before a dormant period. More recent studies of *Anolis* lizards have demonstrated the role of exaptive features in finding multiple routes to relative optimization in a particular evolutionary niche (Poe et al. 2007).

Molecular evolution provides a vast wealth of potential exaptations. Many examples of exaptations at the molecular level arise when proteins take on multiple roles; subsequent gene duplication and further adaptation can lead to divergence between these molecular lineages, but sequence data can confirm their common origin. One such example – which, Gould noted with pride, made explicit use of the term exaptation – was the observation by Wakasugi and Schimmel in 1999 that the carboxyl-terminal domain of human tyrosyl-transfer RNA synthetase has a 49% similarity with a cytokine that recruits phagocytes to the location of apoptotic cells. In other words, the tyrosyl-tRNA synthetase leads what Weiner and Maizels, in an accompanying perspective article, called "a deadly double life", both facilitating tRNA synthesis and alerting the body to incipient cell death. "How and why," asked Weiner and Maizels, "did a tRNA synthetase get involved in the deadly business of apoptosis?"

> The recruitment of tyrosyl-tRNA synthetase as an extracellular death messenger echoes the recruitment of cytochrome c as an intracellular death messenger. Both are essential proteins that serve as harbingers of impending cell death when released from their normal cellular compartments. Release of proteins from their normal locations in the cell may have originally been a symptom of cell death, rather than a cause of it. Evolution… may then have exploited the accidental release of these proteins (and possibly others) to build, amplify, and eventually fine-tune the death circuitry. (Weiner and Maizels 1999)

Crystallin proteins provide transparency in the lens of the eye. They also provide a striking example of exaptation. A vast array of proteins with widely different functions have been co-opted to play the role of crystallins in the lens. These proteins primarily exist in the elongated cells of enucleated lens fibers, and were traditionally viewed as highly specialized, "designed for their ability to confer the required refractive properties onto the transparent lens" (Gould 2002, p. 1242). A series of groundbreaking studies begun by Graeme Wistow and colleagues at the National Eye Institute showed that the crystallins are actually an astoundingly diverse group of proteins, which often play other crucial roles in different tissues as well (Wistow et al. 1987). Wistow and Joram Piatigorsky described as "gene sharing".[11] There are two types of crystallins, so-called "structural" proteins homologous to heat shock proteins, and "taxon-specific" crystallins exapted from other biological roles.

> For example, delta crystallin of chickens is argininosuccinate lyase; epsilon crystallin of ducks is identical with the metabolic enzyme lactate dehydrogenase; tau crystallin of turtles is [the glycolytic enzyme] alpha-enolase; and mu crystallin, found in many marsupials, is ornithine cyclodeaminase…the major lens component of squid S crystallin is related to the detoxification enzyme glutathione S-transferase. (Gould 2002, p. 1243)

As an example of evolutionary convergence by exaptation, species as different as elephant shrews and octopi have, in separate evolutionary events, exapted aldehyde dehydrogenase to serve as a crystallin. Such dual roles occur so often that Piatigorsky called them a "hallmark" of the enzyme-crystallins (Piatigorsky 1993a; see also Gould 2002, p. 1243). While the individual evolutionary history of each of these instances of exaptation may be quite tricky to unravel, the crystallin function clearly appears to have arisen secondarily. Piatigorsky and Wistow (1991) have shown (see Fig. 15.5) that the adoption of this new function typically occurs by "recruitment [or, in Gould's terminology, exaptation] followed by gene duplication and subsequent partial separation of function."

Speculating on why gene sharing is so widespread among the crystallins, Piatigorsky (1993b) suggested "that the properties of proteins required for crystallin function may be less stringent than those for many other biological functions". He proposed that the recruitment of proteins as crystallins may be driven by selective

---

[11] For more detail on this fascinating topic, see Wistow's 1995 book *Molecular Biology and Evolution of Crystallins: Gene Recruitment and Multifunctional Proteins in the Eye Lens*, and Piatigorsky's 2007 book *Gene Sharing and Evolution*.

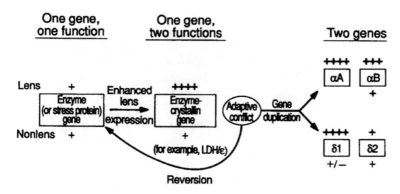

Fig. 15.5 Scheme for the recruitment of crystallins to serve other functions, from Piatigorsky and Wistow (1991). "During crystallin recruitment", they wrote, "a single gene with a single function acquires high expression in [the] lens. If this creates an adaptive conflict, crystallin expression may be lost or gene duplication and specialization may occur." Rectangles indicate genes; + signs above the rectangles indicate expression in lens, while +/- signs below rectangles indicate nonlens expression. From J. Piatigorsky and G. Wistow, The recruitment of crystallins: new functions precede gene duplication. *Science* 252(5009): 1078–1079, 1991. Reprinted with permission from AAAS

pressure on genetic regulatory mechanisms. It is also possible that particular proteins were ultimately recruited as crystallins because of their abundance under the particular conditions of cell elongation, organelle loss, and osmotic stress that prevail in the lens fibers (Piatigorsky 1993b; Wistow 1993; Gould 2002, p. 1244). More recently, there has been a terminological shift in the literature to describing such proteins as "moonlighting"; see Huberts and van der Klei (2010), Monaghan and Whitmarsh (2015), Gancedo et al. (2016), Jeffery (2016), and Min et al. (2016), among many others.

Another example of an exaptation is found in the coiled structure of snail shells. When a tube is coiled around an axis, a cylindrical space is formed in the center, with the tube coiled around it. This space is called an *umbilicus*, and in many species of snail it is filled with calcite. (When filled, it is referred to as a *columella*.) However, in some species, the umbilicus remains empty. Lindberg and Dobberteen (1981) found one such species of snail, *Margarites vorticiferus* Dall, on Attu Island in Alaska, using its open umbilicus as a brooding chamber for eggs and young (Fig. 15.6). Lindberg and Dobberteen noted that a few other species, such as *Munditia subquadrata*, from Australia, and an Indo-Pacific species, *Philippia radiata*, also exhibit umbilical brooding.

The exaptive use of the umbilicus as a brooding chamber is a prime example of a spandrel. In contrast to the cooption of adaptive features for use in a different context, the umbilicus is as clear an example of an architectural byproduct as one could ask for. Use of the umbilicus for this purpose evolved late in the evolutionary history of snails.

Fig. 15.6 *M. vorticiferus* specimen with brood visible in the umbilicus. From D.R. Lindberg and R.A. Dobberteen, Umbilical brood protection and sexual dimorphism in the boreal Pacific trochid gastropod, *Margarites vorticiferus* Dall. Int. J. Invert. Reprod. 3(6): 347-355, 1981. Reprinted by permission of the publisher (Taylor & Francis Ltd, http://www.tandfonline.com)

> The cladogram of gastropods includes thousands of species, all with umbilical spaces (often filled as a solid columella and therefore unavailable for brooding) but only a very few with umbilical brooding. Moreover, the umbilical brooders occupy only a few tips on distinct and late-arising twigs of the cladogram, not a central position near the root of the tree. (Gould 2002, p. 1260)

As another example of a spandrel, Gould cited the enlarged shoulder vertebrae of the Irish elk, *Megaloceros giganteus*, which cave paintings in Spain and France show were covered with distinctively colored fur. The enlarged vertebrae were necessary to support the immense horns of the elk; the hump containing the vertebrae is hypothesized to have played a role in sexual selection (Gould 2002, p. 1261). In its putative sexual selection role, it was a spandrel.

The many different types of exaptations described above, some initially adaptive, and some (informally called "spandrels") not, cry out for careful classification. Gould undertook such a classification in his immense final work, *The Structure of Evolutionary Theory*. As we have seen, Gould and Vrba separated the concepts of adaptation and exaptation in their 1982 paper, and distinguished two types of exaptations (Fig. 15.7). Writing in 2002, Gould proposed a more detailed taxonomy of "The Exaptive Pool", breaking the "Type 2" exaptations shown in Fig. 15.7 into several subsets (Fig. 15.8). To the first type of exaptation, he gave the name *franklins*, in honor of the U.S. dime, with its portrait of Franklin Delano Roosevelt, since it serves as prime example of the concept of something *initially manufactured to serve one purpose, but co-opted for another in a quirky functional shift.* "American dimes," Gould wrote, "are adaptations as money, and exaptations a screwdrivers." (Gould 2002, p. 1278).

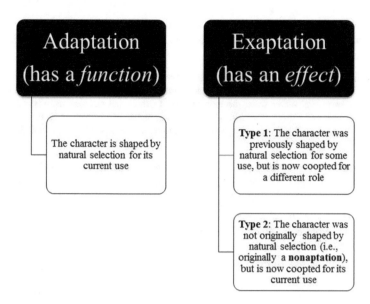

Fig. 15.7 Initial classification of adaptations and exaptations developed by Gould and Vrba in 1982

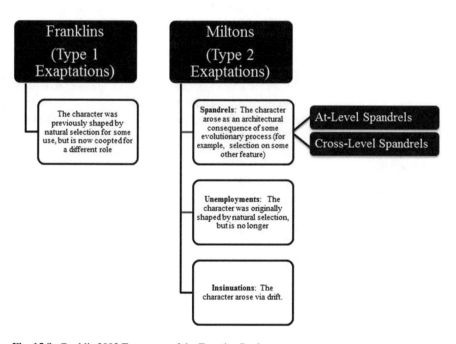

Fig. 15.8 Gould's 2002 Taxonomy of the Exaptive Pool

The second type of exaptation, those that originally were nonaptations, Gould named *miltons*, in honor of John Milton, inspired by his line "They also serve who only stand and wait".

> Miltons break the exclusivity of the adaptationist program by basing a large component of evolvability not upon the potential of already functional (and adaptive) features to perform in other ways, but rather upon the existence of a substantial array of truly nonadaptive features – unused things in themselves rather than alternative potentials of features now functioning in other ways (and regulated by natural selection at all times). (Gould 2002, p. 1279)

However, the category of miltons is not identical with that of spandrels. Miltons come in three flavors. They "can originate in several ways – as nonadaptive spandrels (the most important subcategory I [Gould] shall argue), as previously useful structures that have become vestigial, or as neutral features fortuitously introduced 'beneath' the notice of selection" (Gould 2002, p. 1279).

Gould called vestigial miltons, which "lose an original utility without gaining a new function", *manumissions* or *unemployments*. "Currently nonadaptive as a historical result of their altered status, they fall out of selective control and into the exaptive pool as actual items that must now 'stand and wait' but might serve again in an altered evolutionary future" (Gould 2002, p. 1281). Examples might include genes retained in the genome but unused after a species ceases to pass through a larval stage. The αA crystallin protein in the vestigial eyes of the naked mole rat may offer another case in point. The lens is no longer capable of forming an image, but Hendriks et al. (1987) found that, while the αA crystallin gene accumulates mutations at a much higher rate than crystallin genes in animals with functional eyes, its mutation rate is about one fifth that of "truly neutral pseudogenes". This suggests that αA crystallin still serves some function, and thus is "visible" to natural selection; it is possible that the protein is still involved in perception of light, perhaps in the context of regulating circadian rhythms. Note, however, that it is not definitively known whether there was a period the evolutionary history of the naked mole rat when αA crystallin was truly "unemployed", before shifting to this putative new function.

Miltons that are "introduced beneath selection's scrutiny" Gould called *insinuations*. Insinuations are a result of genetic drift. As an example of an insinuation, in the form of the fixation of random variation via a founder effect, Gould noted the case of the invasive Argentine ant, *Linepithema humile* (Gould 2002, pp. 1282–1283). In its original environment, this species lives in colonies which often engage in mutual antagonism; this keeps the overall population down, and as a result *L. humile* lives in an ecosystem that involves a number of other ant species. *L. humile* presumably migrated to California in a small, closely related subpopulation. As this initial invasive colony grew, its members continued to recognize one another as kin, and therefore did not exhibit antagonism. The colony did not fragment, and continued to grow; cooperation had run wild. This species is now believed to have formed the largest known animal society (Van Wilgenburg et al. 2010).

Gould's Taxonomy of the Exaptive Pool contains a terminological asymmtery. He repeatedly describes franklins as "inherent potentials", specifying that they are "not actual but unemployed 'things out there'. Franklins are alternative potential functions of objects now being used in another way... [they] are inherent potentials, not available things" (Gould 2002, p. 1278). In contrast, Gould specifies that miltons are actually "available things". Dividing exaptations into two categories, one of which is a conceived potential and the other of which is a physical object is logically problematic. This inconsistency is easily remedied, however, by shifting the definition of franklins and miltons so that they each represent the trait or character at issue (each a "thing"), *or* the new function of the character (each a "function"), *or* by defining all items in the taxonomy as *both* their physical presence and their (potential) function(s).

The divisions in Gould's taxonomy represent a personal choice; other divisions are possible. This recalls the issues of taxonomy and classification we discussed in Chap. 2, in the context of the classification schemes of Linnaeus, Adanson, and others. As Foucault emphasized, a classification scheme represents (and formalizes) a style of thought. "I recognize," Gould acknowledged, "that the objective items of the exaptive pool could be parsed in other ways – our decisions about their ranking and secondary ordering require a choice among several logically legitimate alternatives" (Gould 2002, p. 1285). For example, exaptations could be classified according to their "visibility" to natural selection. In this scheme, franklins and miltonic spandrels would be grouped together, while unemployments and insinuations, which are presumed to be currently selectively neutral, would be placed in a second grouping. Another option would be classification according to historical origin, with characters originally arising as adaptations in one group, and those which developed nonadaptively in another. Here, franklins and unemployments would be grouped together in the adapted category, while spandrels and insinuations would reside in the nonadaptive group. Rather than either of these taxonomic options, Gould grouped franklins separately from miltons for a very specific theoretical purpose: to highlight, in a single category, features that "pose a genuine challenge to the exclusivity of adaptationist mechanisms" (Gould 2002, p. 1286). As such, they exemplify the major recurring themes in Gould's work: the importance of structural constraints and historical contingency.

> Franklins enlarge the scope and sophistication of selectionist argument, adding a genuine flavor of formalist limitation and potentiation to an otherwise naively functionalist theory based only upon organic accomodation to selective pressures of an external environment. But miltons emplace a genuinely nonadaptationist component into the heart of evolutionary explanation – for if many features originate as nonadaptations, and if nonadaptations, as material items of miltonic "stuff", stand and wait while occupying a substantial percentage of the exaptive pool, then evolutionary explanations for both the origin of novelties, and for the differential capacity of lineages to enjoy future phyletic expansion and success, will require a revised and expanded version of Darwinism, enriched by nonselectionist themes of a formalist and structuralist research program. I therefore choose...the best taxonomic device for exploring the role of nonadaptation and structural constraint in the exaptive pool of evolvability. (Gould 2002, p. 1286)

One further aspect of Gould's taxonomy remains to be addressed: one that has profound ramifications not just for the roles of contingency and structural constraint, but for the mechanisms of interaction between (and, as I will argue below, the inevitability of) multiple levels of selection. This is the division of the spandrels category into *at-level* and *cross-level* spandrels (Fig. 15.8). The examples of spandrels discussed above have all belonged to the at-level category: characters at one level of biological organization that did not arise adaptively, but were later co-opted for use *at that same level*. However, a character at one level, whether it arose adaptively or not, *will be visible to other levels as a spandrel*, since it has no direct adaptive value at those other levels. *Thus, an adaptation at one level is*[12] *a spandrel at another*.

Gould proposed that *cross-level spandrels provide the source of variability upon which natural selection acts*. Recall that he and Elisabeth Vrba had described exaptations as "an analog of mutation – a source of raw material for further selection". Writing in 2002, Gould saw spandrels as providing the "chance" in Monod's famous mantra of "chance and necessity", the source for evolvability, and the raw material on which natural selection can work. The results of adaptation at the genic level provide a source of variability at the organismal level. For example, "selfish" genes that duplicate themselves in order to increase their presence in the gene pool "may be simultaneously exapted at the organismal level by undergoing a mutational change (only now of potential benefit to the organism as a consequence of the gene's redundancy), and contributing thereby to a new organismal function" (Gould 2002, p. 1287). Similarly, an organismal adaptation to a particular environment may have a cross-level effect at the population level. Marine invertebrates may evolve from producing planktonic larvae, which feed in the water column, to producing non-foraging, bottom-dwelling lecithotrophic larvae, which feed only on the yolk of their eggs. A lecithotrophic larval cycle has adaptive value at the level of organismal fitness (Pechenik 1999). It has no adaptive value, and thus is[13] a spandrel at the species (or population) level, but "may simultaneously impart an exaptive effect to its species by enhancing the speciation rate via the altered demic structure of isolated subpopulations that no longer experience the gene flow previously potentiated by floating planktonic larvae" (Gould 2002, p. 1287).

To the extent that Gould offers the idea of cross-level spandrels as evidence that *adaptationist* arguments are insufficient, philosopher Todd Grantham argues that Gould's apporach is unsuccessful. Grantham (2004) analyzed the case of evolving mollusc life cycles in order to make his point. Suppose, he argued, a species undergoes a shift, driven by natural selection, from producing planktotrophic larvae, which feed on plankton and exhibit high dispersal, to a so-called "direct development" larval form, in which the larvae "crawl away" from the egg mass after hatching. Direct development leads to even less dispersal than lecithotrophy.

> Because direct development decreases larval dispersal, organismic evolution has immediate impact on gene flow among populations (a species-level trait). Suppose further that by decreasing gene flow, direct development elevates speciation rate and therefore becomes more common through species selection. In this scenario, the changes in gene flow count as

---

[12] It might be more precise to say that it *has the potential to be a spandrel*, if a feature is only defined as a spandrel *once has been coopted to serve an exaptive function*.

[13] See note 12; the same qualification applies here.

a spandrel – this change is "injected" at the species level from below and was not introduced by selection *at that level*. It is unclear, however, why this scenario should be thought to challenge adaptationism. The organismic trait (direct development) is an adaptation. This adaptive process has an effect (on gene flow) which initiates another selection process at the species level. The process appears to be driven by selection (albeit at different hierarchical levels). (Grantham 2004)

Grantham presents a strong argument that the idea of cross-level spandrels does not necessarily undermine the validity of adaptationist explanations for evolutionary processes at any particular level. The idea of cross-level spandrels is of far greater importance, he argues – and I agree – for its implication regarding the importance of hierarchical selection. As I will propose in the next chapter, the implications of this idea in terms of multi-level selection are broader and more fundamental than even Gould himself suggested. As Grantham notes, the "only 'moment' of non-adaptationist reasoning [in his hypothetical scenario sketched above] is the original appearance of the new population structure." However, this one moment is perhaps the most important moment of all, not so much because it contains a non-adaptationist process of great evolutionary significance, but because it represents *a step across the bridge between two levels of selection*.

In a particularly beautiful example of an adaptation at the organismal level serving as a cross-level spandrel at the species level, Gould cites the 2001 study by Jeffery Podos (see also Ryan (2001) for commentary) which showed that the beak shape of the Galápagos finches, known to be adapted for food specialization and thus for organismal fitness, influences the vocal signatures of the birds (Fig. 15.9). As Podos summarized,

diversification of beak morphology and body size has shaped patterns of vocal signal evolution, such that birds with large beaks and body sizes have evolved songs with comparatively low rates of syllable repetition and narrow frequency bandwidths. The converse is true for small birds. (Podos 2001)

Podos concluded his study by suggesting far-reaching implications for speciation.

Because song presumably has a central role in reproductive isolation in Darwin's finches..., the linkage between morphology and vocal performance capacities holds important implications for the dynamics of finch speciation. In island songbird clades, including the Darwin's finches, speciation is driven primarily, if not exclusively, by prezygotic isolating mechanisms, as evident from the regular occurrence of viable and fertile hybrids... The effectiveness of prezygotic isolating mechanisms generally depends on the extent to which mating signals among incipient species are distinct, with more distinct signals increasing the probability of 'correct' matings and thus enhancing probabilities of speciation... My data suggest that magnitudes of ecological and morphological diversification among incipient Darwin's finch species will directly determine magnitudes of diversification in vocal features, and thus determine probabilities of speciation (to the extent that trill rate and frequency bandwidth are used by birds in mate recognition). Taking this hypothesis one step further, the high diversity of ecological opportunities for Darwin's finches on the Galápagos Islands may thus have promoted, through extensive morphological adaptation and correspondingly large evolutionary changes in vocal signal structure, conditions suitable for rapid speciation and the marked radiation that has defined the group. (Podos 2001)

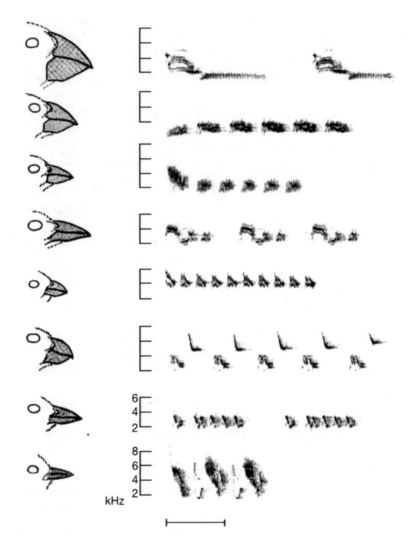

Fig. 15.9 Beak morphology and song structure. Reprinted by permission from Macmillan Publishers Ltd., J. Podos, Correlated evolution of morphology and vocal signal structure in Darwin's finches, *Nature* 409(6817): 185–188, 2001, copyright 2001

Interpreting this through his exaptive lens, Gould wrote that because

> songs function as powerful premating isolating mechanisms, the automatic divergence of song, arising as a side-consequence of ordinary adaptation of bills in feeding, and the different degrees of distinctiveness attached to specific forms of the bill, may have profound consequences in a resulting (and ultimately highly exaptive) differential capacity for speciation among different subclades of this classic group (based upon varying capacities of the resulting song to act as an effective signal for mate recognition). (Gould 2002, p. 1287)

The implications of cross-level spandrels extend far beyond these individual examples, intriguing though they are. Expanding on the suggestion made by Gould and Vrba in 1982, Gould argued that *mutations themselves represent a vast pool of cross-level spandrels.*

> Each mutation arises for a perfectly good reason (usually chemical rather than adaptational in this case) at the gene level. But the effects then imposed upon organismal phenotypes must be designated as spandrels – that is, as nonadaptive side consequences expressed at another level. (Gould 2002, p. 1268)

It could be argued that this interpretation requires expanding the definition of a spandrel to include any feature that arises by random processes (which perhaps overlaps with the definition of insinuations). Perhaps cross-level miltons would be more accurate, given Gould's taxonomy of the exaptive pool.

Gould suggested that the status of mutations as cross-level spandrels (or miltons, in accordance with the taxonomy in Fig. 15.8) fits neatly with the observation that most mutations are deleterious at the organismal level.

> The phenotypic expressions of mutations are spandrels at the organismal level, and we have long recognized the vast majority as deleterious for the organism. But we do not regard this inevitable property…as globally detrimental to organisms. Populations of organisms are large enough, and the generational cycling time of organisms short enough, to tolerate a substantial load of general disadvantage in exchange for the occasional opportunity [of] encountering a favorable spandrel…Thus, the organismal level can usually well afford this carnage of generally deleterious mutational effects in order to win its fuel of positive variants for natural selection. (Gould 2002, pp.1290–1291)

The interpretation of redundant genes as cross-level spandrels, adaptive at the level of the selfish gene, and initially neutral at higher levels, can resolve the supposed "future benefits" paradox in the problem of evolvability, consistent with the arguments discussed in Chap. 14. Natural selection at one level feeds on raw material generated as cross-level spandrels at a lower level. This is the case for selection on organisms. It is also the case for selection on populations. Here, the raw material is variability at lower levels (organisms, genes). The evolution of evolvability can occur *when the group trait that provides a selective advantage is variability itself.* There is thus no magic, no mystery, no seeing into the future. There is simply a store of variation arising at one level, and acting as a substrate for natural selection at another level.

> Traits that confer evolvability upon species-individuals, but arise by selection on organisms, provide a precise analog at the species level to the classical role of mutation at the organismal level. Because these traits of species evolvability arise by a different process (organismal selection), unrelated to the selective needs of a species, they may emerge at the species level as "random" raw material, potentially utilizable as traits for species selection… The phenotypic effects of mutations are, in exactly the same manner, spandrels at an organismal level – that is, nonadaptive and automatic manifestations at a higher level of different kinds of causes acting directly at a lower level. *The exaptation of a small and beneficial subset of these spandrels virtually defines the process of natural selection.* (Gould 2002, p. 1276, my italics)

The importance of this last statement cannot be overemphasized. Indeed, one could ask whether nature would "be Darwinian at all, absent these interesting properties of cross-level spandrels that must supply the fuel of natural selection" (Gould 2002, p. 1291). Yet there is one crucial consequence of these arguments that Gould did not explore, one with far-reaching implications for multi-level selection. It is this consequence – that *evolution cannot occur without the presence of multiple levels* – that we explore in the following chapter.

References

Ahouse JC (1998) The tragedy of a priori selectionism: Dennett and Gould on adaptationism. Biol Philos 13:359–391

Andersson DI, Jerlström-Hultqvist J, Näsvall J (2015) Evolution of new functions de novo and from preexisting genes. Cold Spring Harb Perspect Biol 7(6):a017996

Arnold EN (1994) Investigating the origins of performance advantage: adaptation, exaptation and lineage effects. In: Eggleston P, Vane-Wright R (eds) Phylogenetics and Ecology. Linnean Society of London/Academic Press, London, pp 123–168

Arnold EN (1995) Identifying the effects of history on adaptation: origins of different sand-diving techniques in lizards. J Zool 235(3):351–388

Borgia G (1994) The scandals of San Marco. Q Rev Biol 69(3):373–375

Darwin C (1880) Sir Wyville Thomson and natural selection. Nature 23:32

Darwin C (2009) The Origin of Species by Means of Natural Selection. Modern Library, New York. This edition reprints the second edition of the *Origin*, from early 1860

Dennett DC (1995) Darwin's Dangerous Idea. Simon and Schuster, New York

Di Ieva A, Bruner E, Davidson J, Pisano P, Haider T, Stone SS, Cusimano MD, Tschabitscher M, Grizzi F (2013) Cranial sutures: a multidisciplinary review. Childs Nerv Syst 29(6):893–905

Gancedo C, Flores CL, Gancedo JM (2016) The expanding landscape of moonlighting proteins in yeasts. Microbiol Mol Biol Rev 80(3):765–777

Gould SJ (1997) The exaptive excellence of spandrels as a term and prototype. Proc Natl Acad Sci U S A 94(20):10750–10755

Gould SJ (2002) The Structure of Evolutionary Theory. The Belknap Press of Harvard University Press, Cambridge, MA

Gould SJ, Lewontin RC (1979) The spandrels of San Marco and the Panglossian paradigm: a critique of the adaptationist programme. Proc R Soc B 205:581–598

Gould SJ, Vrba ES (1982) Exaptation – a missing term in the science of form. Paleobiology 8(1):4–15

Grantham TA (2004) Constraints and spandrels in Gould's Structure of Evolutionary Theory. Biol Philos 19:29–43

Hendriks W, Leunissen J, Nevo E, Bloemendal H, de Jong WW (1987) The lens protein αA-crystallin of the blind mole rat, *Spalax ehrenbergi*: evolutionary change and functional constraints. Proc Natl Acad Sci U S A 84(15):5320–5324

Huberts DH, van der Klei IJ (2010) Moonlighting proteins: an intriguing mode of multitasking. Biochim Biophys Acta 1803(4):520–525

Irwin DM, Biegel JM, Stewart CB (2011) Evolution of the mammalian lysozyme gene family. BMC Evol Biol 11:166

Jeffery CJ (2016) Protein species and moonlighting proteins: Very small changes in a protein's covalent structure can change its biochemical function. J Proteome 134:19–24

Lindberg DR, Dobberteen RA (1981) Umbilical brood protection and sexual dimorphism in the boreal Pacific trochid gastropod, *Margarites vorticiferus* Dall. Int J Invert Reprod 3(6):347–355

Mark R (1996) Architecture and evolution. Am Sci 84:383–389

Min KW, Lee SH, Baek SJ (2016) Moonlighting proteins in cancer. Cancer Lett 370(1):108–116

Monaghan RM, Whitmarsh AJ (2015) Mitochondrial proteins moonlighting in the nucleus. Trends Biochem Sci 40(12):728–735

Ohno S (1970) Evolution by Gene Duplication. Springer, Berlin

Okasha S (2006) Evolution and the Levels of Selection. Oxford University Press (Clarendon Press), Oxford

Orr HA (1996) Dennett's strange idea. Boston Rev, Summer Issue: 28–32

Pechenik JA (1999) On the advantages and disadvantages of larval stages in benthic marine invertebrate life cycles. Mar Ecol Prog Ser 177:269–297

Piatigorsky J (1993a) The twelfth Frederick H. Verhoeff lecture: gene sharing in the visual system. Trans Am Ophthalmol Soc 91:283–297

Piatigorsky J (1993b) Puzzle of crystallin diversity in eye lenses. Dev Dyn 196(4):267–272

Piatigorsky J (2007) Gene Sharing and Evolution: The Diversity of Protein Functions. Harvard University Press, Cambridge, MA

Piatigorsky J, Wistow G (1991) The recruitment of crystallins: new functions precede gene duplication. Science 252(5009):1078–1079

Plohl M, Meštrović N, Mravinac B (2012) Satellite DNA evolution. Genome Dyn 7:126–152

Podos J (2001) Correlated evolution of morphology and vocal signal structure in Darwin's finches. Nature 409(6817):185–188

Poe S, Goheen JR, Hulebak EP (2007) Convergent exaptation and adaptation in solitary island lizards. Proc R Soc B 274(1623):2231–2237

Queller DC (1995) The spaniels of St. Marx and the Panglossian paradox: a critique of a rhetorical programme. Q Rev Biol 70(4):485–489

Ryan MJ (2001) Food, song and speciation. Nature 409(6817):139–140

Selzer J (ed) (1993) Understanding Scientific Prose: Rhetoric of the Human Sciences. University of Wisconsin Press, Madison

Van Wilgenburg E, Torres CW, Tsutsui ND (2010) The global expansion of a single ant supercolony. Evol Appl 3(2):136–143

Wakasugi K, Schimmel P (1999) Two distinct cytokines released from a human aminoacyl-tRNA synthetase. Science 284(5411):147–151

Weiner AM, Maizels N (1999) A deadly double life. Science 284(5411):63–64

Williams GC (1966) Adaptation and Natural Selection: A Critique of Some Current Evolutionary Thought. Princeton University Press, Princeton

Wistow G (1993) Lens crystallins: gene recruitment and evolutionary dynamism. Trends Biochem Sci 18(8):301–306

Wistow G (1995) Molecular Biology and Evolution of Crystallins: Gene Recruitment and Multifunctional Proteins in the Eye Lens. R.G. Landes Co./Springer, Austin/New York, Molecular Biology Intelligence Unit

Wistow GJ, Mulders JWM, de Jong WW (1987) The enzyme lactate dehydrogenase as a structural protein in avian and crocodilian lenses. Nature 326:622–624

Chapter 16
The Essential Tension

> *This is about the thirteenth lead I've written for this goddamn*
> *mess, and they are getting progressively worse...which hardly*
> *matters now, because we are down to the deadline again...and*
> *those thugs out in San Francisco will be screaming for Copy.*
> *Words! Wisdom! Gibberish! Anything! The presses roll at*
> *noon... This room reeks of failure once again.*

Dr. Hunter S. Thompson

IN THE preceding pages, we have travelled through crucial periods in the history of science and through a range of topics in the collective dynamics of biological systems. In each of the stops along our way, we have met, in various guises, what I describe as an *essential tension*: a balance between cooperation and competition, a balance between interactions at the local level – between cells or individual organisms, for example – and external pressures originating beyond these local interactions. We have seen how the balance of these apparently opposing drives plays a crucial role in the emergence of an ensemble of elements into a new individual in its own right. This shift is seen in the early theories of crowd formation, and in Durkheim's theory of the origin of the division of labor in human society. It is seen in the balance between alignment with neighbors and collision avoidance that generates an immense murmuration of starlings flying over a field. It is seen in the simple shifts in gene expression that take volvocine algae and their relatives from a single-celled to a multicellular lifestyle. It is seen in the quorum sensing that induces individual bacteria to begin forming a biofilm mat and individual *Dictyostelium* amoebae to form a slug and then a fruiting body. It is seen in the induction of cluster formation by yeast under selective pressure for faster settling in a gravitational field. It is seen in the stretching and folding, in which trajectories exponentially diverge, only to be kneaded back together, that is inherent in the nonlinear dynamics used to model many of these complex biological systems. It is seen in the shift from competition to cooperation via the suppression of conflict that Michod and colleagues define as the key step from selection at the MLS1 to the MLS2 level, which takes an ensemble from being a *collective of individuals* to an *individual collective*. The tension between competition and cooperation manifests itself differently in each instance, and my emphasis on this common theme is meant as anything but a suggestion that

© Springer Science+Business Media B.V. 2018
S. Bahar, *The Essential Tension*, The Frontiers Collection,
DOI 10.1007/978-94-024-1054-9_16

these vastly complex scientific problems can be reduced to a simple formula. Rather, I have focused on this theme in order to highlight its importance as an Ariadne's thread that may lead us to the center of the maze and, with luck, back out again.

While exploring the theme of the essential tension, we have repeatedly encountered the problem of multilevel selection. A unique perspective on this problem is provided by Gould's analysis of the role of cross-level spandrels. In his analysis, features that occur at one level of selection will behave as spandrels when "injected" into another level, and this produces the variability necessary to fuel natural selection at that new level. The variability injected from level n into level n+1 provides an explanation for the false paradox of evolvability. The modularity conferred on organisms by the presence of multiple *Hox* genes, or the vast number of gene duplications in species ranging from *C. elegans* to *Arabidopsis* to *Homo sapiens*, are not retroactively engineered for their possible future benefits; they arise for reasons, such as DNA editing errors or gene selection, that are blind to the collective of cells that they inhabit. At this collective level, they are spandrels: structures without initial adaptive benefit *at this higher level of interest*. They may then be amplified by selection acting at this higher level. The same analysis can be carried out for any pair of selective levels.

Gould concluded his vast *Structure of Evolutionary Theory* by proposing this fundamental role for cross-level spandrels, and suggested that nature might not be "Darwinian at all" without them. For him, cross-level spandrels provided strong support to the themes of *historical contingency, structural constraint,* and *multilevel selection* that he considered the "three legs of the tripod" of an expanded Darwinian theory. Cross-level spandrels provide fuel for, and, in their contingent nature, a complement to, the process of natural selection. Gould argued that

> if cross-level spandrels maintain an important relative frequency among the components of evolutionary change, then these automatic expressions at other levels – introduced separately from, and simultaneously with, the primary changes that generate them at a different focal level – may largely control the possibilities and directions of evolution from a structural 'inside', rather than only from the functional 'outside' of natural selection. (Gould 2002, p. 1294)

This idea fits well with the metaphor of stretching and folding we have drawn from nonlinear dynamics. By providing a source of variability, cross-level spandrels carry a form of sensitive dependence on initial conditions (stretching). An external constraint from the "outside", in this case natural selection, provides the folding.

In formulating his proposal regarding the fundamental role of cross-level spandrels, however, there was one last step that Gould did not take. He described the "inevitability" of cross-level spandrels, injected from one level into another, and providing a fundamental source, perhaps *the only source*, of variation on which natural selection can act. He argued that the role of cross-level spandrels in the evolution of evolvability necessitated a hierarchical approach to evolutionary theory (Grantham 2004). But what are the implications of the role of cross-level spandrels as sources of variability for the problem of multiple levels of selection? The crucial question to ask is this: *where is the level n+1 to receive its fuel of variation, if not from level n?* If level n is the source of variation for level n+1, then the inevitable conclusion must be that *multiple levels of selection (and, by extension, multiple levels of interaction within any complex system of interacting parts) are essential.* The balance and interplay between levels is the last of our essential tensions. Multiple levels of selection are a *sine qua non*: there would be no evolution without them. A set of entities without variability injected from a level below would have no raw material on which natural selection could act. *There is thus no possibility of evolution occurring at only a single level.* A converse argument can also be made. Once variability exists, some balance of cooperation and competition will ensue, and the possibility will emerge for a transition from MLS1 to MLS2, leading to the emergence of a fully realized individual at a new level. One could argue that *multiple levels of selection (and, by extension, multiple levels of interaction within any complex system of interacting parts) are inevitable.*

I suggest that this extension of Gould's cross-level spandrels concept represents the frontier of research at the interface of collective dynamics and evolutionary biology. This book has simply laid the groundwork for the proposal that multiple levels, existing in a perpetual tension, are inevitable. To expand and explore the implications of this idea will demand painstaking thought. Moreover, biology is beautifully messy, and any general statement must entail a wealth of caveats and exceptions. I do strongly contend, however, that exploration of this hypothesis will be a richly fruitful vein to be mined by future research.

$$*** \quad *** \quad ***$$

In exploring the inevitability and essentiality of multiple levels, one important theme must be kept in mind. Nonlinear physics and mathematics have explored the infinity of repeating structures in chaotic attractors and in fractals; the same structures are repeated endlessly, forever, at all scales. This is a theoretical idealization, however, and broad generalizations must be tempered with a strong dose of reality if they are to be applied to the world of actual things. The idea of scale-free behavior in complex systems has made a fundamental impact on scientists' ability to understand various processes: it is found in the complex scaling of critical phase transitions and in the pervasive "rich-get-richer" networks first studied by Barabási and Albert (1999). It has been proposed as a fundamental pattern characterizing "how nature works", in a 1996 book by Per Bak (1996) of the same title. Earthquakes are a classic example of natural phenomena that follow a power law distribution, with very few "Big Ones" of large magnitude; Per Bak described this phenomenon in his classic "sandpile" model.

The idea of scale-free behavior initially seemed so intellectually attractive that scientists rushed to find it everywhere. This was typically done by identifying power-law behavior in a system. A power law distribution is given by $P(x) = x^{-\alpha}$, with x being the measured quantity and $P(x)$ being the number of measurements of that value (or the probability of measuring that value). The log-log plot of a power law is linear, with slope $-\alpha$, so standard procedure has been to look for linear log-log plots in the data. This sort of approach is susceptible to misinterpretation, in part because data must be plotted over many orders of magnitude in order to truly reveal power law behavior, and many real-world data sets barely span one or two orders of

magnitude. Eagerness to identify overarching scale-free behavior has led researchers to cut corners and avoid following up this simple plotting procedure with more
rigorous information-theoretic tests in order to determine what model best fits their
data. This rush to claim evidence of scale-free behavior has been called "log-log
lies", and has done a real disservice to the study of this critically (pun intended)
important behavior.

One recent example of the "over-hyping" of scale-free behavior is in the field of
animal foraging. Various studies had identified a type of scale-free distribution of
step lengths known as a Lévy flight in animal foraging behavior. Here, step lengths
refer to distances traversed between successive movements of an animal, whether
the distance a bee flies between flowers, the distance a *Daphnia* moves in the water
before making its characteristic turning behavior between "hops", or the distance a
mammal travels between successive feeding sites. Lévy distributions can also be
investigated for the time intervals between animal movements, the duration of each
step, hop or flight, etc. Described by a power law with exponent between 1 and 3,
Lévy flight distributions have a characteristic "long tail", with a small number of
large values.

In 1996, Viswanathan and colleagues identified Lévy flight behavior in the flight
patterns of albatrosses, using tracking devices placed on the birds' legs to record
when they touched water (Fig. 16.1). The distribution of flight durations had a char-
acteristically long tail, with a few long intervals spent in the air, and many short
ones. This suggested to the authors that a "Lévy search pattern in animal behavior
may reflect the solution of the biological search problem in complex environments"
(Viswanathan et al. 1996). In a subsequent paper, the researchers demonstrated in an
idealized model that "when the target sites are sparse and can be visited any number
of times, an inverse square[1] power-law distribution of flight lengths, corresponding
to Lévy flight motion, is an optimal strategy" for foraging (Viswanathan et al. 1999).
To confirm their predictions, they digitized data collected by Bernd Heinrich in
1979 on the foraging of bumble-bees in a heterogeneous landscape of clover, and
1996 data from Focardi et al. on foraging deer. When plotted on log-log plots, the
data sets showed slopes consistent with Lévy flight behavior.

More careful analysis of the data, however, soon showed that these conclusions
had been premature (Edwards et al. 2007; Edwards 2011). Firstly, information-
theoretic analysis of the data showed that a power law was not necessarily the best
fit to the data. Indeed, a more careful look at the original data from Focardi et al. and
from Heinrich showed that the measurements did not even represent the time intervals between foraging events. In the case of the deer, the times measured included
the time spent eating at each site, rather than just travelling between sites. For bumble-bees, the data did not represent total flight lengths, but rather the linear distance
between successively visited flowers. And Viswanathan and colleagues, who collaborated with Edwards on the 2007 re-analysis of their earlier work, noticed a

---

[1] Meaning a power law with exponent −2.

Fig. 16.1 Putative Lévy
flight path of an albatross
Reprinted by permission
from Macmillan Publishers
Ltd., G.M. Viswanathan,
V. Afanasyev,
S.V. Buldyrev,
E.J. Murphy, P.A. Prince,
and H.E. Stanley, Lévy
flight search patterns of
wandering albatrosses,
Nature 381: 413–415,
1996, copyright 1996

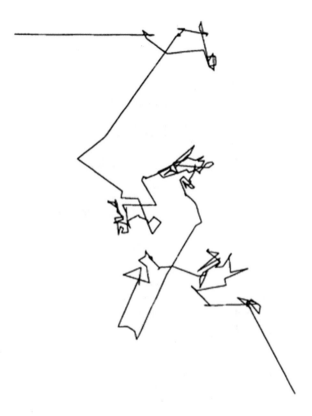

strange pattern in the albatross data. The longest flight times were consistently the
first times in the data set, for every bird studied.

> A salt-water logger only detects whether a bird is sitting on the water or not, and its clock
> starts when it is switched on at a computer. Thus a logger is recording before being attached
> to a bird, and also, crucially, while the bird sits on a nest. The logger is dry, but the bird is
> not flying. So the initial sequence of dry readings includes pre-take-off time plus time spent
> in flight before first landing on water. (Edwards et al. 2007)

The long "flights", then, included time the birds spent sitting on the nest before
beginning to fly (Fig. 16.2). With these spurious data points removed, the long tails
of the distributions were gone, and with them the evidence of Lévy flights. A larger
data set, using more birds, and using GPS trackers to determine when the birds actually left the nests, so that the initial flight time really was a flight time, showed distributions that were not well fit by a power law. Edwards (2011) explored many
other data sets that purported to show Lévy flight behavior, including data from
human fishermen. Re-analysis of this data showed that

> the original power-law Lévy flight model is overwhelmingly rejected for 16 out of the 17
> data sets when tested against three other simple models… Thus, Lévy flight movement pat-
> terns are not the common phenomena that was once thought, and are not suitable for use as
> ecosystem indicators for fisheries management, as has been proposed. (Edwards 2011)

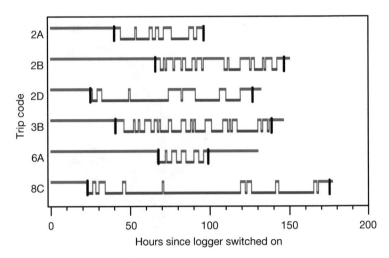

Fig. 16.2 Salt-water immersion logger data from six albatrosses. *Black* lines indicate when the birds left and returned to Bird Island, South Georgia. *Red* lines indicate hours during which the data logger was completely dry; *blue* lines indicate hours during which the logger was wet for some portion of the hour. Reprinted by permission from Macmillan Publishers Ltd., A.M. Edwards et al., Revisiting Lévy flight search patterns of wandering albatrosses, bumblebees and deer, *Nature* 449(7165): 1044–1048, 2007, copyright 2007

Researchers continue to investigate the distribution of animal movement patterns. A recent study of mud snails suggests that Lévy distributions (called "Lévy walks" in this case, more appropriately for snails!) may be innate behavior patterns, independent of interaction with the environment (Kölzsch et al. 2015). A study by Raichlen et al. (2014) of human hunter-gatherers in Tanzania found Lévy behavior in nearly half of all foraging trips. Still, studies suggest that swarming bacteria may follow Lévy flight patterns as they move (Ariel et al. 2015). Meanwhile, other studies, such as one from my own research group on the foraging patterns of two monkey species, show distinctly non-Lévy behavior (Vandercone et al. 2013). Thus, the importance of power law behavior in animal foraging remains an open question; the best one can conclude at present is that this behavior may occur and be useful in some cases, but not all. What is clear, however, is that beautiful concepts of scale-free behavior may sometimes be smashed on the rocks of reality.

Another cautionary tale of hopeful but incorrect inference of scale-free behavior concerns evolutionary dynamics. A simple and elegant computational model developed by Kim Sneppen and Per Bak showed "avalanches" of extinction. The avalanche sizes followed a power law distribution, and the authors suggested that the avalanches represented punctuated bursts of evolutionary change as in Eldredge and Gould's model of punctuated equilibrium (Bak and Sneppen 1993; Sneppen et al. 1995). Another model, proposed by Solé and Manrubia (1996) also suggested critical scaling behavior in extinction dynamics. In Sneppen-Bak model, "species" are arranged in a linear array and assigned random fitness values. The species with the lowest fitness value is removed, along with its nearest neighbors, in order to simulate the cascading effect of local species interactions. Other randomly chosen species are removed as well, in order to simulate the cascading effects of non-local interactions. This resulted in scale-free cascades of extinction "avalanches", a result clearly reminiscent of punctuated equilibrium. This result was particularly intriguing in that it produced "punctuated" extinctions based only on endogenous interactions, suggesting that patterns of macroevolution could be driven entirely "from below". However, Bak interpreted this result as implying that endogenous interactions were the *only* drivers of macroevolutionary processes. This, coupled with Bak opening a book review in the pages of *Nature* with the insulting question of whether "biology is too complicated for biologists" (Bak 1998) understandably alienated a large portion of the biology community. Bak and colleagues drew

> the overextended inference that because such large scale punctuations arise endogenously, the actual mass extinctions of the fossil record therefore need no exogenous trigger of environmental catastrophe, or any other external prod. This claim, emanating from a theoretical physicist with little knowledge of the empirical archives of geology and paleontology, and emerging just as persuasive evidence seems to have sealed the case for a bolide impact as a trigger of at least one actual mass extinction (the end Cretaceous event 65 million years ago), could hardly fail to raise the hackles of observationally minded scientists who, for reasons both understandable and lamentable, already bear considerable animus towards any pure theoretician's claim that success in modelling logically entails reification in nature. (Gould 2002, p. 927)

Spurred by the computational models developed by Sneppen and Bak, Solé and Bascompte (1996) turned to the important question of whether critical phenomena might be relevant to actual macroevolutionary processes. Specifically, they wondered whether the large mass extinction events known to have occurred throughout Earth's history might feature as the "rare, large" events on the tail of a power-law distribution. Looking at extinction data summarized by Raup (1986), they noted the "striking fact" that "the study of the frequency distribution of extinction sizes seem[s] to reveal a continuous connection from small to large events", rather than a bimodal distribution of mass extinctions and smaller extinction events. They noted that other distributions, such as the lifespans of various genera, also were continuous. After interpolating the data with an exponential and a power law function, they investigated whether a power law or an exponential function was a better fit to the data. They wrote that

> [b]oth curves fit well with the data…But the dynamical interpretation of them is far from trivial. If a power law…is involved, then we would conjecture that – at least to some extent – some basic mechanisms operate at different scales. This claim comes from the well-known fact that power-laws are a characteristic of fractal, self-similar objects…A very different situation is found if an exponential distribution is obtained: scale invariance is absent, and particular characteristic scales are relevant. (Solé and Bascompte 1996)

Solé and Bascompte concluded that the extinction and genera lifetime distributions did indeed follow power laws (interestingly, they found that distributions obtained from cladograms did not). They made the connection to punctuated equilibrium, and noted the fractal-like structures observed in various taxonomic data sets. Solé and Bascompte concluded that, while the data was not yet definitive, the study of critical, scale-free phenomena might have significant explanatory power in the investigation of patterns of extinction and diversification.

The same year, Solé collaborated with Manrubia, Benton and Bak to explore scaling behavior using a larger paleontological database. Because of the wide spacing (7 Myr) of some of the data points, Solé et al. again used interpolation in order to fill in these data sets, applying various interpolation methods and finding that each method gave similar results. They concluded that the time series did indeed show power law scaling, and suggested that this shows that the "big five" mass extinctions[2] were not qualitatively different from smaller extinction events, and "could be the skewed end of a continuous distribution of extinction events of different intensity" (Solé et al. 1997). They also suggested that their results strongly argued against Raup and Sepkoski's (1984) suggestion of a 26 Myr periodicity for mass extinction events.

The bubble, however, soon burst. In 1998, Kirchner and Weil published a short letter to *Nature* showing that the power-law behavior found by Solé and colleagues was an artifact of interpolation. They noted that power-law scaling implies some correlation between data points in a time series, and that significant correlations were introduced by the interpolation procedure. Indeed, in the main data set used, Solé et al. had 77 initial data points, spaced at 7 Myr intervals. After interpolating to get data points at every 1 Myr, they had a "time series with 570 points, 86% of which are interpolations rather than real data" (Kirchner and Weil 1998). They noted that Solé et al. had compared their results to white noise as a null hypothesis. But, Kirchner and Weil argued, the appropriate null hypothesis for comparison would be white noise that had been subjected to the same interpolation procedure as the original data. When compared with interpolated randomized values, the interpolated data set showed no evidence of power-law scaling.[3]

$$*** \quad *** \quad ***$$

---

[2] Current data, discussed in Elizabeth Kolbert's excellent 2014 book *The Sixth Extinction*, suggests that we are now in the throes of a sixth mass extinction, resulting from anthropogenic climate change. Indeed, climate scientists have concluded that "humans have changed the Earth system sufficiently to produce a stratigraphic signature in sediments and ice that is distinct from that of the Holocene epoch", and recommend designating the current geological epoch as the *Anthropocene*, beginning with the widespread use of agriculture and the spread of deforestation (Waters et al. 2016).

[3] In a subsequent study, Kirchner and Weil (2000) investigated correlations in rates of extinction and origination of marine families and genera. They found "that extinction rates are uncorrelated beyond the average duration of a stratigraphic interval. Thus, they lack the long-range correlations predicted by the self-organized criticality hypothesis. In contrast, origination rates show strong autocorrelations due to long-term trends. After detrending, origination rates generally show weak positive correlations at lags of 5–10 million years (Myr) and weak negative correlations at lags of 10–30 Myr, consistent with aperiodic oscillations around their long-term trends." Based on these results, they suggested that "origination rates are more correlated than extinction rates because originations of new taxa create new ecological niches and new evolutionary pathways for reaching them, thus creating conditions that favour further diversification."

Fig. 16.3 Scale matters.
Neither you nor I can do
this. Photograph of water
striders by Cory,
reproduced under a
Creative Commons
Attribution-Share Alike 2.1
Japan License

The message to take from these two cautionary tales, of birds "flying" in their nests and interpolations gone wild, is not, I would argue, that we should resist looking for scaling phenomena in the biological sciences. Indeed, a vast wealth of evidence, some of which I have summarized in the preceding chapters, strongly suggests that similar processes of competition, cooperation and selection do indeed operate similarly at different scales. However, one cannot expect these processes to be truly self-similar. Different biological scales have fundamental differences due to the inherent scale-dependence of physical and biological interactions (Fig. 16.3).[4] Interactions at the molecular level are quite different than at the cellular level, interactions at the cellular level are quite different than at the organismal level, and interactions at the organismal level are quite different than at the group, species or

---

[4] For a discussion of how important size scale can be, see Peter Hoffmann's 2012 book *Life's Ratchet*. "Life must begin at the nanoscale," Hoffmann writes. "This is where complexity beyond simple atoms begins to emerge, and where energy readily transforms from one form to another. It is here where chance and necessity meet" (Hoffmann 2012, p. 91). He argues that the reason for this remarkable confluence is the fact that the exchange of energy among various forms (thermal, chemical, electrical, mechanical) takes place with particular ease at the nanoscale, where these forms of energy have similar magnitudes. Hoffmann does a brilliant job of describing the essential tension between "chance and necessity" by which molecules are able to harness the "molecular storm" of thermodynamic fluctuations to ratchet their way up an asymmetric energy landscape, and clearly draws the analogy to the role of genetic noise in fueling evolutionary change. His argument is a perfect encapsulation of the idea of similarities between scales, coupled with the uniqueness of each individual scale, constrained by the physical yardstick of molecular sizes and energies.

Fig. 16.4 Scale matters. A tiny scrap of that red wall hanging would drape quite differently. Judith Beheading Holofernes, painted by Caravaggio (1598–1599). Image in public domain

ecosystem level. Because of the fundamental role of physical scales in nature, some levels of biological interaction are more tightly bound than others. This can lead to discounting interactions at the "looser" levels, as can be seen with the "controversial" history of group selection. However, this is just as misguided as the assumption that all levels are equally tight. Looking for symmetries between levels is a task that must be pursued with infinite respect for the brutal reality of nature. The suggestion that multiple levels are both essential and inevitable must be pursued in that spirit.

In the end, it is perhaps not so much a question of pluralism, but of honesty. Being real, the world cannot be perfectly scale-free. The red velvet cloak that drapes so beautifully over Caravaggio's shoulder hangs stiffly when worn by the inhabitant of a doll house: fibers have a characteristic size scale (Fig. 16.4). I would sink deeper than did ever plummet sound if I tried to compete with a water strider. The Reynolds number of the water means something quite different to a *Daphnia* than to a paddlefish. Multiple scales – and the tensions between them – do exist. They may be inevitable, perhaps even essential. But they don't perfectly mirror each other up and down an infinite chain of being. Like it or not, the world is real.

Discuss.

References

Ariel G, Rabani A, Benisty S, Partridge JD, Harshey RM, Be'er A (2015) Swarming bacteria migrate by Lévy walk. Nat Commun 6:8396

Bak P (1996) How Nature Works: The Science of Self-Organized Criticality. Springer-Verlag New York, Inc. (Copernicus), New York

Bak P (1998) Life laws. Nature 391:652–653

Bak P, Sneppen K (1993) Punctuated equilibrium and criticality in a simple model of evolution. Phys Rev Lett 71(24):4083–4086

Barabási AL, Albert R (1999) Emergence of scaling in random networks. Science 286(5439):509–512

Edwards AM (2011) Overturning conclusions of Lévy flight movement patterns by fishing boats and foraging animals. Ecology 92(6):1247–1257

Edwards AM, Phillips RA, Watkins NW, Freeman MP, Murphy EJ, Afanasyev V, Buldyrev SV, da Luz MG, Raposo EP, Stanley HE, Viswanathan GM (2007) Revisiting Lévy flight search patterns of wandering albatrosses, bumblebees and deer. Nature 449(7165):1044–1048

Focardi S, Marcellini P, Montanaro P (1996) Do ungulates exhibit a food density threshold – a field-study of optimal foraging and movement patterns. J Anim Ecol 65:606–620

Gould SJ (2002) The Structure of Evolutionary Theory. The Belknap Press of Harvard University Press, Cambridge, MA/London

Grantham TA (2004) Constraints and spandrels in gould's structure of evolutionary theory. Biol Philos 19:29–43

Heinrich B (1979) Resource heterogeneity and patterns of movement in foraging bumble-bees. Oecologia 40:235–245

Hoffmann PM (2012) Life's Ratchet: How Molecular Machines Extract Order from Chaos. Basic Books, New York

Kirchner JW, Weil A (1998) No fractals in fossil extinction statistics. Nature 395:337–338

Kirchner JW, Weil A (2000) Correlations in fossil extinction and origination rates through geological time. Proc R Soc B 267(1450):1301–1309

Kolbert E (2014) The Sixth Extinction: An Unnatural History. Henry Holt and Company, New York

Kölzsch A, Alzate A, Bartumeus F, de Jager M, Weerman EJ, Hengeveld GM, Naguib M, Nolet BA, van de Koppel J (2015) Experimental evidence for inherent Lévy search behaviour in foraging animals. Proc R Soc B 282(1807):20150424

Raichlen DA, Wood BM, Gordon AD, Mabulla AZ, Marlowe FW, Pontzer H (2014) Evidence of Lévy walk foraging patterns in human hunter-gatherers. Proc Natl Acad Sci U S A 111(2):728–733

Raup DM (1986) Biological extinction in earth history. Science 231:1528–1533

Raup DM, Sepkoski JJ Jr (1984) Periodicity of extinctions in the geologic past. Proc Natl Acad Sci U S A 81(3):801–805

Sneppen K, Bak P, Flyvbjerg H, Jensen MH (1995) Evolution as a self-organized critical phenomenon. Proc Natl Acad Sci U S A 92(11):5209–5213

Solé RV, Bascompte J (1996) Are critical phenomena relevant to large-scale evolution? Proc Biol Sci B 263(1367):161–168

Solé RV, Manrubia SC (1996) Extinction and self-organized criticality in a model of large-scale evolution. Phys Rev E 54(1):R42–R45

Solé RV, Manrubia SC, Benton M, Bak P (1997) Self-similarity of extinction statistics in the fossil record. Nature 388:764–767

Vandercone R, Premachandra K, Wijethunga GP, Dinadh C, Ranawana K, Bahar S (2013) Random walk analysis of ranging patterns of sympatric langurs in a complex resource landscape. Am J Primatol 75(12):1209–1219

Viswanathan GM, Afanasyev V, Buldyrev SV, Murphy EJ, Prince PA, Stanley HE (1996) Lévy flight search patterns of wandering albatrosses. Nature 381:413–415

Viswanathan GM, Buldyrev SV, Havlin S, da Luz MG, Raposo EP, Stanley HE (1999) Optimizing the success of random searches. Nature 401(6756):911–914

Waters CN, Zalasiewicz J, Summerhayes C, Barnosky AD, Poirier C, Gałuszka A, Cearreta A, Edgeworth M, Ellis EC, Ellis M, Jeandel C, Leinfelder R, McNeill JR, Richter DB, Steffen W, Syvitski J, Vidas D, Wagreich M, Williams M, Zhisheng A, Grinevald J, Odada E, Oreskes N, Wolfe AP (2016) The Anthropocene is functionally and stratigraphically distinct from the Holocene. Science 351(6269):aad2622

Index

© Springer Science+Business Media B.V. 2018
S. Bahar, *The Essential Tension*, The Frontiers Collection,
DOI 10.1007/978-94-024-1054-9

Printed in the United States
By Bookmasters